AN INTRODUCTION TO
Canada's Public Social Services

AN INTRODUCTION TO

Canada's Public Social Services

UNDERSTANDING INCOME AND HEALTH PROGRAMS

second edition

Frank McGilly

Toronto New York Oxford
OXFORD UNIVERSITY PRESS
1998

OXFORD
UNIVERSITY PRESS

Oxford University Press is a department of the University of Oxford.
It furthers the University's objective of excellence in research, scholarship,
and education by publishing worldwide. Oxford is a registered trade mark of
Oxford University Press in the UK and in certain other countries.

Published in Canada by
Oxford University Press
8 Sampson Mews, Suite 204,
Don Mills, Ontario M3C 0H5

www.oupcanada.com

Library and Archives Canada Cataloguing in Publication

McGilly, Frank J., 1929 –
An introduction to Canada's public social services:
understanding income and health programs

2nd ed.
Includes bibliographical references and index.
978-0-19-541232-1

1. Public welfare–Canada. 2. Income maintenance programs–Canada.
3. Medicine, State–Canada. 4. Insurance, Health–Canada. 5. Social
security–Canada. I. Title.

HV105.M363 1998   368.4'00971   C97-932207-3

This book is printed on permanent (acid-free) paper ∞
Printed and bound in Canada.

3 4 5 — 14 13 12 11

For Renée

Contents

Part Three RETROSPECT AND PROSPECT

10 The New Millennium: Sunset or New Dawn of the Welfare State?

Foreword

It gave me great satisfaction, naturally, to be invited to prepare a second edition of this book, but observers, present and future, will probably agree that 1995–7 was a singularly hectic time to have undertaken this revision.

In 1995, virtually every established social program in Canada underwent fundamental reconsideration. All the provinces looked critically at their public assistance legislation, and even when they did not change the basic structure of their programs, they reduced their benefits. Similar re-examination faced medicare and hospital insurance, both of which, according to innumerable public opinion polls, are part of what defines Canada to Canadians (certainly part of what distinguishes us from the United States, widely stereotyped by us as relatively heartless in such matters). The federal government made major changes to Unemployment Insurance, including a change of name. Federal programs that had been repeatedly anointed as 'sacred' were treated quite profanely; for example, Old Age Security was replaced by the Seniors Benefit, featuring an income test that wiped out benefits entirely for millions of better-off Canadians. Federal Family Allowances had already disappeared, to be replaced by the non-universal Child Tax Benefit. For nearly a decade, federal financial transfers to the provinces had been shrinking in relative terms—that is, growing more slowly—but in 1995, for the first time, absolute *decreases* in such transfers were projected for coming years; and furthermore, the links between transfers and the delivery of the provincial health and welfare programs they support, already becoming tenuous, were all but severed, probably undermining the maintenance of national standards.

All of this was taking place in a Canadian political and economic climate dominated by two 'weather systems' either of which was stormy enough to rock the ship of state: the national unity crisis that manifested itself (but certainly did not end) in the indecisive Quebec sovereignty referendum of October, 1995, and the budgetary problems of the federal and all the provincial governments, which had grave consequences for health, welfare, and education programs across the country.

This text does not provide detailed dollars–and–cents information on the workings of our social programs: such information rapidly becomes dated, and, in any case, students of the field owe it to themselves to become familiar with the standard sources of current data (the book will help them to do so). I have tried to show how programs work, and how the programs relate to the policy orientations that underlie them. Dealing responsibly with the changes of the mid-1990s has necessitated many changes; in order to make the changes intelligible, I have felt obliged to devote some attention to the structures of programs as they once were, as well as to what they have become.

A blizzard of details on all the social programs of all the provinces and the territories would blind readers rather than enlighten them. References to the details of specific programs should be taken as illustrating a point, not as providing information for its own sake. In the first edition, many of the illustrations were drawn from the experience of Quebec, both because it is my home province and because Quebec has made many interesting innovations in social policy. This revision pays much more attention to other provinces.

Despite the difficulties of describing the foundations of a structure while it is being rebuilt, the endeavour has a plus side: a cursory glance around the world indicates that things are in pretty much the same state of flux everywhere. If this book succeeds in its goal of introducing readers to the income and health programs and problems of Canada, it will have introduced them at the same time to some of the most daunting domestic challenges now faced by all the industrialized nations of the world.

Frank McGilly

CHAPTER 1

General Introduction

OVERVIEW AND OBJECTIVES

In this text are discussed programs in effect in Canada, under public authority and administration, in two important areas of social policy: income and health services. The three objectives of this book are, in ascending order of importance: (1) to provide basic substantive *information* about programs:

- the statutory base of the program,
- what is provided,
- what part of the population is covered,
- who is eligible to receive benefits, and on what terms,
- who administers the program,
- the resources (chiefly financial) that support the program,
- the scale of the program, its growth or decline over time, and important changes and developments that have taken place;

(2) to acquaint the reader with *authoritative sources* of information and discussion about the programs and the problems; (3) to open the door to an *understanding* of the programs: the precise problem or difficulty each program addresses, or is supposed to address, the institutional framework within which the program operates, how the program is fitted to the problem, criteria for success or failure of programs, and criticisms put forward by various commentators.

Some emphasis is placed on the *information* objective, since students are better prepared to discuss the issues surrounding social programs after they have attained a degree of familiarity with the programs themselves. Certainly the converse is true: students and others are poorly prepared to discuss social policy *without* some familiarity with social programs—though ignorance, as we know too well, has never inhibited the expression of strong opinions.

Information, while necessary, is not sufficient; one could be abundantly informed and still understand little. Information gets its meaning from its context. Attention is given, therefore, to important questions and controversies surrounding social programs, especially to the limits and constraints under which the program operates, or is thought to operate. These may be limitations in terms of *resources* that are available for the program (chiefly but not exclusively financial). The constraints may be *political*, emerging from the fact that people and groups of differing interests and policy preferences have different avenues and degrees of influence and power. Constraints may be imposed by particular social *institutions* that operate in the field of the program. Basically, social policies and programs are seen here as the never fully determinate outcomes of the interplay of three interlinked factors: the requirements of the institutions characteristic of industrial society, the estimation by decision-makers of the pressure of constraints, and the play of relative influence and power among the various interests of society.

As they learn to respect the constraints within which social programs operate, students must not allow their vision of what is necessary and/or possible in social policy to be defined by existing programs. Whatever is is not necessarily right, and is not inevitable; evolution is, or ought to be, continuous. The proper frame of mind is, in Donald Campbell's phrase, to consider programs as hypotheses (Campbell, 1969).

SEVEN MAJOR CHALLENGES

The study of social policy and social services is the study of *actions*. All human actions, individual or collective, are the outcomes of a variety of forces. The roots of social policy are always tangled. Moreover, it is a characteristic of social policy that typically it acts in the face of more or less pressing *problems*: widespread unemployment, poverty among old people, frequent injury and illness among workers. Such a predominant *problem* focus may be an obstacle to the development of clear, coherent, well-coordinated policies: action that quickly fixes today's problem may do nothing to prevent tomorrow's. One may wonder why society does not see the problems coming and make arrangements to deal systematically with them as foreseeable mischances of life, rather than react to problems after they have become urgent. In the vocabulary of students of social policy, it would appear desirable that measures supportive of social welfare become *institutional*—part of the normal fabric of society—rather than, as they so often are, *residual*—designed to repair damage, to rescue casualties, to pick up the pieces.

Whether the social services are designed to forestall damage institutionally or to repair it residually, their development is likely to parallel those changes in social life that have been recognized as posing challenges to society in general. Seven such challenges will be reflected in the descriptions and analyses of Canadian social programs given in this text. The first four arise out of our social and economic environment; the fifth and sixth arise respectively from the system of government to which Canada adheres (with occasional seasons of doubt, it must be acknowledged), and from competing values at work in Canadian society. The seventh is a fiscal issue, temporary in nature, but with repercussions for social policy that will reach far into the future. These seven major challenges are:

(i) changes in Canada's economic context, principally the shift in the composition of economic activity,
(ii) changes in the nature of employment,
(iii) changes in the make-up of the population of Canada,
(iv) changes in the participation of women in social life,
(v) the tension between individualist and collectivist values in Canadian society,
(vi) the system of government in Canada, and
(vii) Canada's public deficits and debt, federal and provincial.

This seventh challenge is less basic than the other six, but it has emerged from their joint impact, and will be at least equally compelling for the foreseeable future of our social policy. This challenge has come to the fore relatively recently in Canada, as it has in almost all industrial nations. Years and years of annual fiscal deficits, at both the provincial and federal levels, have left Canada burdened with a huge accumulated public debt. Payment of interest on this debt eats up an unacceptably large share of each year's revenue—and thus aggravates each succeeding year's deficit. Retirement of the debt and reduction of the annual deficit have therefore moved close to the top of the list of priorities of all governments; and decisions of enormous, long-lasting importance are being made in that light.

In the remainder of this chapter, each of the challenges listed above will be discussed in turn.

(i) The Changing Economic Context

Throughout these pages occur references to the changes that have taken place in the economic context within which our social programs have emerged. *The least misleading simplification of social policy is to define it as society's struggle to keep up with the consequences of advancing industrialization and post-industrialization* (Rimlinger, 1971; Wilensky and Lebeaux, 1965). The most visible evidence of industrial development, besides the increase in goods and services made available, is change in the kinds of work people do, and in the kinds of human relations that prevail in workplaces. In all industrial societies a proportional shift can be observed, over time, in the work people do, from an early concentration on 'primary' activities (based on the extraction of natural resources, like minerals, and Canada's historic four F's: fur, forest, fish, and farm), through the predominance of 'secondary' activities (the processing of raw materials and the manufacture of goods), to a rise in the importance of 'tertiary' activities (the service sector—finance, insurance, trading services, transportation and communications, public services, information processing).

Still further changes can be foreseen. The so-called globalization of trading relations has already induced shifts in the economic structures of nations, and will continue to do so: as producers in emerging countries find their way into world markets (which they must do if their standards of living are to improve), old producers will find it advantageous and/or necessary to move into other occupations, largely those that exploit advanced technologies. This may accelerate another observed tendency in advanced economies, whereby labour markets appear to be shifting in two directions at once: toward better, higher-paying jobs, and toward less rewarding jobs (Banting and Beach, 1995; Economic Council of Canada, 1990).

Such changes are, to be sure, 'economic', but they are far more than that in their effects. They bring in their wake many profound changes in human and social relations, including changes in the role of the family, the extent and nature of urbanization, the typical work career (a large part of the typical life), the relations between employers and employees, the social status of the elderly, the roles of men and women, and patterns of child care. They affect, moreover, the relative strength of the various forces in society; élites rise and fall, yielding to others; and since policy is, in large part, the outcome of the play of social forces, a change in this balance is crucial to policy. Each party among the strong naturally encourages changes that will increase its relative strength; the interests of the weak are always at risk. The social services may be looked upon as society's attempt to deal with at least some of the ramifications of such far-reaching social change.

(ii) The Meaning of Employment

Emerging from the changing economic context is a second theme: the meaning and importance of employment in a maturing industrial society, and, conversely, the meanings of *unemployment* (the situation of someone who is in the labour market, but out of work) and of *non-employment* (the situation of someone who is not in the labour market at all).

In industrial society, the employment relationship is pivotal. Paid work is obviously the most important source of income for the vast majority of the population. Other important material benefits—vacations, pensions, various forms of insurance—are typically attached to employment, although very unequally, as shall be seen. The availability of such benefits has become almost as significant a criterion of a 'good job' as salary or wages. To a remarkable extent, our social programs, including some not essentially related to work, have been built around employment for their financing and their administration.

Many observers see an increasing fluidity in the world of employment. The evidence is ambiguous, but the argument is made that an adult may now expect more job changes in the course of a working life than has been the case in the last several decades. If this trend becomes well established, it will have serious implications for income security. We see around us other great changes: an increase in part-time work, in self-employment, in the numbers of small-scale employers, and most notably in the participation of women in paid work. We see also changes in the labour force participation of men (slightly declining), in the age of retirement, and in the experience of older workers.

(iii) Demographic Change
The fundamental ingredient of Canadian society is the people themselves. Striking changes have taken place, and are continuing to unfold, in the demographic composition of Canadian society.

(a) Age Distribution
The most striking changes, perhaps, are those at the extreme ends of the life cycle: the exceedingly rapid increase in the number and proportion of elderly people, and the sharp decline in the birth rate (a decline that, apparently, may have been arrested in the mid-1990s). Like other industrial countries, Canada is experiencing an inexorable change in the composition of its population by age and gender.

Nature has always provided more male babies than female, so the young are mostly male; but females have lived longer than males, at least since deaths due to childbirth have become infrequent, so the old are mostly female. As our birth rate diminishes and longevity increases, both the age composition and the sex composition of the population are shifting. There are similar changes under way in work force participation, in family composition, indeed in the very nature of families and households—it is estimated, for instance, that upwards of 16 per cent of Canadian households with young children are now headed by single parents, overwhelmingly women, mostly more or less poor (this does *not* mean that most poor children are the children of lone mothers).

These changes have altered the needs to which social policy must respond. Their immediate consequences for social programs (old age pensions, financial assistance to families, job-related programs) are drastic enough to engage the sharpest skills of policy-makers and administrators. Their long-term implications are at least as drastic, not to say frightening.

(b) Ethnic Composition
Equally remarkable is the development of the ethnic diversity of the Canadian population. *No single ethnic group accounts for a majority of the Canadian people.* This should be recognized as a fact of enormous significance. On public occasions, we speak of 'the Canadian mosaic'; the metaphor is more apt than is generally realized. It sets Canada (like the United States) apart from almost all other countries on earth.

This diversity has prevailed in some degree since the eighteenth century, but it became a fundamental characteristic of Canada, whether fully reflected in social mores or not. This came about as a result of the enormous waves of immigration from the 1890s to the outbreak of World

War I in 1914. Most of that early immigration was from Europe, but substantial numbers of persons of Asiatic origins, principally Chinese, also were brought into this country, under explicitly discriminatory conditions. Another huge tide of immigration followed World War II, largely from Europe. More recently, there has been substantial immigration from Asia, the Caribbean, and Africa.

Most of the early settlers of what is now Canada, before and after Confederation, were from France (in Quebec, of course, but also in the Maritimes, Ontario, and Manitoba) and Great Britain, but the picture has changed radically since those early times. The population of the province of Quebec is still dominated by persons of exclusively French ancestry (75 per cent); but in 1991, in Canada as a whole, only 28 per cent reported that their ancestors had been exclusively British. The only provinces in which over half the people professed themselves to be of unmixed British origins were the small provinces of Newfoundland (88 per cent), Prince Edward Island (66 per cent), and Nova Scotia (58 per cent). In none of the larger provinces did people of British origins, including those who identified themselves as *partly* British, constitute as much as half the population. (The term 'British' itself is questionable as an ethnic designation; it lumps together English, Scottish, Irish, and Welsh, groups with distinct ethnic and cultural characteristics, as representatives of any of the four will heatedly insist, and as current political developments in Great Britain confirm.) Close to nine million Canadians consider themselves to have no French or British ancestry at all.

The 1986 census listed seventy-six distinct ethnic strains in the Canadian population. The way in which ethnicity is reported in the 1991 census differs from past practice. Previously people were asked to identify one ethnic origin; in 1991, multiple origins were recorded, and presumably this will be the case in subsequent censuses as well. Since one person may be, let us say, part Irish (considered 'British') and part Italian, that person will show up in both categories. This makes it possible to know what numbers of the population are all or part Italian, Chinese, Caribbean, etc., but reporting on all such combinations is, to say the least, difficult, and will be more so as parentage becomes more mixed. This reporting modification also will make it difficult to be precise in tracing trends from the past.

Members of the various ethnic groups are not dispersed evenly from coast to coast. There are considerable ethnic concentrations across Canada: half a million Ontario residents claimed to be of unmixed Italian origin, for example. Manitoba and Saskatchewan have high concentrations of aboriginal peoples (not, of course, through immigration).

Too many details of this kind obscure as much as they illuminate. The point is simply that Canada is highly diverse from the ethnic point of view; inevitably, there will be significant cultural variations within Canadian society.

That large-scale immigration will have an effect on society is obvious, though just what the effects are remains debatable. With regard to one of the most sensitive aspects—the real effect of immigration on employment opportunities—various observers are far from unanimous. Cultural variations exert subtle impacts on the implementation of social programs. Wherever social policy impinges on the life of the family, as it does, for instance, with respect to the care of the elderly, in the administration of public assistance, and in attitudes toward health care, cultural variations will make a difference. The ethnic and cultural diversity of the Canadian nation is given the highest possible formal recognition in the 1982 Charter of Rights and Freedoms, part of the Constitution of Canada, whose Article 27 requires that 'This Charter shall be interpreted in a manner consistent with the preservation and enhancement of the multicultural heritage of Canadians.' The policy-maker and the administrator ought to be sensitive to such variations, but it is no simple matter to reconcile fairness and equal treatment of individuals with respect for cultural differences, which are necessarily collective phenomena. The courts have recognized this in interpreting the Charter (Magnet, 1987, vol. 2, p. 20; Borovoy, 1988, ch. 11).

The fact that Native peoples, including those of mixed ancestry, account for no more than 4 per cent of the population of Canada, long made it fairly easy for Canadians in general to relegate Native peoples' issues to a low priority. In recent years, Native issues have become much more prominent, due in large part to the Natives' own increasing assertiveness. It would be a miscarriage of intent were this book to appear to collude in this neglect. The initial impact of European settlement on the indigenous peoples of the American continent was simply disastrous. The numbers killed in 'military' and other lethal engagements were considerable, the numbers who died as a result of deprivation following conquest were far greater, but worst of all were the numbers killed by infectious European diseases against which they had developed no immunity. For better or worse, Indians and Inuit continue to receive various other kinds of 'special' treatment.

In Canada, Native Indians have had a special civil status since before Confederation. Some have voluntarily given up their special status, but most have retained it. One aspect of their special status has been that since 1867 the health and welfare services they receive are legislated and administered entirely by the federal government, an exception to the prevailing rule that health and welfare are within the jurisdiction of the provinces. Currently, health and welfare programs for Indians and Inuit are administered by the federal Department of Indian Affairs and Northern Development. The picture is complicated by various intergovernmental arrangements for exceptional cases. It is also complicated by the fact that there are differences between groups of Canadian Indians and Inuit: the Micmacs of Nova Scotia and the Haida of British Columbia are no more similar than one would expect of peoples who developed a continent apart.

In many parts of Canada, the impact on Native peoples of a civilization imported from Europe has continued to be harmful. By conventional measures of social pathology, Indians and Inuit score badly: infant mortality, life expectancy, educational attainment, unemployment, prevalence of infectious diseases, malnutrition, alcoholism and drug abuse, and what the dominant Canadian culture defines as crime (jails in Manitoba and Saskatchewan are occupied by vastly disproportionate numbers of Indians). Facile racist 'explanations' for this state of affairs abound, and their exponents do not yield easily to evidence of Native peoples' capacity to deal with their own well-being when not deprived of the tools to do so. The more sophisticated may perceive that the problems lie with some basic incompatibility between the minority culture and the majority culture. The dominant culture's remedies in matters of health and welfare, which are not uniformly successful even within the majority population, are evidently self-defeating in the widely varying indigenous cultures. But the discussion only begins there.

For the purposes of this text, respect for the special nature of the health and welfare issues of Native peoples, and for the special knowledge required to understand them, leads to the exclusion of those issues from these pages.

(iv) The Participation of Women in Social Life

The fourth major development is the changes that have taken place in the participation of women in the life of society. The machinery of society, including social policy, has lagged in adjusting to the changing participation of women. Writing in the 1940s, Leonard Marsh devoted a substantial portion of his *Report on Social Security for Canada* to 'Women's Needs in the Social Security System'; in the 1990s, many of the needs Marsh analysed are still being met imperfectly, at best (Marsh, 1943/1975, pp. 209–28).

Among the most striking aspects of the change in the position of women has been the aforementioned increase in their participation in the work force (i.e., in the *paid* work force; that women have always done at least their share of the world's work is a truism). Actually, Canada has a long history of special attention to working women—special minimum wages for women (lower than for men), special regulation of their working hours (before wages and hours protection was legislated for men), certain special protections for their health. The special attention has not always been sufficient or salutary. The earlier legislation was of a protective, even paternalistic, character, based on the assumption that women were especially vulnerable to exploitation. (See, for example, the references to 'female labour' in the index of Kealey, 1973, p. 457.) Nowadays, the thrust of advocacy is toward equality of the sexes in social affairs rather than special protection for women. Sometimes overlooked in discussion are the effects such changes must inevitably have on the participation of men.

The significance of this fundamental social change is transparent in the discussion of (Un)Employment Insurance and the Canada and Quebec Pension Plans, where the employment experience of women has led legislatures to pass specific amendments to the relevant statutes. The significance may be less easy to see with respect to some other programs discussed here.

The increase in the number of families headed by women has necessarily had an impact on public assistance programs.

Students will perceive the personal and social costs of unequal treatment easily enough: the injustices and hardships that many women have suffered, the loss to society of contributions that might have been made by women who have been denied the opportunity. The same students will no doubt reach their own conclusions about the effectiveness of the responses of Canadian social policy, actual and proposed, to the position of women.

In some respects, social policy can be criticized for failing to take account of certain aspects of the social roles of women that have *not* changed much. For example, because women remain the primary family care-givers, they are primarily affected by any transfer to 'the family' or 'the community' of the burden of care of elderly, disabled, or chronically ill persons, or patients discharged from hospitals.

(v) The Individualist-Collectivist Tension

A fifth, more abstract, theme is the tension between individualism and collectivism in Canada's response to felt social needs. One current in our prevailing ideology pushes us to rely on the resources of individuals and of primary groups to meet social needs; another current inclines us to seek collective remedies, especially where needs are seen to have been created by the workings of individualism. In our era, 'collective' action more often than not has meant 'state' action, though the state is not our only collectivity. Even after it has been agreed that the state must act in a certain field, the tension persists in the choice of program modes: one current of opinion suggests that a problem such as the relief of poverty be attacked case by case, as with programs of public assistance; the other current prefers to involve the whole population more systematically, as with programs of 'universal' benefits and highly inclusive 'social insurances'.

The waning confidence of many citizens in state solutions to problems, to be discussed later, has given rise to a reliance on other collectivities to assume the responsibility, more or less willingly. One example is the current wave of enthusiasm for community economic development.

The choice of a mechanism to do a certain task may appear to be simply pragmatic, not ideological: we choose the best means to do the job. But ideology has much to do with how one judges effectiveness. And means reflect values as much as ends do. In any case, Canada has always had recourse to collective action at the state level for the accomplishment of economic and social objectives, while at the same time espousing an individualist ethos.

Aside from the question of reliance on collectivist solutions to social problems, the individualist/collectivist tension shows up in attitudes toward the distribution of income. The individualist ethos says that individuals are entitled to all the income and wealth that they can fairly earn in the free market; they may certainly acknowledge obligations to others—most individualist philosophers would insist on that—but such obligations are not, for the most part, decided for them by others. The individualist argues that society as a whole ends up better off through such an arrangement. Adam Smith, considered to be the pre-eminent proponent of economic individualism, entitled his great work *The Wealth of Nations*, not *The Wealth of Individuals*. The collectivist starts at the other end: the inequalities of income that result from the play of the market are unfair, unjust, and socially harmful, and our obligations to each other override any claim to high income and great wealth.

The conflict between individualist and collectivist perspectives will appear also in our discussions of policy related to health and health care. The tension is most clearly apparent, perhaps, in the ongoing disputes over the organization of and payment for medical services, where the medical profession's preference for solo practice and fee-for-service payment confronts the aspiration of provincial health authorities for tighter cost control and a planned match of services to needs. The conflict is less overt, but more significant, with respect to the more basic issue of the allocation of resources to the prevention of illness and the promotion of good health versus the treatment of sickness, discussed in Chapter Nine; that conflict unfolds in the preconceptions and assumptions of people concerning social responsibility and private access to remedial services.

The relationship between individualist and collectivist values in Canadian society is best characterized as one of tension, subject to powerful pulls in both directions, never static for any length of time. Alan Borovoy explores a range of instances of this unsettled tension in his book *When Freedoms Collide* (1988), of which the first chapter is aptly entitled 'The Freedom of Each versus the Welfare of All'.

It is necessary and important to distinguish this individualist-collectivist tension from partisan political positions, 'liberal', 'conservative', or 'socialist'/'social democratic'. Notoriously, many of the largest-scale state-operated collective undertakings in our history, for example, the transcontinental railroads, the generation of hydroelectric power, and publicly operated radio and television broadcasting, have been carried out by small-c and/or capital-C conservative governments, federal and provincial, with the strong, though admittedly not unanimous, support of the traditionally conservative, avowedly individualist elements in society (Finkel, in Panitch, ed., pp. 344–70). On the other hand, many advocates of collective solutions to social problems resist the concentration of collective power in the hands of the state, preferring local and group control (Loney, 1977, pp. 446–72; Shragge, ed., 1993), and most of them argue that *their* main interest is the enhancement of the personal development, freedom, and security of the individual. After all, it was the father of modern Communism, a powerful advocate of collective social action, who foretold the 'withering away of the State' (though Communist state regimes have never shown any tendencies to 'wither away'—not in the sense he meant).

While the convinced advocates of various forms of individualism and collectivism may cling to their positions through thick and thin, the population at large appears to swing from one to the other. Thus, in Canada, as in other industrial countries whose political institutions allow the preferences of populations to be expressed even though imperfectly, the 1960s and early 1970s are seen as an era of confidence in public remedies to social ills, and of consequent expansion of public programs. The middle and late 1980s and the 1990s are seen as an era of loss of confidence in public devices, characterized by contraction, or at least restraint, in public programs and by privatization schemes that reduce the scope of government (Ismael and Vaillancourt, eds, 1988).

(vi) The System of Government
The sixth theme is dealt with in some detail in the next section of this introduction: the interplay between social policy and the federal character of the Canadian state. Canadian federalism is not a mere accident that could have been avoided if people had been more careful; powerful natural and social forces have impelled Canada toward a federal governmental structure (Schwartz, 1974). The student of any aspect of the public life of Canada must pay attention to the influence of federalism in shaping our national experience, the more so because federalism in Canada is constantly evolving. The political tendencies of the 1990s, most prominently but not exclusively the sovereignist movement in Quebec, appear to be propelling that evolution in the direction of the decentralization of powers from the federal level to the level of the provinces. This movement coincides with a powerful tendency to reduce public expenditures, driven by the difficult fiscal situation currently facing Canada and its provinces (and a great many other industrial countries). It is a matter of debate whether the federal system favours the development of programs aimed at meeting social needs. On the one hand, the existence of ten provinces allows for more points of initiative and mobilization of support; on the other hand, the involvement of ten provinces plus the federal government, with complex political processes within each, also allows for more points of resistance (Banting, 1987).

(vii) Canada's Public Deficits and Debt
Since social change shows no sign of abating, the tasks of policy and

administration in the realm of the social services remain endlessly challenging. The greatest error would be to take the *status quo* of any particular moment as securely fixed; certainly, if there ever was complacency about the package of social programs that Canada had developed by the 1980s—the 'social safety net' of popular cliché—it has been thoroughly shaken by the cold winds of the 1990s.

A seventh challenge of a different character currently faces social policy: the financial squeeze felt by the provincial and federal governments. This challenge is of a different character because it has been created not directly by powerful social forces like the six described above, but by tax–and–spend decisions made by governments over the last twenty years or more. Those decisions, to be sure, were made partly in response to serious problems, but governments had some freedom in their choice of responses.

All our governments, provincial and federal, have experienced several years of almost uninterrupted annual deficits—that is, with the exception of a couple of provinces in very recent years, they have consistently spent more than they have collected in taxes and other revenues. The federal government has run overall deficits since 1974. As of the mid-1990s, all are consequently labouring under heavy burdens of accumulated debt.

Governments, like any borrowers, must pay interest on their debt. For much of the period during which public debt grew to its present levels, interest rates have been high; Canada's central bank, the Bank of Canada, deliberately used what discretion it had in the matter to keep interest rates high in order to prevent inflation. (We cannot enter here into the intricate and hotly debated pros and cons of this policy.) By law, the Bank of Canada operates substantially independently of the government of Canada, and certainly of the governments of the provinces. Thus, for many years, the interest payments by governments to their lenders were unusually high. In 1997, interest rates were lower than they had been for many years, but the accumulated debt is so large that the interest on outstanding debt still eats up a large percentage of each year's government revenues, and the repayment of the principal another large slice. The payment of interest and the repayment of principal are known as 'debt service', a dead weight expenditure, that cuts deeply into the ability of governments to pay for programs of all kinds, including social programs.

If incurred for constructive public investments, and if within the state's capacity to repay, public debt is entirely normal and defensible; but to borrow money at interest in order to meet each year's current expenditures is as questionable for a state as for an individual. The governments have been constrained by economic circumstances over which they had limited control: a recession in the early 1980s, a modest recovery, another recession in the early 1990s (not unrelated to the above-mentioned monetary policy of the Bank of Canada), a slow, spotty recovery—but the deficit habit reaches back beyond those recessions. In fairness to the political decision-makers, pressures to keep up spending and to hold down taxes are always strong, during good years as well as bad, but if a government both overspends and undertaxes for years on end, disaster eventually looms.

To appreciate the burden of the interest that must be paid, consider that the federal government has actually had 'operating surpluses' since 1989: the revenues collected each year have been more than enough to meet each year's program and operating expenditures. Even so, the annual interest on the overhanging debt has been so great that federal 'debt service' *has continued to grow* as a percentage of Canada's Gross Domestic Product. The grim outcome is that Canada and its provinces are now in a situation where policy decisions are being driven by the felt need to overcome deficits and to retire debt. That priority is very powerful, even in the face of conditions like the underemployment of youth and poverty among children, to be discussed later. Indeed, similar concerns with the state of public finance have been driving policies in almost all industrial countries, despite the opposition of those who claim that unemployment and poverty ought to weigh more heavily in the scale.

As to deficits, the influence of the ideas of the English economist John Maynard Keynes, enormous in the decades following World War II (and still considerable, if no longer in the ascendant) made deficit financing by governments respectable, even orthodox, as a way of compensating for economic recession. Much simplified, Keynes's idea is that government spending can make up for a slackening of private-sector demand for goods and services of all kinds, thus mitigating the unemployment that insufficient demand brings in its wake. But deficits incurred in periods when there was no such slackening of demand, or incurred because of political reluctance to reduce certain expenditures and/or to collect sufficient revenues in order not to offend certain interests, cannot be justified in Keynesian terms. On the contrary, such deficits play into the deep-seated fear that our governments will use our money irresponsibly to serve their own political interests—to stay in power, to reward their allies, to bribe the electorate with its own money.

At present, then, and for some time in the future, reductions in public program spending appear to be the order of the day in Canada, even among governments of moderate left-wing tendency. There is more than a shade of danger that the fervor for government budget-cutting will be indulged to the point where the country's ability to deal with the basic challenges will be impaired.

Spending on social programs is at least as vulnerable as other public spending—even more vulnerable where the programs' beneficiaries are a politically weak constituency, as are, for example, low-skilled, insecure workers and welfare recipients. Defenders of social programs argue vehemently that the current beneficiaries of social programs can hardly be blamed for the deficits of the past twenty-five years, and that they should therefore not be made to bear the brunt of deficit-reduction measures. Ministers of finance retort that social programs, including income security, health, and education, now account for most, by far, of our public expenditure, and that spending reductions can be sizeable only in fields where the spending is large. The urge to reduce spending has been the dominant factor in most of the changes in income and health programs since the recession of the early 1980s.

As already noted, spending is only one side of the creation of a deficit. The revenue side of public finance is equally important. Deficits can be limited or avoided by collecting sufficient revenues, mainly taxes. The combined rates of personal income tax of Canada's federal and provincial governments are by no means low by international standards, especially not in comparison with the United States. Many observers are highly critical, however, of tax policies that lighten the tax burden on substantial portions of the incomes of corporations and of high-income individuals.

As will be noted later, tax concessions made to certain corporations are often fervently defended, not only by corporation officials, but also by those who speak for the workers employed by those corporations and for the regions and communities in which they do business. Corporation profits constitute a large, visible cash flow. It is appealing to target corporations in general for the arguably insufficient taxes they pay on their profits, both the profits they distribute as dividends to their shareholders and those they retain to 'plough back' into new equipment, etc. Retained (or 'undistributed') profits enhance the wealth of their owners (shareholders), just as dividends increase their incomes. Note, however, that corporations may choose among a number of sources for the money with which to meet their tax burdens: their shareholders, whose dividends they may reduce; their customers, whose prices they may try to raise; their suppliers, whose prices they may try to lower; and their employees, whose total wages and salaries they may try to reduce. (The shareholders, incidentally, include large so-called 'institutional investors', such as pension funds and mutual funds, on whose success millions of people depend.) It would be reassuring if corporations would try to meet their tax obligations solely out of that part of their incomes that goes to the dividends of their wealthiest shareholders and the salaries of their highest-paid executives, but, to put it as gently as possible, that is not their mandate; and the governing bodies of corporations are dominated by the representatives of the major shareholders and the top executives. They are at least as likely to squeeze their workers' wages and the prices they charge consumers.

That said, many fiscal critics do find that taxes upon the profits of corporations are unduly low. They are, in fact, declining as a proportion of total government revenues (Hurtig, 1990). The corporation income tax, by the way, being a tax on profits, is in a different category from such levies as Workers' Compensation premiums and employer contributions to Employment Insurance and the Canada and Quebec Pension Plans, all of which are paid in some proportion to payroll, not profits.

There is criticism also of the overall fairness of *personal* income tax in Canada: not so much the tax structure itself, which is in principle progressive (see Appendix B, 'The Redistribution of Income', pages 283–6), but the existence of concessions that enable a certain number of high-income Canadians to avoid taxes. Every year, taxation statistics are released showing that many Canadians in the highest income strata have, quite legally, paid no income tax at all, and many others very little; the media naturally report these findings with great relish. Whatever their merits, such tax concessions do nothing to help reduce budgetary deficits or accumulated debt.

In their deficit-reduction efforts, governments certainly appear to have focused far more on cutting expenditures than on raising revenues. Since social programs represent, in accounting terms, a very high proportion of total government expenditures, they have borne what many observers regard as an undue share of the burden of deficit reduction. Aided by an unexpected lowering of interest rates in the mid-1990s, both the federal government and the provinces have had some success in reducing (even, in a few cases, eliminating) their annual deficits. Cutting back on debt will take longer, but states have almost always managed to carry a certain level of debt. What will happen in the field of social policy if and when the current wave of budget austerity has passed is, to say the least, an absorbing question.

The challenges identified here do not operate in isolation from each other; they interact. And when measures are put in place to deal with the challenges, millions of individuals and thousands of groups are free to choose how they will react, some, obviously, more so than others: their behaviour cannot be programmed. We must expect a highly complicated interplay between the themes. Action that seems good from one standpoint seems evil from another. At times one theme will dominate, at times another.

In this chapter, the underlying challenges have been expressed as generalities and abstractions, but in the real world they operate specifically and concretely. The student will probably find that the themes mean more *after* the income and health programs have been studied than before. A periodic return to this introductory chapter may prove rewarding.

SOCIAL WELFARE AND THE CANADIAN SYSTEM OF GOVERNMENT

Before we begin looking at public programs in income and health care, a few considerations of a broadly political nature must be mentioned.

(a) Our political philosophy is such that our governments are constrained by *the rule of law*.

(b) Canada is a *parliamentary democracy*, which gives a specific character to the relations between our elected legislators and the executive agencies of government (very different, for instance, from those that prevail in the congressional-presidential system of the United States, with whose processes the media make Canadians so familiar).

(c) The Canadian state operates through a *federal system*, by which governmental powers and resources are distributed among two levels of government; in the present decade, the federation appears to be facing the greatest threat to its survival since its inception more than 125 years ago.

(d) In recent years we have further constrained governments by the adoption of *codes of human rights*, at the provincial and federal levels, that say that certain actions, and actions having certain consequences, are beyond the normal powers of governments.

Each of these characteristics calls for some discussion.

The Rule of Law

Legislated public services in the income and health fields are intended to address problems that are serious and frequently urgent. Sometimes the provisions available under statutory programs do not meet the needs of a given case, and it would appear obvious that something more, or something different, ought to be done. But public authorities may do only what the law allows them to do. At times, they offend by overstepping what the law allows, at other times by failing to do what the law directs. We hold them responsible for acting within their lawful authority, even though something else may look like a good idea at the time; private judgement, however wise, may not be substituted for law. After all, experience amply confirms that public officials are just as likely to overstep their authority in ways we dislike as in ways we like. Besides, the money spent to carry out programs is *our* money, and it is basic to our political system that governments (i.e., public servants) dispose of public resources only as authorized by legislatures, that is, by the elected representatives of the people.

In real life, the rule of law is subject to many far-reaching qualifications, which concern students of politics:

- The laws passed by legislatures typically leave important details to be settled by 'Regulations' that are promulgated by designated authorities, usually the Cabinet. In legislation and in formal documents, the federal Cabinet is called 'the Governor-General in Council'; at the provincial level the Cabinet is 'the Lieutenant-Governor in Council', except in Quebec, where the simple term 'the Government' is now used. Other than the Cabinet, a designated minister of the government, or a specific body, such as the Canada Employment and Immigration Commission or a provincial Workers' Compensation Board, may be given the power to make regulations. Such regulations have the force of law. They are usually written according to the advice of bureaucrats and expert advisers, without much real scrutiny from elected representatives.

- The vigour with which an existing law is enforced may be affected by the political and partisan preferences of governments.
- The administrators who implement the laws are influenced by the organizational interests of the bureaucracies in which they work.
- The great advantage that government officials usually have over the ordinary citizen in terms of access to relevant information, knowledge of procedures, and so on, may mean that the service provided is not exactly what the legislators intended.

Under all this, it may at times be difficult to detect the workings of the rule of law as the 'expression of the will of the majority of the elected representatives of the people'. Even so, the rule of law prevails in principle, as a necessary defence, however flawed, against the dangers of the arbitrary exercise of power by governments and officials. And we cannot have it both ways: to cling to the rule of law when it suits us, and to dispense with it when we find it awkward.

The Parliamentary System

The parliamentary system of government requires that the control of the executive machinery of government be in the hands of persons (the ministers, known collectively as the 'government' or the 'Cabinet') who have the continuing support of a majority of elected legislators. If the government, so defined, is to be held accountable to the electorate, individual members of the majority parties in our legislatures, federal or provincial, have to follow the policies proclaimed by their party leaders as much as, probably more than, they follow their own judgement or the views of their electors. Those legislators who support the government of the day have virtually no freedom to vote or speak against their government's policies; when a member of the government party does so, he or she pays a heavy price for breaking ranks. Elected members can of course influence the policies adopted by their parties, but even opposition party members have limited freedom to disagree publicly with their 'party line'.

Ministers and governments are held accountable both for the content of legislation *and* for its administration. To satisfy this accountability, the government of the day has to rely heavily on experienced bureaucrats for advice, perhaps more than on elected representatives.

These characteristics of parliamentary government affect both the content and the delivery of social programs.

Federal and Provincial Jurisdictions

(i) Canada's Federal Constitution

Canada, a parliamentary democracy, is also emphatically a federal country. This complicates the question of legislation and administration of social programs, because many programs are provincial, some are federal, and some involve both levels of government (Strain and Hum, in Ismael, 1987).

The essence of Canadian federalism is that 'sovereignty' is *divided* among two levels of government; one level is *not* superior to the other. (In time of war and in other crises, the federal government may invoke emergency powers, but that is clearly exceptional.) To avoid chaos, and to maintain the accountability demanded in a democracy, there has to be an understanding as to which level of government may do what. It is seldom easy, however, to draw a clear line distinguishing the responsibilities of the provincial and federal governments; and even if it were possible, a dividing line that at one moment looked clear and reasonable to almost everyone might later look neither clear nor reasonable to anyone.

Our understanding of the division of sovereign powers is based on the words of what we now call our Constitution Act, known until 1982 as the 'British North America Act, 1867, as amended' (the BNA Act). The Constitution Act of 1982 removed the last vestiges of British involvement in Canada's constitutional affairs; the Constitution was 'repatriated'. When enacted by the British Parliament in 1867, the British North America Act listed most of the matters to be dealt with by the federal Parliament in Section 91, and most of those to be dealt with by provincial legislatures in Section 92. This formal division of powers is still the skeleton of Canadian federalism.

An understanding of social policy in Canada requires an understanding of federal and provincial jurisdictions. The picture is quite complex. Among the responsibilities of the provinces, Section 92 lists charities and charitable institutions, hospitals, and municipal institutions. Such institutions administered all the rather scanty existing public responsibilities in social welfare and health care in the colonies that came together in 1867 to form the Dominion of Canada. Accordingly, *'welfare' and 'health', in general, are authoritatively considered to be within the domain of the provinces.*

Likewise, Section 92 gives to the provinces jurisdiction over 'property and civil rights', which is considered to extend to the regulation of most contracts between persons. Employment, a subject intimately related to welfare, is a form of contract. *The regulation of employment is therefore primarily within provincial jurisdiction.* This has important consequences for welfare. (Also within provincial hands is another socially sensitive field, the regulation of housing.) As will be seen, the federal government does have some jurisdiction with respect to employment. And in health and welfare, as in other fields, the federal government contributes financially to provincial programs, and therefore can exert a great deal of influence; but like many things, this too is changing (see the discussion of the 'spending power' on pages 21–3).

Constitutional reform, including agreement on the division of powers, has been a major preoccupation in Canada for the last two decades or so, although both the public and the political actors appear to have lost their appetite for it after two exhausting failed endeavours. The first, in 1990, was an intergovernmental initiative known as the Meech Lake Accord, which needed unanimous approval by provincial governments but failed to obtain the consent of two (Manitoba, by a very close call in the legislature, and Newfoundland). The second, in 1992, was a Canada-wide referendum, in which a reform proposal entitled the Charlottetown Accord was decisively defeated.

Writing in 1997, it is impossible to ignore the threat to the survival of Canada-as-we-know-it posed by the movement for sovereignty in Quebec. Quebec embraces about a quarter of the nation's population; and 75 per cent of its people claim French ancestry (Statistics Canada, 1993). The French-speaking culture of which Quebec is the home put down roots in the 1600s, and was well over a century old before French Canada became British by conquest in 1763. That cultural identity, through all the social and political evolutions since 1763, has remained strong, and since 1867 (and even earlier) has remained central to Quebec's aspirations as part of Canada. French-Canadian leaders have generally been disposed to think of Quebec as representing one of two founding partners in the Canadian state, rather than as one of four (in 1867) or one of ten provinces in the federation. Quebec has always embodied the most serious cultural challenge, therefore, to the integrity of Canada as a nation.

The Quebec referendum of 1995, carried out as promised by the avowedly separatist provincial government that was elected to office in 1994, rejected by an extremely narrow margin a proposal in favour of sovereignty for Quebec. If approved, it would have led, at the least, to a radical change in the relationship between Quebec, as one entity, and Canada as another, or, at the most, to outright separation and independence (not overnight, but eventually). No one expects the issue of Quebec sovereignty to lie quiet. A proposition similar to that of 1995 appears likely to be put before the people of Quebec once again. What is written here about federal-provincial jurisdictional questions, as far as Quebec is concerned, and about Quebec's involvement as a province in a number of financial and other arrangements with the federal government in the field of social policy, must be read against that background.

In the meantime, the strains posed by Quebec have influenced the constitutional positions taken by other provinces and by the federal government, and many observers regard as likely a much enhanced *de facto* decentralization of powers to the provinces. Few provincial (or other) politicians are inclined to turn down the promise of increased powers, but to exercise powers a government must have the necessary resources. Except for Quebec, those provinces that have relied most heavily on the federal system for financial support are rather ambivalent about extensive decentralization; the more financially self-sufficient provinces, which have been less dependent upon transfers from Ottawa, tend to accept decentralization willingly. (In financial terms, Quebec does benefit heavily from fiscal transfers from the federal government, but for the above reasons Quebec is the most 'decentralist' of all the provinces.)

Canadians can anticipate more of the wrangling over federal and provincial responsibilities that has been referred to irreverently as the nation's favourite indoor sport. This is not the place for extensive discussion of this critical issue; however, its relevance to social policy is obvious.

(ii) Amendment of the Constitution

If for some reason some part of the allocation of powers between levels of government appeared to be unsatisfactory, the natural recourse would seem to be to change the Constitution. But until 1982, there was no agreed-upon procedure for such amendments. Despite this lack of clarity, a number of amendments have taken place. (Quebec, it must be noted, never gave its assent to the Constitution Act of 1982.)

The BNA Act was an Act of the Parliament of the United Kingdom and the British Empire. Until as recently as 1982, it could be changed only by an Act of that Parliament. As Canada fairly rapidly asserted its autonomy vis-à-vis Great Britain, the British Parliament came to make changes in the BNA Act only when asked to do so by the Parliament of Canada, and as a rule did so without much further discussion. When the contemplated change involved a change in the powers of the federal and/or provincial governments, it was felt that the Parliament of Canada could hardly request an amendment without some indication of consent by the provinces. Unfortunately, no one had foreseen the need for an amending formula involving the provinces; the BNA Act included no rules by which to proceed (as does, by contrast, the Constitution of the United States). In the face of such uncertainty, the only secure way to proceed was on the basis of unanimity, so all amendments to the distribution of powers to date have had the consent of all the provinces.

It is interesting that the most, if not the only, significant amendments to the distribution of powers have been in matters related to social welfare, notably Unemployment Insurance (Section 91, subsection 2A, added in 1940) and old age pensions (Section 94A, added in 1951 and further amended in 1964).

The procedure for amending the constitution in future is set forth in Part V of the Constitution Act of 1982. This amending process has never been used. As noted, Quebec has never formally consented to it, nor has it enjoyed enthusiastic support across the country. In the aftermath of the 1995 referendum in Quebec, it would be rash indeed to predict what the future holds for formal amendment of the Constitution.

(iii) Judicial Interpretation

When the words of the BNA Act, as amended, left some question of jurisdiction in doubt, Canadians, following the rule of law, took specific cases to court. The leading cases were not typically brought to court on the basis of convictions about the nature of federalism; usually, the litigants were interested only in challenging the constitutional validity of legislation that had affected their interests adversely, whether it was federal or provincial. In our system, the definitions provided by the highest courts in dealing with actual cases are accepted as final. Within the limits of the issues raised in those particular cases, the judicial decisions determine the meaning of the law. In this sense, the Constitution means what the judges say it means.

It is important to understand that a law will be ruled invalid by the courts, no matter how wise they think it is, or how popular it is, if they consider it to be beyond the power (*ultra vires*) of the legislature, federal or provincial, that enacted it. In certain instances where Canadians apparently have not liked the meaning given by the courts to the distribution of powers of government in the Constitution, they have gone through the unwieldy process of amending the Constitution (e.g., in the case of unemployment insurance; see Chapter Three).

In addition to judicial decisions arising out of particular cases, our law allows a so-called 'reference' procedure, by which a *government*, not an individual, may ask a court for a judgment as to the validity of a proposed law, before it is put into effect. A province may refer such a question to its own provincial supreme court or the equivalent; from such judgments, appeals may be made to the Supreme Court of Canada. The government of Canada may refer questions directly to the Supreme Court of Canada. About one-third of significant judicial decisions on constitutional matters have come from 'references', the others from specific cases.

(iv) The 'Spending Power'

It has been stated that the provinces have jurisdiction in the fields of welfare and health. Yet the federal government is involved in many welfare and health programs. In fact, in the popular mind there may even be a tendency to identify our hospital and medical care insurance with the Canada Health Act and our public assistance program, until 1997, with the Canada Assistance Plan (now superseded by the Canada Health and Social Transfer). Both of these impressions would be wrong. Neither of those federal Acts creates a service-providing *program*; without the necessary provincial legislation, no medicare and no public assistance would exist. But the federal government is certainly involved in the financing of health services and welfare provision; and this federal involvement has happened *without* constitutional amendments changing the allocation of powers. How has this come about? The answer may give some discomfort to anyone who expects legal and constitutional doctrines to be clear and straightforward.

Over the years, needs have emerged, and/or policies have been proposed, in such provincial domains as health, education, and welfare, whose financial demands were such that the provinces felt unable to meet them at all, or were very unequally able to meet them. At times the policy needs have been expressed by the provinces, at other times by the federal government. In all cases, so goes the argument, the provinces cannot provide the desired services without the financial help of the federal government;

but the language of the Constitution has stood in the way of federal involvement. To overcome the constitutional barrier, appeal has been made to the implicit power of the federal Parliament to spend its own resources in ways not explicitly forbidden.

This so-called 'spending power', nowhere mentioned by name, is located by sympathetic experts in the wording of Section 106 of the Constitution Act, which allows the 'Consolidated Revenue Fund of Canada' to be 'appropriated by the Parliament of Canada for the Public Service'. This seemingly innocent expression is the indispensable support of some of the most important governmental programs in Canada, including provincial programs in public assistance and in hospital and medical services, and the (now defunct) federal Family Allowances program. The 'spending power' is a vital ingredient in the glue that holds Canada together. The late Frank Scott went so far as to say, 'It is difficult to see how this country could be governed if this power were to be denied' (Scott, 1964, p. 29).

Scott acknowledged that the use made by the federal government of the 'spending power' has given rise to some controversy. In general terms, the question is this: When the federal government gives the provinces money to fund programs that are within provincial jurisdiction, but in doing so imposes conditions that necessarily affect what the provinces are free to do, is the federal government actually imposing its preferences in areas that are the provinces' business? Is it treading on the provinces' field of jurisdiction, under the guise of helping to pay their bills? Scott put it benignly: 'Ottawa learns to induce where it cannot command, *and federal policy* is made by bargains with provincial governments' (p. 28, italics added). Peter Hogg, in his excellent treatment of the subject, concludes that '. . . the sum total of these programmes amounts to a heavy federal presence in matters which lie within provincial legislative responsibility' (1996, ch. 6, section 6.8).

The ordinary citizen may take these constitutional niceties rather lightly, weighed against the palpable benefits of cost-shared programs. A word-mincing legalism certainly ought not to obstruct the pursuit of important national objectives not attainable without some looseness in the constitutional joints. On the other hand, in a constitutional democracy, it is essential that the Constitution be respected. Undeniably, the resort to the vaguely articulated 'spending power' has enabled Canada to mobilize national resources, notably in the creation and delivery of several extremely significant social programs, while preserving the structure of federalism (Strain and Hum, in Ismael, 1987, pp. 358–9).

As social and economic factors and political preferences change, the Canadian federal system adapts. Sometimes the tide of power flows toward the central government, sometimes toward the provinces. In the conventional history, Canada experienced a long centralist wave from the beginning of World War II until the early 1970s, and has been moving in a decentralizing direction since the late 1970s, much accelerated, it would appear, in the 1990s. Provincial reservations about the use of the spending power are expressed more stridently lately than in the past. And, as shall be seen, the evolving pattern of federal cost-sharing (more costs, less sharing) may reduce the federal government's leverage over the programs it subsidizes (Caledon Institute, 1995). Perhaps as a way of recognizing this evolution, the federal government included in the Throne Speech launching the 1996 session of Parliament a statement to the effect that it would not, in future, use the federal 'spending power' to initiate programs in fields of primarily provincial jurisdiction.

In the Constitution Act of 1982 there appears a significant clause, Section 36(2), that recognizes the disparities in material wealth between provinces and purports to commit the federation to the principle of equalization:

> Parliament and the Government of Canada are committed to the principle of making equalization payments to ensure that provincial governments have sufficient revenues to provide reasonably comparable services at reasonably comparable levels of taxation.

The section is phrased so cautiously that a dissatisfied province would probably have great difficulty in making use of it to get any specific relief (Mendelson, 1993a, 1993b). Be that as it may, as a principle it restricts further recourse to the 'spending power'; for if the provinces were in fact provided with 'sufficient revenues' in general, there would be less need for federal cost-sharing in support of specific programs, and less occasion for federal enforcement of common standards.

Great gaps from province to province in service provision might well strain the ties that hold Canada together. Much of the debate about the Constitution concerns the question of whether it is necessary to leave in the hands of the federal government sufficient powers to sustain a unified Canada by assuring adherence to national standards of public services, and, if it is necessary, how it can be done. Among the service standards most often referred to in the debate are those concerned with income security, employment, and health care—because these are now, along with education, by far the largest items in government program expenditures, and because they dominate the field of intergovernmental transfers. The sharing of resources, whether via the 'spending power', equalization payments, or in some other way, is an important part of any constitutional understanding.

Federal and Provincial Taxation Powers

Governments have responsibilities; they need resources with which to carry them out. The most important financial resource of governments is the power to tax. In Canada, the federal and provincial governments have different taxing powers. To simplify greatly a field whose complexities preoccupy legions of accountants, lawyers, and scholars, the difference is this:

the federal government may use *any* kind of tax to raise money; the provinces are limited to so-called *direct* taxation.

By direct taxation is meant a tax that must be paid directly and entirely by the taxpayer from whom it is collected—that is, the taxpayer is not able, in theory, to pass it on to someone else. The personal income tax is perhaps the most obvious example of a direct tax; when you pay your income tax, you have no immediate way to pass on the burden to anybody else. Accordingly, our federal and *all* our provincial governments rely very heavily on personal income taxes for their support. (Many Canadians may be unaware of this, because all the provinces except Quebec let the federal government's Department of National Revenue collect their income tax for them.) Similarly, a tax paid by the consumer of a good or service at the moment of purchase is a direct tax, because there is no one to whom the consumer can shift the burden of the tax; and all the provinces collect such taxes. On the other hand, a sales tax on items used in the course of business is an *indirect* tax: an operator of a business or a professional who pays a sales tax on machinery or services naturally passes on to customers as much of the tax as possible. The provinces may not collect taxes that allow such 'passing on'; the federal government may, and does (the much-maligned Goods and Services Tax).

This division of taxing power limits to some degree the capacities of provincial governments to carry out their responsibilities. Partly to overcome this, important arrangements have been made by which the federal government subsidizes the *general* operations of provincial governments and in addition shares with them the costs of certain *specific* programs (resorting to the 'spending power' discussed above). Entirely apart from this difference in formal taxing powers, the federal government would undoubtedly have been called on to subsidize the provinces, and to share some costs with them, simply to help make up for the inequalities in wealth between provinces. Equalization payments have been part of Canada's public finance system since Confederation, and have been enshrined in the 1982 Constitution Act, as we have seen. The nature of other intergovernmental transfers has changed from time to time; they are undergoing significant change at present as the federal government has shifted from contributing definite proportions (usually 50 per cent) of the costs of specific programs, to a pattern of 'block' grants not tied to provincial expenditures at all.

A government that contemplates a large spending program must also contemplate collecting the revenues to pay for it. The reactions of taxpayers must be anticipated. At the limit, some potential taxpayers may seek to evade the tax burden by leaving the territory. This is a possibility especially for businesses whose capital is fluid, not tied to one location. A provincial government might hesitate to thus put itself at a competitive tax disadvantage with other provinces. Whether the danger of losing taxpayers is serious or not, governments may point to the threat of tax flight as a reason for tax restraint. Obviously, to the extent that a program is financed out of federal tax revenues, interprovincial tax competition is blunted. International tax competitiveness, especially with the United States, is often raised as an issue at the federal level.

Constitutional Restrictions on the Powers of Governments

The Canadian Charter of Rights and Freedoms, which subjects all legislation and governmental administration—federal, provincial and, by extension of the latter, municipal—to possible judicial review, has been entrenched in the Constitution since 1982, which means that it can be changed only by the designedly difficult procedure of constitutional amendment.

The Charter of Rights and Freedoms has now been endorsed as part of the Constitution by all the provinces (including Quebec, *de facto* if not *de jure*). Its terms apply to all governments in Canada, federal, provincial, and territorial, and to their subordinate governments, such as municipalities and school boards. The effect of the Charter has been to add a significant component to the public policy process. The Charter imposes *absolute limits* on the powers of all governments. This is very different from the limits placed on federal and provincial governments by the fact that we have a federal constitution; the federal-provincial division of powers does not assume that there are things that neither level of government may do. In certain areas, there are now some things that neither the provincial legislatures nor the federal Parliament may normally include in legislation, even though supported by the majority of the elected legislators, and, as far as can be told, by a majority of the population (although all legislatures are provided with a notorious escape hatch known as the 'notwithstanding' clause). And in administering laws that are valid in themselves, there are now certain things that governments may not do.

The words of the Canadian Charter cannot possibly cover all cases clearly and unambiguously. Where there is some doubt as to whether some piece of legislation or some administrative action of a government is allowable under the Charter, we resort to the judicial process: we take a specific case to court, and we accept the interpretation of the Charter handed down by the highest courts. To this extent, the judicial process has replaced, or become part of, the political process; within a limited range, the judiciary has been made, in fact, supreme over the legislature. If, after all this, there is widespread dissatisfaction with the way the courts interpret parts of the Charter, it is possible to have the Charter amended. But a change in the Charter would require a constitutional amendment—and, as stated above, amending the Constitution is difficult.

Human rights legislation exists at the provincial level as well. The National Assembly of Quebec (i.e., the legislature of the Province of Quebec), for example, enacted the Charter of Rights of the Person in 1975, and other provinces have similar laws: Alberta's Individual Rights

Protection Act, Ontario's Human Rights Code, etc. Since the federal Charter applies to provincial legislatures and governments as well as to the federal, why do we have charters at two levels? The answer is twofold. First, the provincial rights statutes cover some matters that are specifically within provincial jurisdiction, not covered in the federal Charter. Second, the federal Charter applies to *legislation and to governmental actions*, so its impact on activities in the private sector is highly attenuated; the provincial human rights acts, on the other hand, reach into private affairs. Of course, there are hundreds, if not thousands, of statutes and regulations governing private behaviour, most of which bring some governmental enforcing body into the picture, making the Charter potentially applicable; as Peter Hogg says, 'Without this understanding, the claim that the Charter does not apply to private actions would be grossly misleading', even though literally correct. This second reason is exceedingly important. In the eyes of some, the 'sphere of the private' is precisely 'where all the truly bad stuff goes on'— violations of rights in relationships between private parties, for example, in employment, housing, and access to various facilities [Hogg, 1996, ch. 34, section 2(g)].

Only the federal Charter creates *constitutional limitations* on governmental power. A provincial human rights code does not have constitutional status, being a statute passed by the provincial legislature, subject to ordinary amendment by the same legislature. The provincial codes and charters do have, however, significant political and symbolic force. No provincial government is likely to propose a watering down of its provincial charter except on exceedingly firm grounds.

Thousands of cases involving clauses of the federal Charter have come before the courts. A series of bulky annual volumes is devoted to legal reports of Charter cases. Among them are many cases directly concerned with the administration of welfare and health care programs, and with the content of social legislation. The effect of the Charter cannot be assessed by the number of cases taken to court under its provisions: provincial and federal governments now shape their actions so as to meet the Charter's demands, thereby avoiding litigation in the first place. Predictably, the Charter has become very important in social policy in Canada.

THE INSTITUTIONS OF SOCIETY

Reference was made above to constraints imposed on policy by the major institutions of society. Government necessarily looms large in social policy, but we must constantly remind ourselves that government is only one of the institutions to which society assigns essential functions. The word 'institutions' in this sense does not mean identifiable corporate bodies like the University of Toronto, the United Church of Canada, Le Club de Hockey Canadien, the Canadian Congress of Labour, or IBM Inc.; it means the *patterns established by our society for the carrying out of socially important functions*, for example, the family in the care and socialization of children, parliamentary democracy in government, the relationship of employee to employer in the field of work, the relationship of professional to client in the field of certain expert services, notably medical and legal. Among the most significant institutions of our society is the corporation, which has become our principal vehicle for the ownership of the means of production (not only the 'private ownership of the means of production'—the corporation form has been greatly favoured in Canada as a vehicle for public ownership as well, notably in the communications, transportation, and public utilities industries). There are other forms of ownership of business enterprises—sole ownership, partnerships, co-operatives—but the corporation form is most important. The organizations that employ people play a large role in the delivery of many services that are crucial to social policy in Canada, and their role seems destined to expand in future as governments appear willing to let their roles diminish.

Other focal institutions in society are the institution of private property, including private ownership of much of the means of production; the market, meaning the way in which we allow most economic activity to be determined by more or less free exchanges among buyers and sellers (this includes the *labour market*, in which work is exchanged for incomes); and organized religion, still enormously influential despite recurring reports of its decline. Institutions of this kind largely constitute our social environment; they direct our attention to certain problems rather than others; they influence our choices among available solutions; they bias our judgement of how large a part of our resources we should, can, or will devote to the improvement of particular conditions (Armitage, 1996, ch. 7).

To illustrate the importance of social institutions in social policy, consider the way in which our understanding of 'the family' and of 'employment' has affected the ways we look at and deal with some of the burning social questions of the day: unemployment, the socialization of young people, the way formal education is conducted, the experience of women at work, poverty and deprivation, the determination of eligibility and benefits in public assistance, retirement, the financing and the benefits provisions of public pension plans, the care of young children. It is all too easy to take our particular social patterns as the *natural* way of ordering behaviour, and to forget that all such patterns are social artifacts, to which there are real alternatives.

Members of a society operate within its institutional framework much as fishes swim in water—barely aware that it is there, but unlikely to survive if pulled out of it. Our social institutions lead us to take for granted certain things as relevant, and certain things as good. It is important for the student of social affairs to challenge the conventional wisdom as to both relevance and goodness. Real understanding requires that one step back and take a careful look at the standards of right and wrong, of important and unimportant, that underpin the society in which one has grown up. Some will see the need for much change, some for little. What matters is that one exercise the responsibility to look critically.

As was true of the 'major challenges' raised previously, such general considerations as these probably will mean more after the reader has looked at some concrete instances to which they apply, whereupon a re-reading of this section may be advisable.

THE UNDERLYING POINT OF VIEW

The Acton Ideal

Intended primarily for the use of students who are approaching the study of social policy and social programs perhaps for the first time, this volume has been written to be compatible with a wide range of points of view about the nature of society. The author has not set out to defend any particular point of view; on the contrary, the conscious objective has been to push the student to make his/her own analysis and to arrive at his/her own position. But three caveats must be recorded.

The first is that the bulk of the text concerns specific programs, all of which are *explained*. There are reasons for the specific features of programs (the content of legislation, revenue sources, rules for eligibility for benefits, administrative structures), and these reasons are presented so that one may understand why the programs are the way they are. Critical questions are raised here about each program, but if the explanations given are persuasive, readers might be led to believe that the programs are being defended rather than explained. Such is not the intention.

The second cautionary note has already been sounded: that social programs are most often conceived in response to recognized problems, with some degree of urgency. There is a considerable difference between a *remedy* to alleviate the symptoms of a problem, and a genuine *solution* to the problem. (Un)Employment Insurance is no doubt of great help to many people who lose their jobs; but it is not, cannot be, and was never envisaged as a solution to the problem of unemployment. Workers' Compensation clearly helps many workers who suffer injuries on the job, but it does not contribute directly to the improvement of work safety. The reader should be alert to this distinction between remedies and solutions.

The third caveat has to do with the possibility or impossibility of a bias-free presentation of such matter as is dealt with in this book. It is generally agreed that no writer can be entirely free of preferences and inclinations, even, perhaps, of bias and prejudice. But there is much disagreement as to what to do about it. We will leave out of consideration those who would wilfully distort the truth in order to advance a cause. One camp argues that the only honest thing to do is to declare one's position, however partisan, and then give it free rein. The opposite camp, defined by Lord Acton, the great nineteenth-century English liberal, estimates a writer 'very much less by his own ideas than by the justice he does to ideas that he rejects—not for his national, his religious views, but for his appreciation of nations, religions, parties not his own' (Acton, 1961, p. xix). In a textbook, Acton's ideal is the more appropriate, even if imperfectly realized. It is hoped that the reader will be equipped both factually and conceptually to reach an independent opinion.

The assumptions and predilections underlying this text may be characterized as *liberal, pluralist,* and *interventionist*—liberal in the sense that individual freedom is given primacy as both morally sound and socially efficacious, while recognizing that significant collective (mostly state) interventions are necessary to protect the moral primacy and the welfare of individuals. The reader may judge whether the presentation of material has been biased thereby.

Canada's economic system is a mixed capitalist one. There is considerable public enterprise, and considerable public regulation of private enterprise (despite recent professions of enthusiasm for 'deregulation', often opposed by those who are about to be deregulated). But the private ownership of enterprise remains central, and public intervention has been intended at least as much to enhance the performance of the private sector as to restrain it. Thus the public programs of income maintenance and security and health care provision reviewed in this volume are seen as complementing rather than as subverting the prevalent economic system. The object of this text is to be equally useful to all students, whether they judge the mixed capitalist system to be on the whole good, bad, or somewhere in between.

The 'explanations' of social policy and programs in this book are based on a mixture of theories.

(a) The minimal maintenance of an industrial society absolutely requires some provision for repair of the human wear and tear that inevitably takes place; therefore, a good deal of weight is given to so-called 'functional' or 'convergence' theories, according to which all modern societies (capitalist, socialist, totalitarian) do what they have to do in order to keep moving. In observable practice, they do many similar things, despite their ideological differences (Burawoy, 1985; Rimlinger, 1971).

(b) 'Convergence' theories cannot explain everything. Decision-makers in any society may see the 'functional', 'convergent' demands differently. Leaders must make choices as to which of the demands are to get higher priority, and also as to the means to be chosen to pursue them. These choices are influenced by values derived from a great number of influences—religious, ethnic, cultural, historical, even climatic—whose interacting impacts are problematic.

(c) The variety of these influences poses a severe challenge to theories of economic, or any other, determinism. In a determinist perspective, some identifiable crucial variable is seen as determining history: not every detail, but certainly the broad sweep.

Among determinist ideologies, the most widely received in intellectual circles are the several world views grouped under the label of Marxism. Whatever their differences, Marxists are united in asserting that the class relations that ensue from private ownership of the means of production ('capital') largely determine the course of history in capitalist societies, up to the predicted evolution/revolution which leads them into socialism/Communism. (Marxists are not alone, of course, in considering the ownership of the means of production to be important.) The histories of capitalist nations have, however, been exceedingly diverse: some have become and remained democratic, some not; some (few) have experienced genuine socialist or Communist revolutions, most not (and in recent decades, some of those revolutions have been reversed, with mixed results); some capitalist nations have proceeded relatively quickly towards planned economies, some have continued to rely more on the processes of free markets.

These differences in the experiences of capitalist societies may be explained largely in terms of the different degrees of power exerted by actors and interest groups espousing different preferences. If any group or class dominates, it does so from issue to issue in such different fashions that variables other than economic group or class must be allowed much weight.

The ideologue of left or right argues that such differences are a superficial political façade that hides the actual motives and the truly important doings of the powerful. A pluralist view of politics puts more faith in the observable evidence, and less in assumptions about what is *really* going on behind the scenes.

To illustrate the difference between the determinist and the pluralist views, consider the question of the origins of such social reforms as unemployment insurance, old age pensions, and medicare. They all represent the outcomes of lengthy political struggles, including some episodes of violence and some surprising use of the state's power to compel, for example, compulsion imposed upon the medical profession in Saskatchewan and Quebec (Taylor, 1988, pp. 307–27, 409–10). A Marxist is committed to the belief that capitalist society is dominated, by and large, by one particular class, the owners of capital. Logically, then, these and other reforms must have been at least grudgingly accepted by that dominant class, if only for strategic reasons. But many elements of the presumed dominant class stubbornly resisted, even though the reforms can be and were argued to have been in their interest. At the same time, the subordinate classes often have failed to unite behind such reform measures, which *apparently* are favourable to them. Can this only be because the dominant class succeeds in deceiving the subordinate classes as to where their own true interests lie (the Marxist concept of 'false consciousness')? Or did they have good reasons of their own?

(d) To questions of this order, a reasonably coherent ideology strives to provide answers. But the answers are seen to vary sufficiently that any ideological camp sooner or later takes on a pluralist appearance, in spite of itself. This is certainly true of Marxism, thanks in part to the richness and variety of the analyses Marxism has evoked from intellectuals world-wide; but it also is true of fascism, less respectable intellectually, but politically alive and well throughout our battered century. For another thing, doctrinal answers to awkward historical questions invariably require elaborate rationalizations of such evidence as we have. That these rationalizations are convoluted does not mean that they are wrong. But a lot is asked of the believer, who must proceed on strong, usually unflattering assumptions about the consciousness and the motivations of various people who appear to be on the other side, whether they be politicians, business managers, public administrators, judges, journalists, labour leaders, even professors— assumptions that may be *plausible* but are not *demonstrable*.

Less strenuous assumptions about the consciousness of protagonists in social issues are required by a point of view that sees outcomes as much less rigidly determined by any single factor, such as the class structure. A pluralist power analysis readily acknowledges that the forces in any society are unequal, and that some interests will have more power, and will get their way more often, than others; but it expects that the powerful will often differ among themselves (as they appear to do), that the less powerful can sometimes actually win a round or two by uniting and/or by forming alliances with some of the powerful (again, as they often appear to do), and that, in any case, power itself is multidimensional—power in one arena may not be portable to another.

Likewise, pluralist assumptions about human nature and motivation are not strained. As a general rule, the less one must assume, and the more one can show, the better. It is assumed, for instance, that people of all classes are motivated largely by self-interest—but self-interest, like power, is multidimensional, having different meanings in different circumstances. Perceived self-interest sometimes motivates self-sacrifice.

(e) Finally, for the purposes of an introductory text in the field of social policy, the compelling argument for a pluralist critical stance is one made familiar a generation ago by such writers as Robert Dahl, Charles Lindblom, and Karl Popper (Dahl and Lindblom, 1953; Dahl, 1967; Lindblom, 1965, 1977; Popper, 1961, 1966). If any particular determinist picture of society is in fact correct, then the evidence gathered on pluralist principles ought to show it to be correct; pluralist assumptions about the nature of knowledge do not rule out determinist conclusions, if the evidence is sufficient.

The opposite is not true. Determinist assumptions *do* rule out a pluralist conclusion. In politics, for instance, the pluralist will acknowledge that the more powerful interest is always more *likely* to prevail if it really tries, but can be persuaded of the contrary by contrary evidence in a given instance; the ideologue would insist that the social structure dictates whose interests will prevail in a given social contest—and evidence seemingly to the contrary must be re-interpreted to conform with the determinist assumption.

Charles Lindblom himself illustrates this point well. Proceeding on liberal pluralist assumptions about human nature, society, and the interpretation of evidence, he reaches many conclusions agreeable to most Marxists, as in his *Politics and Markets* (1977); but he regards the epistemological assumptions of many Marxists (e.g., 'false consciousness') as so much excess baggage.

Students are better served by an analysis that leaves room for a broader rather than a narrower range of conclusions and encourages them to judge for themselves.

Approaching the Subject From the Operating End
Something remains to be said about the value to students of a treatment of social policy that focuses, as this book does, on the operating details of programs—the nuts and bolts of social policy. It is more common to approach the subject with a thesis in mind about the structure of society. The program details are then seen as relatively less significant and interesting in themselves, analytically useful principally for the support they lend, or the challenge they pose, to the underlying thesis. For purposes of this text, the angle of vision is almost the opposite of this.

It is assumed that the character of a program can be meaningfully deduced from its own internal structure. If unemployment insurance does, in the final analysis, serve more than anything else to smooth out some rough spots in the functioning of the capitalist labour market, students will be able to see that for themselves from the structure of unemployment insurance and the way it has changed over the years, so long as the program details are seen in their context rather than as a tangle of disconnected rules and procedures. That some characteristics of unemployment insurance may appear clearly to fill that system-maintaining function, while others may appear to have an opposite tendency, is both an illustration of the complexity of social policy *and* a test of various possible points of view about the real functions served by social programs. The nuts and bolts can tell a great deal about the machine of which they are part and about the society that needs and/or develops such a machine.

Second, programs may be thought of as experiments, and program changes as changes in the variables (Campbell, 1969). As in other fields, the examination of specific evidence—what happens when you do this? and what happens when you do that?—is one good way to test general ideas about social policy and about society.

Finally, assuming we did locate and remedy the flaws in our society, it is too much to hope that the need for social programs would disappear. New programs would be needed, and they would require many of the same tools as do our current social programs.

BIBLIOGRAPHY

Since this book is primarily addressed to students, all bibliographies are briefly annotated. It is not suggested, of course, that they are exhaustive.

Acton, Lord (1961). *Renaissance to Revolution: The Rise of the Free State.* Introduction by Hans Kohn. New York: Schocken Books.

Alberta. *Act for the Protection of Human Rights.*

Armitage, Andrew (1996). *Social Welfare in Canada Revisited.* 3rd edn. Toronto: Oxford University Press. A helpful review of relevant concepts and concrete issues, strengthened by the author's extensive experience in government. See especially Chapter 7, 'Politics, Power and Organizations', which gives a broader and more opinionated view of the whole range of organizations, public and private, active in the field of welfare in Canada, than is given here, and Chapter 8, 'The Discipline of Social Policy', in which the spectrum of viewpoints on social policy—liberal, conservative, Marxist, feminist—is critically reviewed.

Banting, Keith (1987). *The Welfare State and Canadian Federalism.* 2nd edn. Kingston: McGill-Queen's University Press. This book is by far the most careful extended political analysis of the influence of Canada's federal political structure upon Canada's welfare policies. Good for both novices and initiates.

—— (1985). 'Institutional Conservatism: Federalism and Pension Reform', in J. Ismael, ed., *Canadian Social Welfare Policy* (cited below).

Banting, Keith, and Charles M. Beach, eds (1995). *Labour Market Polarization and Social Policy Reform.* Kingston: Queen's University School of Policy Studies. Is the Canadian labour market becoming polarized into better jobs and worse jobs, as economies around the world globalize, ours included? And if so, what can be done about it?

Battle, Ken, and Sherri Torjman (1993). *Federal Social Programs: Setting the Record Straight.* Ottawa: Caledon Institute of Social Policy. A factual review of the nature and the scale of federal programs.

Blache, Pierre (1993). 'Le pouvoir de dépenser au coeur de la crise constitutionnelle canadienne', *Revue générale de droit* 24/1.

Borovoy, Alan (1988). *When Freedoms Collide: The Case for Our Civil Liberties.* Toronto: Lester & Orpen Dennys. Borovoy is general counsel of the Canadian Civil Liberties Association.

Bryden, Kenneth (1974). *Old Age Pensions and Policy-Making in Canada.* Kingston: McGill-Queen's University Press. Before launching into his history and analysis of policy in relation to old age pensions, Bryden provides an enlightening description of the Canadian political system.

Burawoy, Michael (1985). *The Politics of Production: Factory Regimes under Capitalism and Socialism.* London: Verso. A leftist writer examines capitalist and socialist practice in the very heart of industrial society—the factory—and finds many important convergences.

Caledon Institute of Social Policy (1995). *The Dangers of Block Funding*. Ottawa: The Institute.

Campbell, Donald (1969). 'Reforms as Experiments', *American Psychologist* 24.

Canada. Economic Council of Canada (1990). *Good Jobs, Bad Jobs: Employment in the Service Economy*. Ottawa: The Council. Explores the issue of labour-market polarization, currently causing much anxiety.

Canada. Economic Council of Canada (1976). *People and Jobs: A Study of the Canadian Labour Market*. Ottawa: The Council. Many of the trends in employment mentioned in this chapter—the shift toward tertiary industries, the participation of women— were already well recognized twenty years ago, when this report was prepared. See especially Chapter Four, 'Recent Developments in the Canadian Labour Market'.

Canada. Human Resources Development Canada (1997). *Social Security Statistics: Canada and the Provinces 1970–71 to 1994–95*. An annually updated twenty-five-year summary of data on all social security programs. The most complete record available.

Canada. Statistics Canada (1993). *1991 Census: Ethnic Origins*. Catalogue No. 93-315.

Canada. *Statutes of Canada, 1982*. The Constitution Act. Part I, 'The Charter of Human Rights and Freedoms'. Every responsible Canadian should read the Charter attentively.

Canadian Council for Policy Alternatives (1997). 'The Great Deficit Hoax', *Monitor* (April). Attacks the use of current public finance deficits to justify cutbacks in social programs. The *Monitor* can be relied upon for criticism of the too-close links it sees, as in the deficit issue, between the interests of large corporations and the policies of governments.

Dahl, Robert A. (1967). *Pluralist Democracy in the United States: Conflict and Consent*. New York: Rand McNally.

Dahl, Robert A., and Charles E. Lindblom (1953). *Politics, Economics and Welfare*. New York: Harper and Row. Widely regarded as a definitive statement of the liberal pluralist position in political science and social science generally; as such, both hailed and condemned.

Djao, A.W. (1983). *Inequality and Social Policy: The Sociology of Welfare*. Toronto: John Wiley and Sons. Djao briskly and critically reviews various analytical perspectives on social welfare in capitalist society: 'Rugged Individualism', 'Modified Individualism', 'Reluctant Collectivism', 'Social Democracy', and 'Marxism' (cf. Mishra, and George and Wilding, cited below). The author concludes that Marxism provides the best theoretical explanation for the development, and the shortcomings, of state welfare programs.

Finkel, Alvin (1979). *Business and Social Reform in the Thirties*. Toronto: Lorimer.

——— (1977). 'Origins of the Welfare State in Canada', in Leo Panitch, ed., *The Canadian State: Political Economy and Political Power* (cited below). In his book (cited above) and, more briefly in this article, Finkel presents evidence to show that powerful business interests have, on the whole, supported rather than resisted Canada's welfare-state legislation, albeit not always with enthusiasm, because much of it has been essential to the maintenance of modern industrial capitalism.

George, Vic, and P. Wilding (1985). *Ideology and Social Welfare*. Rev. edn. London and Boston: Routledge and Kegan Paul. George and Wilding classify writers on social welfare on a sort of right-to-left dial ranging from 'rugged individualists' to 'Marxists', as is done in many other books and articles (cf. Djao, cited above, and Mishra, cited below). This approach helps clarify the differences one encounters in social welfare literature. As always, the student will want to look critically at the fairness and completeness with which the authors present points of view that are different from their own.

Guest, Dennis (1997). *The Emergence of Social Security in Canada*. 3rd edn. Vancouver: UBC Press. A workmanlike history, covering both income and health care programs. As the title implies, more satisfactory for the period of 'emergence', particularly from the 1940s to the early 1970s.

Harrington, Michael (1976). *The Twilight of Capitalism*. New York: Simon and Schuster. While the title does not look infallibly prophetic twenty years after publication, this book is still interesting for its analysis of developments in the capitalist world, and for its treatment of Harrington's intellectual hero, Karl Marx, whom the author does not see as a deterministic totalitarian.

Hogg, Peter W. (1996). *Constitutional Law of Canada*. 4th edn. Toronto: Carswell. The standard legal textbook on the topic, both authoritative and highly readable. See especially Chapters 2 and 3, on the BNA Act, the Constitution Act, the distribution of powers between federal and provincial levels of government, and federal-provincial financial relations, and Chapter 34, on the Canadian Charter of Human Rights.

Hum, Derek P.J. (1983). *Federalism and the Poor: A Review of the Canada Assistance Plan*. Toronto: Ontario Economic Council.

Hurtig, Mel (1990). *Mel Hurtig on Corporate Taxes: An Outrage and a Scandal*. Ottawa: Canadian Council on Policy Alternatives.

Ismael, Jacqueline, ed. (1985). *Canadian Social Welfare Policy: Federal and Provincial Dimensions*. Kingston: McGill-Queen's University Press.

————, ed. (1987). *The Canadian Welfare State: Evolution and Transition*. Edmonton: University of Alberta Press.

————, and Yves Vaillancourt, eds (1988). *Privatization and the Delivery of Social Services in Canada*. Edmonton: University of Alberta Press.

Kealey, Greg, ed. (1973). *Canada Investigates Industrialism: The Royal Commission on the Relations of Labour and Capital, 1889 (Abridged)*. Toronto: University of Toronto Press. From this early expression of public concern with working conditions in industry, Kealey has selected the testimony of witnesses on a range of issues, including the working conditions of women and children. Much of it is appalling.

Lindblom, Charles E. (1965). *The Intelligence of Democracy: Decision-Making through Mutual Adjustment*. New York: Free Press. In this book and the one cited below, Lindblom compares the outcomes of authoritative co-ordination ('politics') and of 'mutual adjustment' among more or less free actors (akin to markets).

———— (1977). *Politics and Markets*. New York: Basic Books.

Loney, Martin (1977). 'A Political Economy of Citizen Participation', in Panitch, ed., *The Canadian State: Political Economy and Political Power* (cited below).

McGilly, Frank (1995). 'The Knowledge Base of Social Welfare', in Joanne Turner and Frank Turner, eds, *Canadian Social Welfare*. 3rd edn. Toronto: Allyn and Bacon. One view of the range of matters that are relevant in the study of social welfare. Compare with Armitage (cited above), ch. 8.

Magnet, Joseph E. (1987). *Constitutional Law of Canada*. 3rd edn. Toronto: Carswell.

Marsh, Leonard (1943; republished 1975). *Report on Social Security for Canada 1943*. Toronto: University of Toronto Press. This *Report*, written by the Director of Social Research of McGill University, was commissioned by the House of Commons Special Committee on Social Security in 1943 as part of Canada's preparation for post-war reconstruction. It is a must-read for the serious student of Canadian social policy.

Mendelson, Michael (1993a). 'Fundamental Reform of Fiscal Federalism', in Sherri Torjman, ed., *Fiscal Federalism for the 21st Century*. Ottawa: Caledon Institute of Social Policy, pp. 22–4.

———— (1993b). *Social Policy in Real Time*. Ottawa: Caledon Institute of Social Policy.

Mishra, Ramesh (1981). *Society and Social Policy: Theories and Practice of Welfare*. London and New York: Macmillan. Mishra classifies theoretical approaches to welfare on a different basis from that of George and Wilding and Angela Djao (cited above). His classification ranges from those 'functionalists' who see welfare as an inevitable component of any industrial society, irrespective of political regime, meeting needs that cannot be ignored if the system is to function, to those who see it as part of the class struggle, not inevitable at all. To locate writers on the ideological spectrum will enhance understanding of their work. Major contributors to the discussion of social policy will not always agree with the way they are classed by scholars; and besides, the leading writers on these matters do not always maintain one ideological stance in all their writings.

Ontario. *Human Rights Code. Revised Statutes of Ontario, 1990*, Chapter H-19. Originally enacted, *Statutes of Ontario, 1981*, Chapter 53.

Pal, L. (1985). 'Federalism, Social Policy, and the Constitution', in Ismael, ed. *Canadian Social Welfare Policy: Federal and Provincial Dimensions* (cited above), pp. 1–20.

Panitch, Leo, ed. (1977). *The Canadian State: Political Economy and Political Power*. Toronto: University of Toronto Press. A collection with a pronounced left-wing slant.

Polanyi, Karl (1957). *The Great Transformation*. Boston: Beacon Press. What the freeing of the markets for labour and land, necessary adjuncts to the Industrial Revolution, meant to the world, for better and for worse.

Popper, Karl (1966). *The Open Society and Its Enemies*. 5th edn. Princeton, New Jersey: Princeton University Press.

———— (1961). *The Poverty of Historicism*. 3rd edn. London: Routledge and Kegan Paul. Popper criticizes theories that treat history as in any sense the inevitably determined unfolding of a discernible process. Marxism is his principal target, but only because it is the most powerful contemporary variant of historicism.

Prince, Michael J. (1987). 'How Ottawa Decides Social Policy', in Ismael, ed., *The Canadian Welfare State: Evolution and Transition* (cited above), pp. 247–73.

Quebec. An Act respecting the Rights and Liberties of the Person, 1975.

Rimlinger, Gaston (1971). *Welfare Policy and Industrialization in Europe, America and Russia*. New York: John Wiley and Sons. This is one of the best-known elaborations of the idea that social welfare measures are best explained as functional necessities of industrial society, whatever the political structure in any given country.

Saskatchewan. Bill of Rights Act, 1947.

Schwartz, Mildred (1974). *Politics and Territory: The Sociology of Regional Persistence in Canada*. Montreal: McGill-Queen's University Press. How ethnicity, geography, climate, natural resources, economic activity, history, and other factors contribute to the distinct vitality of the regions of Canada—with profound consequences for our political life.

Scott, Frank (1964). 'Our Changing Constitution', in W.R. Lederman, ed., *The Courts and the Canadian Constitution*, pp. 19–34. Toronto: McClelland and Stewart. The Carleton Library series.

Shifrin, L. (1985). 'Income Security: The Rise and Stall of the Federal Role', in Ismael, ed. *Canadian Social Welfare Policy: Federal and Provincial Dimensions* (cited above), pp. 21–8.

Shragge, Eric, ed. (1993). *Community Economic Development: In Search of Empowerment and Alternatives*. Montreal: Black Rose.

Simeon, Richard (1972). *Federal-Provincial Diplomacy: The Making of Recent Policy in Canada*. Toronto: University of Toronto Press. Another political analysis of how Canada's federalism works in concrete instances. One of the cases Simeon examines is the development of the Canada Pension Plan.

Stevenson, G. (1977). 'Federalism and the political economy of the Canadian state', in Panitch, ed., *The Canadian State* (cited above). A left-wing view of the functioning of federalism in Canada.

Strain, F., and D. Hum (1987). 'Canadian Federalism and the Welfare State', in Ismael, ed., *The Canadian Welfare State: Evolution and Transition* (cited above), pp. 349–71.

Taylor, Malcolm (1988). *Health Insurance and Canadian Public Policy*. 2nd edn. Toronto: University of Toronto Press. As does Bryden (cited above), Taylor introduces his topic with a brief and very clear description of the system within which policy emerges in Canada: federal and provincial governments, political parties, bureaucracies both public and private, interest groups.

Titmuss, Richard (1963). *Essays on the Welfare State*. 2nd edn. Boston: Beacon Press. The chapter entitled 'Social Welfare and the Family' is an exploration of social needs created by industrialism, calling for a collective response.

Webb, Steven (1995). 'Social Security Policy in a Changing Labour Market', *Oxford Review of Economic Policy* 11/3 (Autumn).

Wilensky, Harold, and C.N. Lebeaux (1965). *Industrial Society and Social Welfare*. 2nd edn. New York: Free Press. The 'Preface to the Second Edition' is an excellent introduction to the study of contemporary social welfare, even though written some time ago.

Yelaja, Shankar, ed. (1987). *Canadian Social Policy*. 2nd edn. Waterloo: Sir Wilfrid Laurier University Press. Several articles on both specific issues and general orientations in Canadian social policy.

PART ONE

Income Programs

CHAPTER 2

Income: Basic Concepts and Issues

THE CONCEPT OF INCOME

The idea of earning an income through employment—that is, through working for some person or some organization in return for *wages* (typically, so much per hour) and *salaries* (typically, so much per year or month)—is so familiar to us that it is easy to lose sight of its significance. It is one of those pervasive social institutions referred to in the previous chapter, so much a part of our everyday life that we take it for granted. But in fact, income through employment is a concept specific to a certain type of society, and there are signs that in our society the employment relationship is now undergoing change.

In the first place, there are in the world today, as there have been in the past, many societies in which income from employment is *not* all-important; people live their lives without much exchange of cash at all; they depend for their livings largely on what they produce themselves and on the barter of goods and services. They are not paid a money income for the work they do. In the contemporary world, the societies that work like that are not the ones that we are pleased to call 'advanced'.

In the second place, in our society there are other kinds of incomes. There are incomes without work and, of course, there is plenty of work without income. Self-employed people certainly work, but the incomes they earn are not salaries or wages, they are *profits*, as are those portions of the surpluses of corporations that are distributed to their owners, or shareholders, as 'dividends'. (Also, some self-employed people incorporate their businesses, from which they receive profits, then pay themselves salaries as executives.) It is salutary to bear in mind that the self-employed and the corporations may have losses as well as profits. Many people receive incomes in the form of *interest* on money they have loaned, and in the form of *rent*. Incomes from employment, profits, interest, and rent are known to economists as 'factor' incomes, being incomes paid *in exchange* for some factor that contributes to production.

Everybody is familiar with other kinds of income, received by people from government on the basis of some status or as a consequence of the occurrence of some contingency. These are known as 'transfer' incomes:

they are 'transferred' from some source to certain recipients who do not provide anything explicitly in return for the incomes. In Canada, the familiar Old Age Security benefits and the more recent proposed Seniors Benefit and Child Tax Benefit are examples. Through these programs, incomes are transferred from the whole body of Canadian taxpayers to the mothers (usually) of children under a certain age, and to Canadian residents over a certain age. Many of the recipients are also taxpayers, but that does not affect the 'transfer' character of the incomes thereby distributed. 'Public assistance', or 'welfare', is a 'transfer' from the taxpayers to recipients who qualify principally by being poor; here the recipients are not substantial taxpayers, so the transfer is mostly from one large group, taxpayers, to a small group, welfare recipients.

Other programs to which this book shall give close attention are programs that have a 'social insurance' character. In each such program, people in an income-earning category make (compulsory) contributions, and if and when some contingency strikes that deprives them of their incomes from earnings, the program provides a replacement income. In one sense, these too are 'transfer' incomes in that the recipient does not have to *do* anything specific in return; in official statistical breakdowns, incomes of this sort are classified as transfer incomes. But they are different from the first class of transfer incomes, in that one must first make sufficient contributions while earning to qualify for an income when earnings are interrupted or terminated, and one must have suffered the occurrence of the defined risk.

In modern society, income from employment is by far the largest income source for by far the most people. This crucial fact looms in the background of *all* our social policy related to incomes—and not very far in the background. It turns up under the heading of the 'incentive to work' in discussions of income programs where recipients of income benefits may conceivably be seen as having a choice between working and not working. There are different points of view as to how much transfer income it takes to dampen an individual's incentive to work, and as to how important money is compared to other values associated with work; but an income program that actually did reduce total work effort sufficiently to lower appreciably the national production of goods and services would be costly for society beyond the amount of money transferred. On the other hand, most would agree that certain people *ought* to be well enough provided for that they should feel little necessity, or incentive, to work: for example, the single parents (usually the mothers) of children who are young and/or who need special care, or persons whose age or health is such that they would probably not be able to maintain themselves by working. Allowing for such exceptions, the effect on the incentive to work is always an important criterion of income programs.

It can be helpful to think of income programs as being of three kinds: (1) those restricted to people who *do* normally have incomes from work;

(2) those whose benefits extend to income earners and non-earners alike; and (3) those intended for non-working people. The student approaching the subject for the first time may be surprised to see how few programs there are in this third category, and equally surprised to find that few even of the programs in the third category are aimed directly at the poorest among us.

OBJECTIVES OF INCOME PROGRAMS

The direct, overt objectives of income programs may be stated as *adequacy, security*, and *equality* or *equalization*. The proponents of programs are likely to have other objectives in mind as well; and whatever the intent, a program may well have outcomes very different from or even contrary to those it ostensibly pursues.

Adequacy

Conceptually, the objective of *adequacy* speaks for itself. If a person is working, it is socially as well as personally valuable that his/her income be adequate; this may require special measures if his/her work income is judged to be less than adequate. If a person is prevented from working, by our social values he/she ought still to be provided for adequately.

When it comes to determining exactly what level of provision is meant by 'adequate', we may expect to find differences of opinion. Whatever we mean by 'adequacy', obviously the amount required for a person with dependants is larger than the amount required for one person alone. Largely because of the issue of incentive to work, the kinds of instruments we use to assure adequacy for employed or employable people are different from those considered appropriate for people who are separated from the labour market (such as the seriously disabled, lone mothers of young children, or older people).

Security

For most people who do earn an income, and for their dependants, the interruption or cessation of that income is a serious matter, perhaps a catastrophe. In their individual capacities, people may try to save for the proverbial rainy day, but not many people could provide, on their own, for all their needs, should their income be cut off for any length of time. With a moment's reflection, we can perceive that the loss of income for income earners would rapidly do harm not only to those workers and their dependants, but to other people, too—such as the owners and employees of businesses that depend on them to buy their products. 'Security' is therefore security not just for individual income recipients, but for the system of which they are a part.

Social programs have been developed to provide a measure of security in the face of some of the occurrences that regularly threaten the continuance of incomes in industrial society: unemployment, aging, some special kinds of injury or illness—not, be it noted, sickness or injury in general, for Canada, like many countries, has no public programs to cushion the impact of loss of incomes through illness as such (but see below, under 'Add-ons', in the section devoted to unemployment insurance).

All industrial societies are money societies: people spend money to obtain their material satisfactions. In a money society, the natural response to a widely distributed risk, whose probability can be estimated, and whose money costs can be assessed, is to *insure* against the financial consequences of the occurrence of the risk. Most insurance is voluntary and private; but some of the risks associated with insecurity of income are such that they could not be met by reliance on voluntary insurance. One reason for this is that the most insecure are usually the least able to afford the insurance. Another is that insurance against certain risks inherent in the functioning of society will not work unless everyone concerned is taking part in the insurance scheme. Typically, therefore, such insurances become *social* insurances, covering whole sectors of society, and are necessarily compulsory.

One unavoidable practical question concerning income security programs directed at the risk of loss of work income is this: should the income provided through the program bear some relation to the employment-related income that the person was receiving before misfortune struck? The answer could easily be 'No'; such benefits could be paid at a flat rate to all who suffer the occurrence of the risk. This was the recommendation of the historic Beveridge Report, *Social Insurance and Allied Services* (1942), considered to be the foundation document of the British welfare state. In Canada, we have said 'Yes', income benefits should be related to prior income, because that is the income that is being insured. The next question is, what relation? One third, one half, three quarters? To that question our answers vary from program to program.

Equalization

Finally, no one disputes that incomes are very unevenly distributed in our society. Within the labour market, there are exceedingly broad inequalities, from the lowest-paid, irregularly employed worker to the highest-paid executive, and lately the inequality has been widening. Rewards in the other streams of income are even more unequally distributed—for example, income from the profits of enterprises or income from interest on bonds and mortgages. The defenders of capitalism do not deny that inequality of income is characteristic of capitalist societies. Quite the contrary: they insist that the inequalities of reward that prevail are socially useful, even necessary, because they channel resources into the production of the goods and services that society evidently prefers. If society has a preference for entertainment, sports, and computers, among other things, so be it: certain entertainers and athletes and computer wizards will be exceedingly well rewarded, as will some of the business men and women who promote entertainment, sports, and computers. And so on. (Pro-capitalists also point to substantial inequalities of material reward in non-capitalist industrial societies—but that is another issue.)

In practice, while the extremes of inequality frequently are denounced, there are very few voices that clamour for anything resembling real equality of material standards for all people. It is more realistic to speak of an objective of *limited equalization* of incomes: a little taken off the top and the middle, a little added to the bottom. At times even the equalization that is actually brought about is an implicit outcome rather than an explicit objective. Obviously, if the state gives to somebody who would otherwise have zero income an income of $800 a month, some equalization takes place, since the $800 is provided largely by taxpayers whose before-tax incomes are a good deal more than that. Similarly, certain measures are intended to shore up the incomes of workers, either by assuring them a certain rate of earnings higher than they might otherwise have obtained, or by supplementing their work incomes if these are very low. These measures also have an immediate equalizing effect, though there is some disagreement about their longer-term consequences.

In fact, in Canada we probably achieve as much equalization through tax measures as through direct income provision. Tax measures that tend to have an equalizing effect include exemptions of very low incomes from income tax, low rates of income tax on low incomes, progressively higher rates on higher incomes, refundable tax credits, and other devices. As we shall see, the federal government's benefits for older people and for parents of dependent children have now become income-tested and are closely meshed with the federal income tax system, entirely replacing the 'universal' programs for children and the elderly that until recently were pillars of Canada's social security system (see below, 'The Money Must Come from Somewhere: Taxes, Contributions', and 'Universal and Selective Transfers', and Chapter Six, 'The "Tax Expenditure" Approach'). By concentrating benefits on lower-income people, these changes will tend to equalize incomes.

Income and other taxes are not used solely to collect revenues. Income taxes in particular are administered to achieve objectives of economic and social policy and for other purposes. The manipulation of income tax for policy purposes can have the greatest impact when applied to people whose incomes are high enough to be most affected by it. For example, a reduction in the rate of income tax intended to encourage some type of behaviour will matter little to persons whose incomes are so low that they pay little or no income tax, but will make a great difference to higher-income people. Unfortunately for the attainment of even 'limited equalization', a modest-looking tax concession to a high-income earner will often be of greater dollar value than an impressive-looking cash benefit to a low-income individual.

For example, Canada has an economic policy objective of encouraging people to save, both to increase investment and to dampen inflation. The government has an interest in inducing people to provide for their own post-retirement incomes, in order to reduce their reliance on income transfer programs like the Seniors Benefit and public assistance. These are legitimate goals. To meet them, as we shall see in Chapter Five, an income-receiving Canadian may put aside each year 18 per cent of income, up to a maximum of $13,500 (18 per cent of $75,000), for contributions to a registered pension, deductible from income for income tax purposes. The 18 per cent maximum is allowed to all income recipients, but (a) obviously, 18 per cent of a low income is less than 18 per cent of a higher income; (b) in any case, few low-income or even average-income people could afford to put aside 18 per cent of their incomes; and (c) even if they did, their tax savings would be much less than those of high-income people, because they pay income tax at a lower rate. For a high-income earner, the pension tax break will be worth $6,500 per year or more—and the record shows that higher-income Canadians, being rational, take full advantage of it. That compares with the transfer of $1,020 per child per year that a low-income family will receive through the Child Tax Benefit, discussed in Chapter Six. (A note of caution: the calculation of the dollar benefit of the pension deduction is more complicated than has been shown here, but the more complicated calculation would probably show a still greater benefit to upper-income people.) Thus, in this manipulation of the income tax system, other policy objectives take precedence over the objective of equalization.

This is perhaps the most transparent of many tax breaks available to wealthy people, most of them more arcane, that considerably dampen the equalization of incomes achieved by the combination of the tax system and the income transfer measures in place in Canada. Hence our opening assertion that Canada's commitment to the equalization of incomes is *very* limited.

Two Words of Caution

(1) The life of a society consists of many complicated interactions among people and processes. Any income program is therefore sure to have effects other than those that are its overt purposes. Much analysis of social policy concerns the tracing out of the real effects of programs and the real intent behind them. The difference is that reliable evidence about intent is obviously more difficult to collect than evidence about effects. This does not deter the more severe critics of social policy from concluding that the 'real' purposes of social programs are very different from the purposes declared in the laws that create them.

(2) This text will cover certain social programs of income security and income maintenance. Our governments rely far more on other branches of policy than on income programs to promote the general economic well-

being. They try to stimulate certain kinds of investment (such as the elusive 'Research and Development'); they attempt to influence the flow of money between savers, financiers, and entrepreneurs ('monetary policy'); they try to control the balance between government revenue (mostly taxes) and expenditure ('fiscal policy'); they manipulate specific taxes to influence the behaviour of taxpayers; they regulate Canada's international trading; and so on. In focusing our attention on income-related social programs, we are not questioning the crucial impact that policy in those other fields may have on the realization of social objectives. *That impact is far greater than what is achieved through public income transfers.*

THE MONEY MUST COME FROM SOMEWHERE: TAXES, CONTRIBUTIONS

Our concern will be mostly with programs that pursue the goals of adequacy and security quite directly, by providing cash incomes. As we have said, the goal of equalization is seldom explicit. The money to provide such transfer incomes must come from somewhere. Almost totally, it comes from other ('factor') incomes. The only exception may be the financing of the Canada and Quebec Pension Plans, where some of the pension money comes from interest paid by institutions to whom the pension fund managers lend their surpluses, and some possibly from dividends paid on shares owned by the pension funds. Since the flow of incomes earned through employment is the largest, most transparent, most easily accessible money flow in the industrial economic system, it is constantly tapped to finance government programs, through *taxation* and through the collection of special *contributions*. Other taxable money flows are *consumption* and *value added* in the production of goods.

Broadly speaking, there are two ways to mobilize the money needed to finance income-providing programs: (1) by the tax-and-transfer process, where the money comes from the general taxes and other government revenues; and (2) by the contributory process, where the potential beneficiaries of programs make special contributions (or have special contributions made on their behalf) exclusively to support specific programs. The latter approach is often conceptualized as 'social insurance', because all the people considered to be vulnerable to a certain risk, damaging to society as well as to themselves, are obliged to pay a certain amount—a sort of 'premium'—in order to have money available should they fall victim to the risk. Sometimes a program basically supported by contributions is subsidized by government out of tax revenues as well. It is estimated that payroll-related contributions now actually *exceed* income taxes for Canadian employees, taken all in all.

In assessing the net effect of a program, its tax and/or contributions base (who pays, and how much) must be taken into account as well as the benefits it provides (who gets, and how much). It is possible for the collecting and the paying to work against each other, in terms of redistribution.

Tax-supported programs will be seen to differ from programs supported by contributions in virtually all important respects. Tax-supported programs are part of the overall responsibility of the elected government, taxation being the share of its money that society allots to government. Oliver Wendell Holmes called taxes 'the price we pay for civilization'. Contributory programs are paid for by specific contributions from specific groups of people. Not being part of the general revenues of government, the contributions are not available to the government for its general purposes (but see the discussion of the financing of the public pension plans in Chapter Five, Discussion Questions). Tax-supported programs are administered by government departments that are properly subject to the political control of the government of the day, while contributory programs are administered, as a rule, by more or less autonomous special bodies created for the purpose. And the income benefits paid through contributory programs generally vary in accordance with the contributions paid in by individuals, and those contributions, in turn, generally are proportionate to individuals' incomes. Higher earned incomes yield higher benefits, lower incomes lower benefits. The benefits paid through tax-supported programs, on the other hand, tend to bear some inverse relation to the beneficiary's means, so that those in the lowest income strata receive the largest benefits. Across-the-board flat-rate benefits, once common, have all but disappeared.

The taxpayer-contributor may not feel much of a difference between paying his/her income tax and paying his/her contributions to Employment Insurance or the Pension Plan. Some writers go so far as to speak of EI and C/QPP contributions as taxes. Both in principle and in practice, however, there are fundamental differences in the way tax-and-transfer and contributory programs work.

BENEFITS AND ELIGIBILITY FOR BENEFITS

The legislation and the regulations governing each program must say which persons are *covered* by the program, and under what conditions those who are covered become *eligible* to receive the benefits provided, and how those *benefits* are to be determined.

Benefits Related to Contributions

In the case of contributory programs, where the focus is normally protection against loss of income on the occurrence of a particular risk, the pattern is what you would expect: the contributors are the ones covered by the program (or, to put it the other way around, everyone covered by the program is obliged to contribute), and anyone who is covered and who falls victim to the risk becomes eligible for the benefits. Contributions normally bear some proportionate relation to earned income—the more you earn, the more you contribute, within minimum and maximum limits; and benefits typically also bear some relation to income—the more you have been earning, the more you receive, again with minimum and maximum limits.

Universal and Selective Transfers

In the case of tax-and-transfer programs, there are two ways of distributing benefits: universally and selectively. A *universal* program gives a certain amount of money to everybody in the population who meets some straightforward personal criterion that has nothing directly to do with income level; a well-known example was the former federal Family Allowances program, where benefits were paid to all Canadians with children of certain ages. It must be said that, to the dismay of many, we no longer have any universal income transfer programs in Canada. All tax-and-transfer programs at both federal and provincial levels are now subject to income limits, in such ways that upper-income people do not receive any benefits. A *selective* program, as the name implies, selects its beneficiaries on the grounds that their incomes are more or less low. Usually, people with very low incomes receive larger amounts than people with not-so-low incomes, but there is always an income limit above which no benefits are paid at all.

Universal programs were vulnerable to the obvious criticism that they benefited the wealthy as well as the poor. In our history, this has been dealt with in steps. First, the benefits were treated as subsidies to the recipient, pure and simple, and were not taxed. Part of the rationale was that, once the money had been transferred from the government to the individual, at some administrative cost, it made little sense to collect some of it back in tax, at some further administrative cost. In the second stage, presumably when the expenditure on benefits was high and it was clear that low-income people were receiving only a small part of the transfer, the benefits were *taxed* as income, like any other kind of income. In this way, a recipient whose total personal income was at or above the average returned anywhere from about 30 per cent to a little over 50 per cent of the transfer in taxes, depending on income level and province of residence. A very wealthy person still kept about half of the benefit. In the third stage, benefits were not only taxed back; they were subjected to a special recovery process if the recipient's income was above a certain level (generally around the population average), at such a rate that the benefit disappeared entirely at a higher income level. This process is graphically called the 'clawback'. At that point the benefit ceased to be universal. This was the history of Canada's Old Age Security program, up to 1997. The fourth stage was the elimination of the 'universal' program altogether, replacing it with something else. That has been done both with Old Age Security, in the process of transition to the Seniors Benefit (see Chapter Five), and the federal Family Allowances program, now replaced by the Child Tax Benefit (see Chapter Six).

Tax Expenditures

Over the last twenty years or so, Canada has made increasing use in income security programs of a technique long used to encourage particular kinds of investment by individuals and corporations, namely, the technique of 'tax expenditures'. This means an amount of tax that, under specified conditions, is not collected, or is even reversed and paid *to* the taxpayer. Either way, a tax expenditure is a cost to the public treasury, even though it shows up as less revenue rather than as more expenditure. In the field of income support, there are three main classes of tax expenditure: (1) Tax *exemptions* allow taxpayers a certain amount of their income free of income tax. (2) Tax *deductions* allow taxpayers to deduct certain expenses from their incomes before calculating their taxes. (3) Tax *credits* accord taxpayers a stated credit in their income tax account, reducing their taxes payable by the amount of the credit; the credit may be made *refundable*, in which case a person who owes little or nothing in taxes will get a calculated amount back from the government. Of all of these, only a refundable tax credit is of any help to persons whose income taxes payable are zero, and so it can be effectively targeted towards low-income people.

The rationale for tax expenditures is that it is more efficient to leave income in people's hands rather than collecting taxes with one hand and then, probably much later, giving money back as benefits. In a manner of speaking, the tax expenditure technique collapses the taxing and the transferring into one transaction between the individual and the government.

Income Tests, Means Tests, Needs Tests

Selective programs make demands of administrators and beneficiaries that universal programs do not. In a selective program, a decision has to be made in each case as to whether an individual is eligible to receive benefits. The most important factor in the eligibility decision is the person's income, and possibly his/her other resources. A program the amount of whose benefits is based entirely on the *income* of the prospective recipient is called an *income-tested* program: the higher the recipient's income, the lower the benefits. An *income test* has the great advantage of being, for the most part, easy to administer: most people declare their incomes every year when they pay their income taxes. That some income may be concealed is well known; the detection of the concealment of income takes up a large part of such effort as is required in the administration of income-tested transfer programs, as it does in the administration of income tax collection.

A person's income is not a fully reliable indicator of his or her material situation. Individuals with modest incomes may have at their disposal substantial *means*, that make life more comfortable than their incomes might suggest—means such as ownership of their homes or other useful possessions, the resources of immediate family members, money in the bank, etc. A program that takes such matters into account in determining the benefits to be paid to recipients is categorized as a *means-tested* transfer.

Programs that call for a means test usually go a step further, and require an assessment of the person's *needs* as well, following some definition of needs established by law; benefits are scaled according to the apparent gap between needs and ability to meet those needs. Such programs are called *needs-tested* programs. Looked at with attention, the terms *income-tested*, *means-tested*, and *needs-tested* are self-explanatory. Among other differences, it should be immediately clear that needs-tested programs will call for more investigation of the lives of potential recipients, and will give more discretion over benefits to administrators, than do means-tested programs.

After we have applied these descriptive categories—'contributory', 'tax-and-transfer', 'universal', 'selective', 'income-tested', 'means-tested', and 'needs-tested'—to specific programs, their implications will be seen more clearly.

Legislation, Administration, Finance, Benefits

We shall be talking entirely about public programs. The most basic information about each program is information about the problem it is intended to meet, its authorizing legislation, its intended beneficiaries, the body responsible for its administration and the kind of authority that body possesses, the benefits provided, and the financing of the program. These elements are interrelated: understanding of one will contribute to understanding of all the others. A good grasp of these basic facts concerning social programs, and the way these characteristics are tied together, will enable the student to interpret various analyses and criticisms of the programs, and to form responsible opinions about them.

The first program to be examined is one of the most familiar: Employment Insurance, or, as it was known until 1995, Unemployment Insurance.

STUDY QUESTIONS

The foregoing material has all been rather general in nature. No doubt it will mean more after there has been some opportunity to study concrete programs and to see particular instances of these general notions. But before proceeding, the student should be able to answer the following questions.

1. In the context of industrial society, exactly what does 'employment' mean? What makes 'employment' so important a concept in thinking about social policy?
2. Can you think of an instance where our reliance on the family as a social institution affects the benefits paid out under any income program of which you have some knowledge?
3. Various commentators on the federal Charter of Rights and Freedoms have said that the Charter, to a greater or lesser extent, takes powers

away from the elected legislature and gives them to the unelected judiciary. What does that mean?

4. In Canada, which level of government has jurisdiction over social welfare legislation and health legislation in general? Why (that is, on what authority do you base your answer)?

5. What is the basic difference between the taxing powers of the two levels of government?

6. What difference does it make whether some policy field is within federal or provincial jurisdiction?

7. What is meant by 'security' and 'adequacy' as goals of social policy in relation to income?

8. How does a 'tax expenditure' program differ from a 'transfer' program?

CHAPTER 3

Employment Insurance

WHAT IS EMPLOYMENT INSURANCE?

Over a period of fifty-five years, Unemployment Insurance (UI) became one of the best-known social programs in Canada, a pillar of Canada's social security edifice. It remains such in its new guise of 'Employment Insurance'. All employed people pay Employment Insurance premiums out of every paycheque; all their employers add still larger premiums on their behalf. Every year, millions of Canadians receive Employment Insurance benefits; during the last few years preceding 1996, the average number of recipients in any month consistently exceeded one million. For reasons presented below, the number of beneficiaries has decreased sharply. Still, it is estimated that about one third of Canadian workers receive EI benefits at least once in their working lives.

First enacted in the Unemployment Insurance Act of 1940, and first implemented in 1941, unemployment insurance has undergone many changes. The most recent have been those of the legislation of 1995, entitled, in a striking piece of political rhetoric, the *Employment* Insurance Act. Though it may be some time before this new name replaces the old in common usage, this text will use the now-current term Employment Insurance (EI), except where the reference is clearly to years prior to 1995. Whether it is called UI or EI, the program is one of insurance against part of the loss of income through unemployment, a concept that is best expressed by the generic term 'unemployment insurance'—lower-case 'u', lower-case 'i'—so this term is used here in that generic sense.

The 1995 Act introduced some fundamental innovations, notably, discrimination between infrequent and frequent recipients of benefits, a shift from *weeks* worked to *hours* worked as the basis for entitlements, and a much-enhanced concentration on so-called 'Employment Benefits' intended to better equip the unemployed person to return to work. The Act also made some less basic but nonetheless important changes, notably a reduction in the dollar benefits payable, a reduction in the premiums to be paid into the EI fund by employees and employers, an increase in the length of time in work required before workers become eligible to receive benefits, and a shortening of the maximum periods for which benefits are paid. The changes will be phased in over a period of years, so some pre-1995 characteristics of EI will continue in effect for some time. The following discussion takes into account the 1995 changes, in the context of the development of unemployment insurance over nearly sixty years.

As the name implies, EI is an insurance program. Beneficiaries receive a certain cash benefit when they fall victim to a certain risk. The risk in this case is the *temporary, involuntary interruption of earnings from employment.* As already noted, 'employment' means work as a paid employee of someone else, the most important source of income in our society. A person has to have earned an income through employment for some time before his/her income can be considered to have been interrupted; a person still looking for a first job has not suffered an 'interruption' of employment; and a person who has worked at a number of jobs a few days or weeks at a time, too sporadically to have accumulated sufficient hours of work within the qualifying period, is also not considered to have suffered such an interruption. And since the risk insured against is defined as only *temporary* involuntary unemployment, the situation of the unemployed person who remains unemployed for a very long time calls for some other remedy than Employment Insurance. So does that of the person who leaves a job voluntarily, or who terminates a job and is not interested in finding another.

We have already emphasized the importance in industrial society of *employment*; in such a society, employment is the most important vehicle for the distribution of incomes to people. A break in the flow of salary or wages to an individual has serious consequences, and not only for the individual.

Unemployment is a fairly characteristic phenomenon of industrial society—at least of capitalist industrial society. The moral and economic waste brought on by recurring waves of unemployment has always been one of the principal reproaches directed at the capitalist system by its critics. (How non-capitalist societies deal with variations in demand for labour is an interesting topic, too, but not one upon which we shall dwell.) In their pursuit of profits, private-sector companies discharge workers permanently or lay them off temporarily for various reasons. A major determinant is the overall economic performance of the country as a whole. There are 'up' times when investors are investing, producers are producing, merchants are selling, in a self-reinforcing process, and employment is consequently at a high level; and there are 'down' periods when the indications are that a limit has been reached, that further activity will be unprofitable, and investment, production, sales, and consequently employment decline. One of the major objectives of the economic policies of government—monetary, fiscal, and trade—is to smooth out the ups and downs of this cycle while still maintaining long-term growth.

Even in 'up' times, individual companies may discharge workers temporarily or permanently because their sales slow down, so they reduce production; because they fare badly in competition; because the tastes of the consuming public change, so they stop producing certain things altogether; because a new technology changes the labour demands of an industry, ending the need for the services of some workers, perhaps requiring the services of other workers, perhaps not. Certain businesses have seasonal ups and downs regardless of the business cycle, so that fewer workers are needed at predictable times of year. Foreign competitors of Canadian firms are sometimes able to put their products on the Canadian market at attractive prices, making it difficult for Canadian firms to compete. (The reverse is also true.) Some of this foreign competition comes from developing countries, whose industries are rapidly mastering modern technology.

Most of the foregoing analysis of the causes of unemployment applies to the private sector. The public sector, including health and education, now employs close to three million Canadians. Traditionally jobs have been more secure in the public sector, but in recent years many public-sector workers have been vulnerable to unemployment as needs have changed and as some public services have been reduced. Partly in recognition of this fact, Canada's Unemployment Insurance program, which at first did not cover public-sector employees, was modified in 1971 to include them.

It will be evident that a program of unemployment insurance, which provides an income to workers only after they have become unemployed, can have little effect on the above-mentioned sources of unemployment. One of the intended effects of Employment Insurance, to be sure, is to stabilize the flow of incomes to workers through periods of unemployment and so to enable them to continue their consumption of goods and services; in this way the contagion of unemployment from firm to firm is slowed. Associated with Employment Insurance are programs of relocation and retraining of unemployed workers; if successful, these associated programs will improve the employability of people and will reduce unemployment. But Employment Insurance cannot and does not solve the problem of unemployment, and its most fervent advocates never claimed that it would.

Considering that unemployment is a perceptible risk with clear financial consequences, could not individuals insure themselves against that risk privately, as against other risks? Historically, trade unions and workers' associations did this—and faced some of the same administrative problems that confront unemployment insurance today. See, for example, the account of such trade-union-based programs in Great Britain in the famous *Minority Report* of the Poor Law Commission of 1909, Part ii (National Committee to Promote the Break-up of the Poor Law, 1909, pp. 166–7, 288–93). In fact, a certain amount of something resembling private unemployment insurance does exist, in the form of termination agreements bargained for by upper-income executives who operate at levels where there is indeed much job insecurity. Also, a few firms, mainly large ones, provide laid-off employees with so-called Supplementary Unemployment Benefits, paid for out of the companies' own resources, but this is not done on a true insurance basis. An insurance company, interested in making a profit in a competitive setting, could not possibly extend affordable unemployment insurance to workers of modest incomes in insecure jobs—the very ones who need it the most urgently. If unemployment insurance covering the whole labour force is to work at all, it cannot be competitive; it must cover entire classes of workers; and within the classes of employments covered, participation must be compulsory.

The Risk

The risk insured against is a *temporary, involuntary* interruption of income due to unemployment. An interruption due to some cause other than loss of job, for instance, a disabling illness, is equally painful, but does not fit the definition, strictly speaking (but see below, under 'Add-ons', for ways in which the strict definition has been relaxed). The point is that a person who is *unable* to work is not a victim of short-term loss of job so much as of loss of ability to work.

No insurance program will insure against a condition that one can inflict upon oneself, so the unemployment has to be involuntary. A person who leaves a job voluntarily is not a victim of the risk that is insured against, and is therefore not entitled to EI benefits immediately; if unable in the course of time to locate another job, however, such a person may be considered to have become involuntarily unemployed. To be eligible then for benefits, the insured person must show willingness to work, as defined in regulations and interpreted by administrators. The discretion of administrators in judging the eligibility of applicants is limited by law and regulation, but some discretion remains. What would be insisted upon by an administrator as the appropriate effort to find a job when jobs are readily available might be acknowledged by the same administrator to be quite pointless when employment is scarce.

The main burden of unemployment falls, of course, on unemployed workers and their dependants. It may be thought, then, that the workers are the ones at risk, that they are the ones being 'insured', and that they ought therefore to pay the premiums out of which the benefits will be financed. But since a person with no income ceases to be a significant consumer of goods and services, substantial unemployment is bad for business. It is especially bad for business if many workers—possibly hundreds—lose their jobs simultaneously in one location, as can readily happen in one-company or one-industry communities or regions. Thus, employers as a class are somewhat at risk, as is society as a whole; and both employers and society benefit along with laid-off workers when cash benefits are paid to the unemployed. Furthermore, employers are, after all, the ones who decide whether and when workers' employments are to be interrupted; the design of Employment Insurance gives employers a certain financial incentive to limit unemployment, thereby reducing the total EI benefits paid and, consequently, the premiums that have to be collected. In the Canadian EI system, employers are obliged to share in paying the cost of the insurance: in fact, they pay 1.4 times as much as employees do. 'Society', through the federal government, did share in the direct costs until 1989, but no longer does so.

Coverage

Coverage under EI is very simple, in principle. Practically all employed persons in Canada are covered. The few exceptions are not very significant. 'Employed' means 'employed by someone else'; the self-employed are excluded. This is consistent with the principle that you cannot be insured against something over which you have control. The self-employed are subject to the same changes in economic conditions as are workers employed by others. The difference is that the self-employed may decide to give up their businesses at the moment they think in their own best interest. It is difficult, therefore, to insure them against unemployment. This is so even though the self-employed may feel they had no more choice over shutting down than workers had over being laid off.

For insurance purposes, there must be some minimum measure of what constitutes 'employment'. From the inception of the plan, this was defined in terms of a minimum number of hours of work *per week*. In recent years, there has been a considerable increase in the number and proportion of people working part-time and/or working irregularly. The hours-per-week eligibility rule has accordingly been replaced with a rule that dictates that a worker is covered by EI once he or she has accumulated a minimum number of hours within a year, and remains covered no matter how few hours are worked in any given week. Many part-time workers are thus now eligible for EI. The former hours-per-week minimum made it fairly easy for employers to avoid the payment of unemployment insurance premiums, with or without the collusion of the workers, by simply employing workers for fewer hours per week than the minimum. Such avoidance will require greater ingenuity now that the minimum is stated in *total* hours in a year, not hours per week. Incomes of part-time workers are better protected. It is feared, however, that some employers may be deterred from hiring part-time workers because they will be obliged to make EI contributions on their behalf.

Until 1989 a worker ceased to be covered by UI at age sixty-five. If the worker continued to work, no contributions were collected, and if the worker became unemployed, no benefits were paid. The rationale for this cut-off was fairly evident: relatively few people continue to work after sixty-five; only a few of those who do are the principal supports of households; employed or not, a Canadian at age sixty-five begins to receive Old Age Security or Seniors Benefits and if he or she has been employed, is probably eligible for benefits through the Canada or Quebec Pension Plan and possibly other programs. It might be thought, therefore, that older workers do not need the protection of EI. Besides, many people are persuaded by the argument that encouraging older people to stay in the work force may aggravate the difficulties of younger workers finding jobs.

In 1988, a sixty-five-year-old worker challenged the validity of the age cut-off in the Unemployment Insurance Act, alleging that it violated that clause of the Canadian Charter of Rights and Freedoms that renders unconstitutional any law that discriminates on the basis of age alone. The challenge was upheld on appeal to the Federal Court of Canada. Rather than appeal further to the Supreme Court, the federal government chose to modify the Act so as to allow older workers to participate in Unemployment Insurance.

This, by the way, neatly illustrates the effect of the Charter. Whatever may be the reasons of policy underlying a piece of legislation, it is vulnerable to challenge if it appears to violate any article of the Charter. If the courts decide that the legislation indeed discriminates on one of the prohibited grounds, the defendant government may seek to show that the discrimination is within 'reasonable limits' that can be 'demonstrably justified in a free and democratic society'. If the courts decide that the government has failed to show this, the legislation is declared invalid.

The near-universal coverage of the work force under UI was a late development. When UI was initiated, in 1941, less than half of Canadian workers were covered. Until 1970, coverage was limited to employments where the risk of unemployment was considered to be relatively high, and to workers whose regular annual incomes were less than a certain amount. The purpose of those limitations, presumably, was to restrict UI to those people most likely to be unemployed and to those likely to suffer most from even brief periods out of work. The amendments of 1970 broadened coverage to include virtually all employed people. As noted above, the exclusion of most part-time workers has been corrected by the amendments passed in 1995. Estimates of the proportion of employed Canadians covered by unemployment insurance have been as high as 93 per cent. The coverage of part-time workers will probably raise the percentage to even higher levels.

The Canadian fishing industry has been given special treatment under UI since 1956. The structure of the industry is unusual in that a high proportion of fishers are technically 'self-employed'; they own their own boats and gear, and they are paid for their catch, not for their hours of work. But many fishers sell all of their catch to one buyer, with whom their relationship is rather like that of employee to employer. (The traditionally used word 'fishermen' has been officially superseded by the gender-neutral, if uncolloquial, 'fishers'.)

Most of the fishing industry is highly seasonal, and in some fishing regions, particularly in eastern Canada, there is little alternative employment for fishers during the off-season. It is probable that a number of fishers will therefore become eligible for benefits predictably and repeatedly, compromising somewhat the character of EI as a protection against the *risk* of 'temporary, involuntary' unemployment. At the same time, without EI thousands of people in the fishing industry would face hardship year after year. Inclusion of fishers via a special section of the Act was one of the significant modifications made in 1970; and the contraction of benefits to workers in the fishing industry has been one of the most hotly contentious parts of the changes announced in 1995.

The treatment of seasonal workers has been a continuing subject of discussion and controversy (see 'Criticisms' below). It was a major focus of the 1986 Report of the Commission of Inquiry on Unemployment Insurance, the Forget Commission (Canada, Commission of Inquiry, 1986). In 1995, radical changes were introduced that will alter EI's coverage of seasonal workers (under the classification of 'repeated claimants').

As with any insurance, there is an important difference between being *covered* by Employment Insurance and being *eligible to receive* the benefits it provides. Over 90 per cent of workers are covered at a given moment. Because a certain quantity of work is required before one becomes eligible for benefits, not all covered workers are, at any one moment, eligible. And since there is a time limit on the duration of benefits, there are at any moment a number of unemployed workers who were covered but who have exhausted their benefits. The tighter the eligibility rules at both ends—the qualifying end and the duration end—the fewer the workers eligible to receive benefits. Thus the numbers of beneficiaries and the total amounts of benefits are certain to decline in coming years, even if unemployment remains at current levels.

Eligibility for Benefits

In principle, a person is eligible to receive benefits: (1) if covered by EI in the first place; (2) if employed for the minimum required number of hours in the preceding year, earning a certain minimum total income—in other words, if the worker has a current real attachment to employment; and (3) if unemployed—that is, does not quit, or is not fired 'for cause'; in both those cases, theoretically, the person's problem is not the disappearance of employment, and in both cases eligibility for benefits is delayed; and (4) if 'able and willing' to work. If one is not able to work, then *that* is the problem, not the unavailability of a job (but there is provision for benefits during a short-term job loss due to illness). Similarly, if a worker does not demonstrate the necessary willingness to work, then his/her unemployment is not considered involuntary. Assessment of 'willingness to work' is difficult, is partly discretionary, and is the root of much contention.

The detailed rules governing eligibility have been manipulated many times in the history of UI/EI. They were made more restrictive in the 1995 Employment Insurance Act. The minimum period of time that a worker must have worked, and consequently the minimum contributions that employer and worker must have made to the EI fund in order to qualify for benefits, vary according to the level of unemployment in the worker's region. If regional unemployment is relatively high, the number of hours of work needed to qualify is lower, and vice versa.

It is possible for a worker to work too few hours in the year to be eligible for benefits at all—yet to have made contributions to EI for every hour worked. As this would be patently unfair, a worker whose total work income for the year falls below a certain level will have all the employee's share of EI contributions returned through the income tax system.

To reinforce the earlier point that eligibility must be distinguished from coverage, a Statistics Canada report released in March, 1995 estimated that barely half of the 1.5 million Canadians then unemployed were receiving unemployment insurance benefits, though virtually all of them had been covered by UI. By contrast, in 1990, under the looser eligibility rules then in effect, 77 per cent of the jobless were receiving UI benefits (Statistics Canada, 1995).

Once the unemployed worker has been judged eligible to receive benefits, according to the criteria of the region in which he/she works, the amount and duration of the benefits payable must be determined.

Benefits

The employee, while working, is being insured against an interruption of income; also protected, implicitly, are all others who depend on that employee's income. The worker and the employer each pays a premium that is a certain percentage of the worker's weekly income, up to a ceiling. *What is insured is the worker's income under the ceiling*; income above the ceiling is not covered. The ceiling has tended to be somewhere around the national average annual income from employment, so necessarily many workers have incomes that go well above the EI ceiling. Since premiums bear a certain arithmetical relation to insurable incomes, it is logical that benefits should also bear some arithmetical relation to insurable incomes. Hence, weekly EI benefits are a certain percentage of normal weekly income, up to the ceiling. This percentage has varied over time; lately it has been 55 per cent in the simplest case, with possible embellishments. A 1994 change, retained in the 1995 Act, allows 60 per cent for workers with modest incomes and with dependants. (A similar provision in the original Act, by which lower-income earners got a higher percentage while higher-income earners got a lower percentage of insurable income, was eliminated many years ago; in spirit, that feature has returned.) Under the 1995–6 amendments, frequent claimants will get benefits at a reduced rate—50 per cent.

Benefits are based on the *weekly* income of the individual worker, not on *annual* income, nor on family or household income. That explains why a certain percentage of total UI/EI benefits has gone to people with annual incomes well above average, and even more so, to persons in families with quite substantial household incomes. In 1992, 40 per cent of claimants lived in families with total incomes over $50,000, which was then very close to the national average household income (Human Resources Development Canada, 1994, p. 47).

Conceptually, this is consistent with the *insurance* character of EI: in a pure insurance model, there is no concern with the circumstances of insured persons at the time they receive benefits to which they are entitled. But since 1990, benefits have been subjected to an after-the-fact income test. Recipients of EI benefits whose total income, for a year in which they receive benefits, exceeds 1.5 times the maximum insurable, have their benefits 'clawed back' through the federal income tax system, at the rate of 30 per cent. Under the new 1995 Act, the clawback thresholds have been sharply reduced. A substantial measure of *equalization* has thus been added to the *security* objective of Employment Insurance: EI has been made more redistributive downwards than it was previously. (Since it was hotly controversial at the time, it is worth noting that the government's original proposal to base the clawback on *family* rather than *individual* income was beaten back, largely by the opposition of women's groups; it would have had an adverse impact upon the incomes of families with second income earners, and a majority of those considered to be second earners are women.)

There is also a time dimension to benefits. Benefits do not begin immediately the day a worker is laid off—there is a waiting period. The duration of benefits depends on the number of hours (formerly weeks) the claimant has worked in the past year or since the last claim for benefits, and on the regional unemployment rate. The maximum duration of benefits corresponding to a given number of hours of work is longer in regions of high unemployment, where it is harder to find a new job. The duration of benefit periods, having been lengthened under the Unemployment Insurance Act amendments of 1970, has been more and more severely limited by successive amendments since the mid-1980s, and was limited further still by the 1995 Act.

Finally, for years there have been grumblings that certain classes of workers, including but not limited to workers in seasonal industries, and not all poor, have drawn benefits repeatedly, establishing a pattern of working a little more than the weeks required to qualify, and drawing benefits for nearly as long as possible. The Forget Commission Report provides evidence that a large proportion of *claims* was being made by a small proportion of *claimants*.

Many employers also behave in ways that exploit the program for their own advantage, or so it appears. It has been pointed out that 12 per cent of Canadian firms, with 14 per cent of the employees, accounted for 38 per cent of the benefits paid (Cox/Canadian Press, 1995).

There is no necessary implication of fraud or abuse in the analysis of the evidence; a great many people are caught in situations that leave them little choice but a succession of irregular jobs interspersed with periods of unemployment. But in a good deal of the discussion of this issue, there are definite implications of abuse of the system by both employers and workers; examples have always been highlighted of workers who earn high incomes in brief periods of intense activity, then coast for months on EI, often with the complicity of their employers, and of employers who synchronize their production schedules and layoff periods with EI benefit periods.

In response to this evidence and these criticisms, the Employment Insurance Act of 1995 created the beginnings of a two-tier Employment Insurance system. Regular claimants are to have one scale of eligibility rules and one scale of amounts and duration of benefits. Frequent claimants, defined as persons who have made three claims in the past five years, will have another scale, with longer qualifying periods, reduced benefits, and shorter benefit periods. At the same time, training, rehabilitation, and reinsertion programs are to be expanded, ostensibly to improve the labour market chances of frequent claimants.

The controversy surrounding these and other issues, brought to public attention by the Employment Insurance Act, are reviewed below (see 'Criticisms').

Add-ons: Sickness, Maternity/Adoption, Age, Employment Benefits

(1) *Sickness.* What if someone loses a job because he/she becomes sick? Or loses a job and then, while unemployed, becomes too sick to work? That person is not 'able to work', and thus might be thought not to fit the definition of 'unemployed' for purposes of EI. Yet, there is nowhere any kind of public program giving cash benefits to people on account of illness. (Workers' Compensation sickness indemnities are limited to workers who suffer from work-related illnesses.) There are employee benefit programs and private insurance programs that provide indemnities during periods of illness, but no public programs. To fill this gap partially, provisions for time-limited sickness benefits were added to UI early in its history. Unlike ordinary unemployment benefits, sickness benefits are means-tested, that is, they take into account certain other income the beneficiary may be getting, for example, from a company sickness benefit plan.

(2) *Maternity and Adoption.* An employed woman is unable to work for a period of time before and after the birth of a child. Being unable to work, she falls outside the narrow definition of 'unemployed'. Furthermore, an insurance program normally insures only against involuntary occurrences, and there is a presumption that maternity is voluntary. Despite these reservations about their appropriateness as a component of EI, maternity benefits are paid to an eligible woman off work for a limited period of time around the anticipated date of the birth of the child. A similar benefit is paid to a worker, male or female, who takes time off work upon the adoption of a child.

(3) *Age.* Until 1989, a worker ceased to be covered by UI at sixty-five, whether still working or not. UI provided a lump-sum termination benefit at age sixty-five, to tide a worker over until old age benefits from other programs began to arrive. This benefit became available at age sixty-five, however, not specifically at retirement. In a 1988–9 court case, referred to earlier, the exclusion of workers over sixty-five was ruled to violate the clause of the Canadian Charter of Rights and Freedoms that prohibits discrimination in any law on the basis of age. As a consequence, workers over sixty-five are treated the same as workers under sixty-five and there is no more termination benefit.

Some critics ask whether UI is the right vehicle for these added-on programs, however badly needed they may be. If sickness, maternity and adoption, and age benefits are needed, it is asked, what sense does it make to tie them to the coverage, eligibility, and benefit limitations of Unemployment Insurance, and to pay for them out of a payroll deduction? For since its inception, UI/EI has been financed wholly (as now) or mostly (until 1989–90) by a special payroll levy.

(4) *Employment Benefits.* A new category of benefits was added in 1995–6, entitled 'Employment Benefits', to distinguish them from the above cash benefits. The category is new; most of the benefits included are not entirely new, though they are expanded. These changes too are to be phased in over a period of years. They are directed especially at frequent claimants, whose work experience indicates that they could benefit from some help in finding and holding jobs. These include greatly expanded training and education programs (to be operated in close collaboration with the provinces, out of regard for constitutional sensitivities), assistance to unemployed persons creating small businesses, and a modest program of time-limited wage subsidization to encourage employers to hire unemployed people.

Financing

At first, from 1941 to 1971, financing of UI was simple: equal contributions were collected from covered employees and their employers, then the federal government contributed, out of its tax revenues, 20 per cent of the total contributions collected. The principle was the one proposed by Joseph in Egypt (Genesis 41): in 'fat' years, more was collected than was paid out in benefits; a 'fund' was thereby built up, which could be loaned out to make additional revenue from interest; then in lean years, when benefits paid exceeded contributions plus the government's 20 per cent, the saved-up fund was drawn upon. Over the cycle of good years and bad years, the fund was supposed to remain solvent. In fact, it did not. The deficits of the bad years greatly exceeded the surpluses of the good years, to such an extent that the federal government on occasion felt obliged to subsidize the fund out of tax revenues.

After the 1970 amendments to the Unemployment Insurance Act, financing grew more complicated. There was no longer a fund expected to grow in good years and to dwindle in bad; Unemployment Insurance was put on a year-to-year pay-as-you-go basis. The employers' contributions were raised to 1.4 times the employees'. The employee plus employer contributions were calculated to meet the costs of 'regular' benefits, to be paid when unemployment was at a level that was 'normal' or lower; when unemployment was higher than 'normal', the federal government contributed enough out of taxes to cover the consequent 'extended' benefits, as they were called. The government share each year depended upon that year's unemployment experience, and was accordingly unpredictable. Unfortunately, unemployment was invariably higher than 'normal', and the 'government share' higher than had been hoped. This arrangement made UI a substantial and rather volatile item in the annual budget of the federal government, far more so than when the government's share was limited to 20 per cent of employee and employer contributions.

Another turn was taken in the 1990 amendments to the Unemployment Insurance Act. The distinction between the financing of 'normal' and 'extended' benefits was abolished; all cash benefits were now to be paid entirely out of employer-employee contributions, and the Joseph-in-Egypt principle was resurrected, i.e., the fund was recreated, intended to grow in years of prosperity, as a cushion against years of recession. The 'government share' was now to be used exclusively to finance retraining and job development projects, not to pay UI benefits.

In the latest version (see below, '1995–6'), cash benefits are still financed wholly by contributions; the 'fund' is retained; and the fund itself is to pay for most of the cost of employment development and training activities, categorized now as 'employment benefits' (see above). The government will finance, out of taxes, a certain share of the so-called 'employment benefits'.

It is important to grasp the difference between 'employee and employer contributions' and 'government share'. There is a natural tendency to regard unemployment insurance as a government program like any other, and to think of the money involved as taxation, since it is collected from workers and employers in much the same way as taxes are collected; but unlike taxes, UI contributions are not available to the government for any other purposes than those of UI. The distinction is significant in the heated discussion of whether UI is really 'social insurance', designed to meet a certain risk, for which earnings-based contributions are appropriate, or an 'income transfer', intended to supplement low incomes, for which taxation is appropriate. This brings up important issues of justice, efficiency, intergovernmental finance, and integration with other social welfare measures.

Administration

The job of running UI has three major components: (1) collecting contributions and paying out benefits, in a manner somewhat similar to that of any insurance business; (2) making judgements about claims and entitlements and in dealing with appeals from initial decisions, a quasi-judicial function; and (3) making regulations that have the force of law when proper approvals from Cabinet and/or Parliament have been obtained, a delegated law-making function. It is generally agreed that this combination of administrative, quasi-judicial, and quasi-legislative functions calls for a special kind of responsible agency, different from an ordinary department of government headed by a minister who is a member of the governing political party. Accordingly, in 1940, the Unemployment Insurance Commission was created. Its name was recently changed to the Canada Employment and Immigration Commission.

Like other Boards and Commissions with special functions, this Commission is a relatively small body of persons selected by the government on the basis of their qualifications, broadly representative of the principal interests involved in the issue of employment. They are given secure terms of appointment so as to make them independent of the government of the day. Besides these formal protections of its independence, the Commission is assured by the legislation of sufficient financing to allow it to operate independently of the political preferences of the government.

Since 1977, however, dating from a time when the financing of UI called for a relatively large input from the tax resources of the government, the Canada Employment and Immigration Commission has been located more closely within the embrace of the Department of Employment and Immigration, reduced to three members, and chaired by the deputy minister of that department. In the eyes of some, this has severely compromised its supposed independence (Commission of Inquiry, 1986, pp. 253–5).

As with any important public program, the current shape of Employment Insurance can best be understood with a knowledge of its origins and some of its history.

BACKGROUND TO THE UNEMPLOYMENT INSURANCE ACT, 1940 AND AFTER

Employment and welfare are both matters that are supposed to be within provincial jurisdiction. Yet here, in Employment Insurance, we have a program clearly related to both employment and welfare, authorized entirely by federal legislation, run by a federally appointed Commission, with no direct provincial involvement at all except in workforce training. How did this apparent constitutional anomaly come about?

1940. The explanation goes back to the 1930s and the severe economic depression that then afflicted virtually all industrial countries. In Canada, every indicator of economic health was at an alarmingly low point. Unemployment was severe for years on end. The only available public programs for income support of unemployed people and their dependants were relief programs for the indigent, paid for and run by the provinces and municipalities. These programs, not having been designed for a massive unemployment crisis, were swamped by the unprecedented numbers of people in extreme need.

Widespread unemployment created a problem of social order as well, for the large numbers of able-bodied unemployed men began to make their discontent and anger unmistakably clear. Partly to isolate these transient unemployed men, partly presumably to give them some material support, the federal government created 'work camps' in remote areas where the men were fed, housed, and kept more or less occupied (Struthers, 1983). The fear of disorder certainly spurred the search for solutions—some writers go so far as to say that the fear of social unrest was the principal motivation for the social reforms of the 1930s, including unemployment insurance.

In Canada, there was wide though not universal agreement that the problem was too broad in scope to be successfully met by action at the provincial level. The provinces, whose taxing powers are limited by the Constitution, had their hands more than full coping with the swollen costs of relief of the indigent. Some of the poorer provinces and many munici-palities technically went bankrupt during the 1930s, meaning that they were unable to pay all their bills and to repay money they had borrowed. This was obviously serious; it threatened public confidence in the legiti-macy of the institutions of government.

The provinces with the heaviest welfare burdens were naturally the ones with the least resources. Now, the federal government had wider taxing powers than the provinces, and only the federal government, with its nation-wide base, was in a position to equalize the burdens as between the regions worst hit and those faring less badly. But in the face of these real-ities, the division of powers in the British North America Act was gener-ally understood to stand in the way of federal legislation to deal with the crisis. The provinces had the power, but not the resources; the federal gov-ernment had larger resources, but not the power.

In 1935, the federal government of the day (a Conservative Party gov-ernment) charged into this constitutional dilemma, and pushed social leg-islation through Parliament, including an act creating an unemployment insurance program. It was immediately challenged in the courts as being beyond the powers (*ultra vires*) of the federal Parliament. And the courts agreed—first the Supreme Court of Canada and then the so-called Judicial Committee of the Privy Council, the highest court of the British Commonwealth and Empire, at that time the final appeal tribunal in mat-ters relating to the Canadian Constitution.

The only way around this was to change the Constitution. A later federal government (Liberal this time) first got the agreement of all the provinces to a change in the BNA Act to permit the Parliament of Canada to legislate in the field of unemployment insurance, next secured the amendment through the then necessary Act of the Imperial Parliament in Great Britain, and then had the Parliament of Canada pass the Unemployment Insurance Act. (We will see this process again when we look at programs for older people—agreement on an amendment to the BNA Act, followed by the amendment, followed by desired legislation.)

The Unemployment Insurance Act was passed in 1940 and came into effect in 1941, a tremendous landmark in the history of social policy in Canada.

1971. After thirty years of experience, the Act was amended in 1971 in a number of basic ways:

(1) *The number of workers covered was greatly increased.* As we have seen, the original program had excluded high-income earners and those in some employments in which unemployment was not seen as a serious risk, such as public service and teaching. On the other hand, certain highly seasonal and some highly unemployment-prone employments, like fishing and most logging, has also been excluded. In fact, in 1941 less than half the total work force had been covered. Since 1971, just about every kind of employment has been included, presumably on the ground that everyone is now subject to some risk of loss of job, and perhaps also on the ground that all should contribute, whether likely to be unemployed or not, and there has been no income limit for coverage under UI. As discussed above, there *is* a ceiling on the level of income that is covered.
(2) *The method of financing was changed.* Not to repeat all that was said above under 'Financing', there was no longer to be a fund. Employer contributions (now 1.4 times as large as employee contributions) and employee contributions together were supposed to cover benefits for moderate unemployment; the federal government was henceforth committed to pay benefits attributable to abnormally high unemployment.
(3) *Benefits were increased*, especially in terms of their duration. Of course, the great increase in numbers of people covered meant that the revenues of the Commission increased correspondingly—but changes in benefits increased expenditures even more. It became possible to receive benefits after a very short 'qualifying' period of work. The concept of 'extended' benefits was introduced, allowing for extra periods of benefit in exceptionally depressed regions, or even nation-wide under certain conditions. Under the new provisions, benefit outlays increased so much more than expected that the rules were tightened up in the mid-1970s: a longer period of work was required to qualify for initial benefits; the maximum duration of benefits was reduced and was more closely linked to the number of weeks of work prior to loss of job. The benefit rate—the percentage of insured income paid to unemployed persons who qualify for benefits—has been reduced a number of times as well.

1990. Another sharp turn was taken at the beginning of the 1990s. Amendments enacted in 1990 threw all costs of cash benefits upon the fund generated by employer and employee contributions; the government thereafter paid only the costs of associated programs, such as training, relocation, job creation (supplemental income to UI beneficiaries who take part in projects 'of service to the community'). In other words, *the govern-*

ment withdrew from the financial support of unemployment insurance as such. The amendments restored the 'fund', expected to rise and fall with the business cycle. And a variant of income testing was introduced, that reduces the net benefits paid in a year to a high-income worker who experiences a period of unemployment: if the worker's total income for the year exceeds a certain amount (1.5 times the maximum insurable earnings), some portion of the UI benefits received will be 'clawed back'; this is in addition to the obligation to pay normal income tax on UI benefits received.

1995–6. The Canadian economy went through a serious recession in the early 1990s. The federal government slowed the drain on the UI fund by reducing benefits in a couple of steps—again increasing the qualifying period, again shortening the maximum duration of benefits, and tinkering with the benefit rate. While there has been an economic recovery according to many economic indicators, unemployment has not been reduced as much as expected. The factors mentioned in our first chapter under the headings '(i) The Changing Economic Context' and '(ii) The Meaning of Employment' were clearly affecting unemployment in Canada. Above all, perhaps, the pressures upon the tax system created by the accumulated public debt of Canada and the provinces were taken by the federal government to call for restrictions in all programs.

Despite some trimming of benefits, total expenditures on UI had grown sharply—'Regular Benefits' alone had jumped from $10 billion in 1989-90 to $15 billion in 1991–2; all UI expenditures had peaked at over $20 billion in 1991, when the average number of regular benefits claimants throughout the year was well over one million. Total benefits fell to $17 billion in 1994-5. (One must not be misled; total expenditures may fall because there are fewer unemployed, or because eligibility restrictions mean that fewer of the unemployed are eligible for benefits, or because benefits to individuals have been reduced—or, as in 1994–5, for all of those reasons.) Even though UI technically does not form part of the government's revenues and expenditures, the money flow in UI is significantly large; it affects the tax capacity of all governments; and the incentives it creates arguably affect the performance of the Canadian economy. In 1995 the federal government argued that the way benefits were structured acted as a brake on adjustments that had to be made to cope with prevailing economic trends.

After a much-touted Social Security Review and public consultation process, the government introduced, late in 1995, the Employment Insurance Bill, incorporating the changes to the system that have been described above: a two-tier system of benefits, one for infrequent claimants, a less generous one for frequent claimants; a series of 'Employment Benefits', aimed at improving the employability of the frequently unemployed, to be paid for mostly by the UI fund; a shift to hours of work rather than weeks of work in the determination of eligibility to receive benefits, thus encompassing the inclusion, for the first time, of part-time workers as both beneficiaries and contributors; further raising of the quantity of work required to qualify for benefits, and further reductions in the permitted duration of benefits; and the basing of the clawback provision on *family* income where applicable.

The details of these provisions of the new Employment Insurance are of course important: as the saying goes, 'The devil is in the details.' The contribution rates, maximum insurable earnings, even the benefit rates, etc.—are, however, subject to change from year to year; any presentation here of such details would rapidly become outdated. What matters for present purposes is that the variations in the basic components of EI—financing, benefits, etc.—be seen in relation to each other and in relation to their economic and social context.

LEGISLATION

The current EI legislation is the Employment Insurance Act of 1995. As noted already, certain parts of this Act are quite new; other parts (contribution rates, benefit rates) represent further amendments to the Unemployment Insurance Amendment Act of 1989 and subsequent amendments up to 1994. The 1995 legislation represents a sharp turnaround from the previous major Amendment Acts of 1971 and 1976. As with any large and complex piece of legislation, amendments have been and will probably continue to be frequent.

While very little if any remains of the wording of the original 1940 Unemployment Insurance Act, its place in the history of Canadian social policy is assured.

CRITICISMS

As befits a social program that has worked its way into the social fabric of the country, (Un)Employment Insurance has been abundantly discussed, criticized, and defended. The references provided at the end of this chapter give a sampling of this discussion. Some main themes of the various critiques will be identified here.
• Predictably, some critics of the program say that benefits are too low. The benefit rate is now not much more than half of the unemployed person's regular weekly income, up to a maximum—and less for a frequent claimant.

Over UI/EI's fifty-seven years of existence, benefits have sometimes been higher for beneficiaries with dependants, sometimes not; currently they are slightly higher. Limiting the duration of benefits in accordance with the duration of previous consecutive work time is favourable to those who are rarely unemployed and who have no trouble building up maximum entitlement, but it does much less to soften the hardships of those who are particularly vulnerable to unemployment. In a sense, this is a basic criticism of the insurance approach, which naturally tends to tighten up on those who make the most claims, and which, in its pure form, takes no account of need.

The maximum insurable income is in the neighbourhood of the national average industrial income, so naturally there are many workers whose regular incomes are above that maximum insurable; for all of them, EI benefits replace less than half of their customary income. By the same token, these upper-income workers pay a smaller percentage of their earnings in EI contributions than workers whose incomes are at or below the maximum insurable income; hence, the criticism that EI is financed by a levy that is regressive above the maximum insurable income, a criticism that unfortunately is not always accompanied by the acknowledgement that benefits, too, fall off as a proportion of income.

• There is a problem in terms of congruity between federal policies with respect to EI and federal contributions to provincial public assistance programs. In prosperous times fewer people will need welfare, and in times of recession more people will need welfare. Since the late 1970s, eligibility for UI/EI benefits has been tightening up from the 1970 standards—longer qualifying periods, shorter benefit periods. This means that fewer people will qualify for EI benefits, and of those who do, more will exhaust their EI benefits. An unemployed person who does not qualify for EI, or who has used up all the EI benefits available, has little alternative but to apply for public assistance. Unfortunately, while Ottawa has been cutting back on EI, it has also been reining in its financial support of provincial public assistance programs (as will be seen in more detail in Chapter Six). The provinces are in charge of public assistance, and are free to expand their programs if they wish, but since they are subject to the same fiscal constraints as the federal government, their recent tendency has been to restrain all their expenditures, including expenditures on welfare. The outcome is very distressing for those poor people whose poverty is caused by their long-term unemployment. Advocates for welfare recipients denounce the combination of restrictions on EI benefits and reduction of federal cost-sharing in public assistance as the dumping of a grossly unfair portion of the deficit-and-debt burdens of the federal government onto the poor.

• As noted above, the experience of the UI *fund*, before 1970 and since 1990, has been rather chequered. Beginning approximately with the 1990 amendments, however, the fund has functioned somewhat as designed. Unemployment insurance spending has been heavy, especially throughout the recession of 1991–3 and its lagging after-effects, but fund revenues have exceeded fund outlays for a few years. This is due, in large part, to the restrictions on benefits; but it is also due to the somewhat paradoxical fact that total employment in Canada has been *increasing* even while unemployment has been high: 7.2 million jobs in 1966, 12.4 million in 1993 (Human Resources Development Canada, 1994, pp. 13–14). The steadily increasing numbers of employed persons have generated a steadily increasing level of contributions into the UI/EI fund. Since the economic recovery of the mid-1990s, the fund has been in surplus.

It has struck many observers as ironic that the government should proceed with severe restrictive measures while the EI fund was sitting on a surplus of approximately $5 billion, built up over three years or so. There is no doubt that the reform has been driven by the felt imperative for overall deficit and debt reduction. The federal government willingly acknowledges that the state of the public finances has indeed influenced EI reform, but argues that the real objective has been to improve the long-term employability of the Canadian work force. As to the surplus, the government maintains that the EI fund is *supposed* to generate a surplus at times, and that to tamper with the surplus would compromise the fund's capacity to meet its obligations the next time economic activity begins to droop—and past experience lends some credence to this argument.

The point here is not to lean to one side or the other of the debate. But the proper use of surpluses in the EI fund, and the management of deficits in years when they occur, are issues that will remain relevant and controversial.

• The administration of UI/EI is often criticized for being bureaucratic and unfeeling. CEIC employees themselves express frustration over the limits within which they must work (Commission of Inquiry, 1986). The requirement that a recipient be actively looking for work is administratively unworkable at times of high local unemployment. The onus thereby placed on the unemployed person to prove his/her entitlement is seen as a variant of 'blaming the victim'.

• Initially, UI provided more generous benefits for unemployed workers with dependants. The 1995 Act envisions a slightly higher benefit rate for infrequent claimants with dependants. Arguments pro and con must bear in mind that the amounts formerly paid to families with dependent children through Family Allowances had been greatly increased over time, and that the Child Tax Credit that replaced the Family Allowances in 1993 is designed specifically to be of greatest assistance to lower-income families.

• Some critics allege that the system is too vulnerable to rip-offs, not only by less scrupulous recipients of benefits, who have been the usual suspects, but also by self-seeking employers, who can lay off workers temporarily, at little risk of losing their experienced work force, because the availability of benefits (paid for by all employers and employees) will keep them close to home. It should be borne in mind that UI was *intended* to make the plight of the unemployed less desperate, thereby enabling them to stay in their preferred line of work and in their own localities, rather than having to jump around from industry to industry and from place to place. In the consultation process preceding the 1995 Employment Insurance Act, there was much debate over employer exploitation of UI. Commented Minister of Human Resources Development Lloyd Axworthy, in his preliminary Green Paper, *Agenda: Jobs and Growth—Improving Social Security in Canada* (1994, p. 42):

A recent study by Statistics Canada reports that some businesses have structured their basic hiring and compensation practices around the UI program—for example, planning layoffs to coincide with UI qualification periods, and recalls with the end of UI benefits.

The automobile industry was singled out for special attention. Another much-cited abuse is collusion among employers and workers, whereby workers allegedly take extended periods off work while their employers attest that they are 'actively looking for work'.

In 1996, Quebec's Conseil de la santé et de la bien-être claimed that over 80 per cent of layoffs in that province fixed a date for rehiring; this and other evidence led the Conseil to conclude that the unemployment insurance system directly subsidizes employers who adapt their labour practices to its provisions for entitlement to benefits (Quebec, Conseil de la santé et du bien-être, 1996). In the nature of things, the extent of this manipulation is difficult to ascertain.

• Accepting that the very purpose of UI/EI has always been to make a period of unemployment tolerable, some critics argue that the trade-off between staying on EI benefits and looking for a different job is such that many will choose EI, at least for a time. Thus, they say, EI 'induces unemployment'. They argue that it also inhibits the mobility of workers to a damaging degree, since it incites them to stay where they are, rather than move to locations where their services might be more productive. Paradoxically, this may be most important among well-paid workers, especially those in certain seasonal industries, whose total annual incomes are fairly good. Employers in these industries can lay off their workers with full confidence that they will be available when recalled to work, and also with a good conscience, knowing that their workers' hardships while unemployed will not be intense. 'Induced unemployment' would be expected to be much less important among poorer-paid workers, whose total incomes are low, whose EI benefits are at the minimum levels, and who need every dollar they can get. Certainly EI cannot induce unemployment among people who lose their jobs before they have qualified for benefits, nor among those who have exhausted their benefits, and obviously the numbers of those increase with each extension of the required qualifying period and each shortening of the maximum duration of benefits. The extent of induced unemployment is admittedly difficult to estimate. Also, some participants in the debates about EI consider it to be a distraction to fuss about a marginal amount of induced unemployment when Canada is facing a serious structural economic problem of long-term unemployment.

• EI does not treat industries differently according to how high or low their unemployment experience has been. (As will be seen in the next chapter, this differentiation is standard practice in Workers' Compensation.) The idea of 'experience rating'—charging higher premiums in the industries and/or firms that generate the most unemployment—has often been proposed. It is argued that such a premium structure would be fair: employees most likely to draw EI benefits would contribute most to EI, as would their employers, while those least likely would contribute least—and it would give employers an added incentive to keep workers at work as long as possible. The counter-argument is that most of the unemployment-prone industries, and certainly most of the unemployment-prone workers, have enough trouble without increasing their obligatory EI contributions. Moreover, in the worst-affected industries, such experience rating would add greatly to the cost to the employer of each additional worker hired, and might indeed induce unemployment by motivating employers to hire fewer workers, and/or by worsening the industry's competitive position (Commission of Inquiry, 1986, pp. 124–5).

• While it is hardly fair to criticize UI/EI for failing to do what nobody ever said it would do, it is wise to recognize its limitations. EI will not lift a poor person, or a poor family, out of poverty. In fact, it is something of a cliché that very few poor people ever get anything from EI, and that only a few EI recipients are really poor. Of course, EI prevents many a person of modest income from falling into poverty, and that is a great benefit to society. But if a worker whose income maintains the family at the level of poverty loses her job, EI pays benefits that are lower still. (Poverty in Canada is proportionately much more frequent among households headed by females than by males.) Many relatively poor people do work whenever they can, but if they work only sporadically or for short spells, they will at best be eligible for only limited EI benefits. The lengthening of the qualifying period means that it may be even more true in the future than in the past that very little in the way of EI benefits will go to poor people.

• Finally, we may situate Canada's experience with unemployment insurance within a debate now occurring in all industrial countries concerning the relative merits of 'passive' and 'active' social security measures, including EI. At the outset, the reader should be on guard against taking 'active' and 'passive' as pejorative terms: 'active' probably sounds better, but that does not mean that any 'active' program is inherently better than any 'passive' program.

The basic proposition is that the chronically unemployed may indeed need income support, such as EI provides, but that they really need something else as well, something that will change their relationship to the labour market. Few will challenge the statement in that simple form, but there are considerable differences in the kinds of public interventions recommended by different authorities.

In this context, 'passive' means, by and large, programs that give people money under stated conditions; 'active' means programs like vocational training, job creation, and so forth, that try to change the situation of the beneficiary. The benefits provisions of EI would be regarded as 'passive', while the training, job-finding, relocation, and enterprise-encouraging provisions, now called 'Employment Benefits', would be deemed 'active'. Students of social policy generally agree that the countries of continental

Western Europe invest much more heavily in active programs than do Britain, the United States, and Canada, and at the same time countries in Western Europe tend to have more generous financial benefits (Muszynski, 1994).

Advocates of active programs base their position on the need to do something about the *causes* of unemployment and unemployability, particularly the problems of the individual worker facing structural factors in the labour market, rather than simply palliating the symptoms by distributing money, whether generously or meanly, to the unemployed, who then stay trapped as a class in an endless cycle with no real hope of change.

Advocates of passive programs, who are, of course, disinclined to use that label themselves, point to the spotty record of a great many job training and job creation programs. They observe that over the last quarter-century total employment has increased absolutely and relatively far more in North America than in Europe, with all its emphasis on 'active' programs, and if many (but by no means all) of the North American jobs, regrettably, have been at low wages, that still may be better for the individual and for society than no jobs. They make the standard small-l liberal philosophical argument that 'passive' income distribution enhances individual choice, whereas 'active' programs inevitably impose controls, and depend for their success on the wisdom and foresight of planners and bureaucrats. Finally, they point out that 'active' programs cost so much to run—they must pay for teachers or trainers, job counsellors, other staff, facilities, etc.—that if they 'pay off' at all, it can only be in the very long run.

While by no means an extreme statement of the 'active' position, the government Green Paper *Agenda: Jobs and Growth—Improving Social Security in Canada,* which launched the public consultation process leading to the Employment Insurance Act, tends more in that direction than any previous federal government policy statement; and the inclusion in the Employment Insurance Act of a new section on 'Employment Benefits' (few of which are really new), indicates movement along that track.

OTHER ASPECTS OF THE ISSUE

The student should remember that this text gives no attention to a number of programs intended to deal with the consequences of unemployment: job creation programs, tax incentives to employers to provide their own supplementary employment benefits, etc. It must not be inferred that these issues and programs are regarded as unimportant.

Similarly, this text does not attempt any analysis of the responsibility of government for the overall levels of employment and unemployment in Canada, or of the various instruments of policy that are used in the pursuit of high employment and economic objectives: fiscal policy, monetary policy, industrial policy, regional development policy (intended to improve the economic prospects of regions in difficulty), and so on. What is done under these policy headings helps determine the terrain within which social policy must operate (Mendelson, 1993). These topics cannot be treated fully here, but their importance must be recognized.

Nor, finally, is there any discussion of just where Employment Insurance fits into the fabric of a society whose economic character is basically capitalist or mixed capitalist. For example, we do not discuss the proposition that the real purpose of Employment Insurance, as of most government programs, is to enhance the process of accumulation of capital, most of which is in private hands. The objectives of this text will be satisfied if the reader becomes equipped to understand the discussion of such broader issues.

PROBLEMS FOR DISCUSSION

1. Some commentators question whether EI has become an income maintenance program for certain classes of frequently unemployed workers rather than an income security program for all workers. They argue that you can say beforehand which groups of workers, at which income levels, in which parts of the country, are going to contribute much and receive little, and which are going to contribute little and receive much, and in practice, this had made UI more of a transfer program than an insurance program dealing with an unpredictably distributed risk. Through its tighter eligibility rules, the 1995 Act will reduce this redistributive effect, but this problem will probably remain on the agenda for discussion. The issue is far from a matter of words ('insurance', 'transfer'): if one is trying to redistribute income to the relatively poor, it is questionable to do it through a program that is financed exclusively by a non-progressive levy upon the incomes of workers, and that designedly pays higher benefits to well-paid than to poorly-paid workers. This issue is debated in the majority and minority reports of the Forget Commission.

2. One school of thought argues that certain policy decisions of governments, such as those affecting interest rates and the international trading value of the Canadian dollar, have among their known and foreseeable consequences a certain level of unemployment. Seen in this light, Employment Insurance is not so much insurance as compensation to certain workers for the burden that they bear as a result of those policies (Soderstrom, 1984).

3. From the very first, critics have questioned the foundations of the insurance principle and its financing by premiums. In the absence of anything like EI, financial support of the needy was a burden on the general taxpayer, or, if you wish, on the whole society. These critics say that the premiums structure places the burden of much of this support on the workers themselves, with little or no help from the general taxpayer, notably, for purposes of the argument, the taxpayer who is a wealthy owner of capital.

The very large 'employer's share' of EI contributions does not necessarily refute this criticism, for it is seen by employers as simply part of their labour costs. (From this point of view, it would not matter whether employees paid all, employers paid all, or employers and employees split the premiums between them: it comes to the same thing in the end. Of course, if employees paid all, it may be questioned whether employers *would* include in the wages they paid amounts sufficient to cover EI premiums, but that does not affect the point that EI premiums are a payroll cost.) In this view, EI premiums improperly shift the burden of support of the needy unemployed from the general taxpayer to the employees.

The other side of this argument is that EI benefits are *not* limited to the 'needy unemployed', let alone the indigent. On the contrary, the indigent get little from EI; the burden borne by EI is the support of the workers, not the support of the indigent; and therefore the criticism of the insurance principle is misdirected.

4. The other side of the question concerning financing should also be raised. The employer-employee contributions amount to an additional charge on payroll. The wage or salary that the employee is paid must include the employee's contribution to EI, then the employer must tack on a still larger contribution of his/her own. As we shall see, many other programs are financed by employer levies based on the incomes of employees. The employer naturally takes account of all this as part of the expense that each employee represents. At the margin, these extra levies may deter the employer from hiring more employees, or may encourage the employer to invest in machinery for which he/she will not be taxed extra for every hour it works. Some spokespeople for labour argue that the financing of programs out of premiums based on payroll amounts to a discriminatory tax that falls on employment cost alone and spares all the other factors that measure the ability to pay of businesses or, *mutatis mutandis*, of public-sector organizations.

5. The fragility of the independence of the Canada Employment and Immigration Commission from government control has been commented on above. One may ask whether, in the public mind, the Commission's 'independence' is clearly understood, or regarded as particularly important. It is possible that many Canadians think of Employment Insurance as entirely a government responsibility, financially as well as politically, and regard EI premiums as taxes. Ever since the 1989 amendments terminated the government's tax-supported share of financing of EI benefits, this view of UI/EI has been logically untenable, but logical incoherence does not prevent its wide acceptance. And the way in which the 1995 Act brought EI into line with the government's debt- and deficit-reduction policy further watered down any sense of EI's independence from the government.

6. With reference to what we have called 'add-ons', the question has already been raised as to whether the EI program is being pushed out of shape, and

therefore rendered vulnerable to political attack, by being asked to take on burdens for which it is not designed.

STUDY QUESTIONS

The above outline of Employment Insurance contains very few specific details about the program. It refers to the employee's contribution rate, and to the employer's considerably larger rate, and to maximum and minimum income levels upon which benefits are calculated, without saying what those rates and levels are. One reason for omitting these figures is that many of them change at regular intervals. Sources of detailed data of this kind are listed in the following bibliography. One objective of the bibliographies in this text is to direct the reader to answers to such factual questions as the following:

1. How much does an employee contribute to EI? How much does an employer contribute?
2. What percentage of insurable income is paid as the weekly EI benefit? What is the maximum weekly income on which employee and employer contributions and benefits are based? What, then, is the maximum weekly benefit? What is the minimum insurable income? the minimum benefit?
3. In some recent year, how much money was paid out in UI/EI benefits? By your calculation, how much of this was paid for by employees? by employers?
4. How many people received UI/EI benefits in the latest year for which you can find the answer? How has that number varied in recent years?
5. What was the average weekly benefit paid in the latest year for which reports are available? How close was the average to the maximum payable? What was the average duration of benefits? Did many people receive benefits for the maximum period possible? How long must a person be out of work before receiving benefits?
6. How much was spent on sickness and maternity benefits in the latest reported year? On what terms are sickness and maternity benefits given?

In addition to being readily able to locate information, the reader should be able to respond to questions such as the following:

7. What is the logical linkage between the financing, the administration, and the benefits of EI?
8. How does it happen that an employment-related welfare program like EI is operated at the federal level?
9. Some people say that EI is a massive income transfer program masquerading as a social insurance. What basis is there for such a perception?
10. Some years ago, the chairman of the then Unemployment Insurance Commission said, 'I think it is only fair to say that the program can

only have an indirect effect on the poverty situation.' Is that not an alarming admission about a program that distributes billions of dollars annually?

11. One can understand workers being obliged to contribute to EI; after all, they get the benefits. What justification is there for making employers pay any contribution, let alone a much larger one?

12. For years, many observers complained that UI provided no protection whatever for the increasing number of part-time workers in Canada. Now that part-time workers are covered, some observers are complaining that part-time workers will be hurt because now they have to pay premiums and employers have to pay EI premiums for them. Whose complaint is more justified?

BIBLIOGRAPHY

Bailey, D.J., and M. Naemark (1977). 'A Note on the Transfer Payment Implications of Benefit and Contribution Operations under the UI Act', *Canadian Statistical Review* (November). Crunches some numbers to assess the income-transferring propensities of UI in the 1970s.

Caledon Institute of Social Policy (1995). *Critical Commentaries on the Social Security Review*. Ottawa: The Institute.

Canada. The Unemployment Insurance Act. *Statutes of Canada, 1940*, Chapter 44.

Canada. Act to Amend the Unemployment Insurance Act. *Statutes of Canada, 1971*, Chapter 48.

Canada. The Employment Insurance Act, *Statutes of Canada, 1995*.

Canada. Commission of Inquiry on Unemployment Insurance (Claude E. Forget, Chair) (1986). *Report*. A 'must read', at least in part. This exceedingly thorough Report proposed a number of controversial reforms to UI—so controversial that two members of the Commission submitted their own, diametrically opposed Minority Report. The stated guiding principle of the majority Report was to limit UI to the maintenance of income through periods of involuntary unemployment, and to shift to other welfare programs burdens not properly borne by Unemployment Insurance. The Commission did *not* recommend simply cutting off benefits without proposing an alternative form of relief. The Report was given a generally frigid reception, but the 1990 amendments followed some of its recommendations, notably the withdrawal of the government's tax-based contribution and a certain emphasis upon employment retraining for unemployed people, but Mr Forget himself dismissed the 1990 amendments as superficial. The Report recommended coverage of part-time workers; this was included in the 1995–6 Employment Insurance legislation.

Canada. Department of Labour (1970). *Unemployment Insurance in the 1970s*. The government White Paper announcing and explaining the reforms presented to Parliament for UI reform in 1970–1. Historically important, because the changes made were very extensive; from 1971 to 1990, UI had a character unlike what had gone before or what has followed.

Canada. Economic Council of Canada (1976). *People and Jobs: A Study of the Canadian Labour Market*, pp. 143–70. Ottawa: The Council. This excerpt considers whether UI adds to the number of unemployed people in Canada. (See also Green and Cousineau, cited below.)

Canada. Employment and Immigration Canada (1981). *Distributive and Redistributive Effects of the UI Program*. Task Force on Unemployment Insurance, Technical Study 10.

———— (1981). *Income Distribution through UI: An Analysis by Individual and Family Income Class in 1977*. Task Force on Unemployment Insurance, Technical Study 11.

Canada. Human Resources Development Canada (1996). *Basic Facts on Social Security Programs*. Exceedingly useful compendium of federal programs, published at irregular intervals.

Canada. Human Resources Development Canada. *Inventory of Income Security Programs in Canada*. Comprehensive descriptions and statistics on federal and provincial programs. Published at irregular intervals.

Canada. Human Resources Development Canada (1994). *Agenda: Jobs and Growth— Improving Social Security in Canada*. The Green Paper in which the federal government announced in broad terms its intentions for social security reform in coming years.

Canada. Statistics Canada regularly publishes a series of reports on all aspects of unemployment insurance, formerly entitled *Unemployment Insurance Statistics*, now entitled *Employment Insurance Statistics*.

Cox, Wendy/Canadian Press (1995). 'Ottawa to unveil UI reforms in Commons this week', *The Gazette*, Montreal, 27 November. Includes data on exploitation of UI by employers.

Cuneo, Carl (1979). 'State, Class and Reserve Labour: The Case of the 1941 Canadian Unemployment Insurance Act', *Canadian Review of Sociology and Anthropology* 16. UI is analysed as an instance of the Canadian government acting to enhance the process of the private accumulation of capital.

Dingledine, Gary (1981). *A Chronology of Response: The Evolution of Unemployment Insurance from 1940 to 1980*. Supply and Services Canada. An authoritative summary of the 1940 Act and amendments to 1980.

Finkel, Alvin (1977). 'Origins of the Welfare State in Canada', in Leo Panitch, ed., *The Canadian State: Political Economy and Political Power*. Toronto: University of Toronto Press. Finkel espouses the 'social control' and 'legitimation' theories of the welfare state: the principal reasons for social reform in the capitalist state are to keep people in order and to do just enough good to allow the state to look 'legitimate' as a promoter of the public interest. (See Pal and Struthers, cited below.)

Green, Christopher, and J.-M. Cousineau (1975). *Unemployment in Canada: The Impact of Unemployment Insurance*. Ottawa: Economic Council of Canada. Do the very existence of UI, and its particular characteristics at the time, affect the unemployment rate in Canada? The conclusions are necessarily tentative, but the analysis is thorough enough to let the reader come to an independent opinion.

Mendelson, Michael (1993). 'Fundamental Reform of Fiscal Federalism', in Sherri Torjman, ed., *Fiscal Federalism for the 21st Century*. Ottawa: Caledon Institute of Social Policy.

Muszynski, Leon (1994). *Passive to Active! Rhetoric or Action?* Ottawa: Caledon Institute of Social Policy.

National Committee to Promote the Break-up of the Poor Law (1909). *The Minority Report*, Part ii, 'The Unemployed'.

Pal, Leslie A. (1985). 'Revision and Retreat: Canadian Unemployment Insurance 1971–1981', in Jacqueline S. Ismael, ed., *Canadian Social Welfare Policy: Federal and Provincial Dimensions*. Kingston: McGill-Queen's University Press. Pal reviews some 'models' that seek to 'explain' UI theoretically, notably the 'social control' and 'legitimation' models, finding none completely persuasive. He sees the developments in UI in the 1980s as attempts to promote more productive employment of the labour force. He calls this a 'labour management' model, in which experts and bureaucrats are seen as having great influence on policy.

People's Commission on Unemployment: Newfoundland and Labrador (1978). '*Now that We've Burned our Boats . . .*'. St. John's: Creative Printers and Publishers Ltd. In the 1970s, plagued by continued high unemployment in their province, some labour and community groups organized their own 'commission of inquiry' composed of four individuals who heard testimony from all kinds of people in Newfoundland and Labrador about their experiences—generally devastating—with unemployment. Though now twenty years old, the report they produced is a remarkable account of the human costs of extreme unemployment, as pertinent today as when it was written.

Québec. Conseil de la santé et du bien-être (1996). *L'harmonisation des politiques de lutte contre l'exclusion: Resumé*.

Riches, Graham, and Gordon Ternowetsky (1990). *Unemployment and Welfare: Social Policy and the Work of Social Work*. Toronto: Garamond. This collection of nineteen articles is less concerned with unemployment insurance than with policy and service responses to the problem of unemployment.

Shragge, Eric, ed. (1993). *Community Economic Development: In Search of Empowerment and Alternatives*. Montreal: Black Rose. Dissatisfied with and suspicious of the performance of corporations and governments in fighting unemployment, a movement has developed to promote local enterprise under local control. Some experiences are described here.

Soderstrom, Lee (1984). 'Unemployment Compensation: An Alternative View', in *Unemployment Insurance*. Ottawa: Canadian Centre for Policy Alternatives.

Struthers, James (1983). *No Fault of Their Own: Unemployment and the Canadian Welfare State 1914–1941*. Toronto: University of Toronto Press. A history of developments leading up to the UI Act of 1940. Struthers assigns much weight to the 'social control' motive, but he gives more credence than does Finkel to other roots of social reform.

CHAPTER 4

Workers' Compensation

WHAT IS WORKERS' COMPENSATION?

Workers' Compensation (WC) is a social insurance program—in many industrial countries, including Canada, the first social insurance to be implemented. It does exactly what insurance is supposed to do: it protects a defined population against some part of the financial consequences of the occurrence of a strictly specified risk whose probability can be fairly well estimated from experience. The price of the insurance is scaled to the varying levels of risk in different parts of the population.

Since the regulation of employment and welfare are both within the constitutional responsibilities of the provinces, every province has its own Workers' Compensation program, under the direction of a Workers' Compensation Board or Commission (in what follows, the most common acronym WCB shall be used to signify all such bodies). The older term 'Workmen's Compensation' has been replaced by the gender-neutral 'Workers' Compensation'. In Quebec the responsible body is the Commission de la santé et de la sécurité du travail (CSST). There is a federally legislated program for workers in those exceptional industries that fall under federal jurisdiction. Employees of the federal government itself are covered by the Government Employees Compensation Act, claims under which are administered, however, by the WC program in the province where they work. Most of what follows applies to all such programs. Interesting experiences in many provinces will be referred to, but the programs of the largest provinces, Ontario and Quebec, will be described in somewhat greater depth.

Risks

The risks are narrowly defined. The first is the risk of incurring costs for the health care and rehabilitation of a worker who is injured at work or who contracts an illness on account of the work. The second is the risk of interruption of income or impairment of earning capacity, temporary or permanent, resulting from an accident or an illness *arising out of work*, including the consequences to dependants in the case of the worker's death. The words 'arising out of work' are key words and must be kept in mind throughout this discussion. WC deals with a strictly limited range of injuries and illnesses. We have seen that Employment Insurance provides limited, short-term income replacement for a worker who is out of work on account of other kinds of illness. The Canada and Quebec Pension Plans provide long-term pensions for contributors (i.e., income earners) who become disabled through any kind of illness or injury. These are the only public programs that help replace income lost due to illness. The harmonization of work-related disability benefits under Workers' Compensation with generalized disability pensions under the pension plans is a source of concern that will be explored in Chapter Five.

The income-related risks met by WC are by all measures major ones. In round figures, about 800 Canadian workers die each year as a result of work accidents. Over 800,000 cases are compensated each year, over half of which tend to be cases of disablement, temporary or permanent. Total benefits paid annually in all the provinces and territories approach $6 billion (Human Resources Development Canada, 1994, Table 1). While our focus is on the financial consequences for workers and dependants and the remedies provided by WC, lost income is only one facet of this problem. Injuries and illnesses are by far the most important causes of work time lost in Canada. Work time lost means income loss for the worker, and lost production for society.

Coverage

Logically, the program might be expected to cover all employed people who are subject to risks of work-related injury or accident. Up to about 1980, programs tended to exclude people in safe work environments, such as office workers and university professors. As an example of what is now the common pattern, a major revision of the Quebec law in 1979 brought almost all employed people within the coverage of WC. Part-time workers are covered. Coverage still varies widely among provinces, from 60 per cent to 90 per cent of the work force. In Manitoba, for instance, the program does not cover agricultural workers, many of whom are engaged in family enterprises, nor workers in financial services, a relatively low-risk industry; consequently, coverage extends to about 63 per cent of employed Manitobans.

Eligibility for Benefits

A person covered by the program becomes eligible for benefits if he/she has suffered an injury in the course of work that has caused at least some specified minimum interruption of ability to work, and/or has required medical attention. A worker qualifies for benefits on similar grounds for work-related illness. Conditions in certain industries and occupations are known to be linked to certain illnesses. WC Acts include lists of industries and the illnesses presumed to be related to them; a worker who has been employed in a given industry in the immediate or not-too-distant past, and who contracts an illness associated with that industry or occupation, has a *prima facie* eligibility for benefits, almost never contested. If the worker's illness is *not* among those presumed to be associated with his/her employment, the burden of proof is on the worker to show that the illness *did* arise out of work.

As a general rule, the question of fault or responsibility is not raised; all that matters is the determination that the injury or illness was work-related. That determination itself may be contested. In the rare case where an injury can be shown to be clearly the result of the worker's own wilful negligence, the worker may be denied benefits. When the worker dies as the result of a work-related injury, leaving dependent survivors, the question of his/her possible contributory negligence is not pursued.

While the defining expression 'injury or illness arising out of work' may seem clear, the actual determination of whether or not an accident occurred at or on account of work may not be simple. In all WCBs, a considerable jurisprudence has been developed, and considerable time is taken up, making fine discriminations about such things as the conditions under which travel to and from work is considered 'work' for purposes of determining the right to compensation for an accidental injury.

Benefits

First, medical and other health care costs are covered, including costs of rehabilitation therapy. These services are often provided directly through facilities operated by the Workers' Compensation program.

Second, where there has been an interruption of work and/or an impairment of earning capacity, income benefits are provided. These are scaled according to two factors:

(1) the worker's normal income, up to some limit. Two patterns prevail for computation of income benefits: most provinces, including Quebec and Ontario, allow for benefits of up to 90 per cent of net insured income during disability, others for up to 75 per cent of gross insured income. Still others fall in between, some providing one rate at first, then another rate after a period of time. For example, New Brunswick pays 80 per cent of net income for two years and 85 per cent thereafter if the disability continues. 'Net income' means income minus income tax and compulsory contributions (Employment Insurance, Canada/Quebec Pension Plan, etc.);

(2) the extent of the disability, increasingly being interpreted in terms of *reduction in the worker's earning capacity*, which need not be the same as reduction in bodily or mental ability (see below, on 'deeming' awards). The extent of the disability is subject to change for better or worse over time, so the loss of earning capacity may be reassessed on occasion. Benefits continue as long as the disability is judged to last; there is no time limit. Thus the payment of benefits may be temporary (as in most cases) or permanent; and the proportion of maximum benefits actually paid depends on whether the disability is partial (as in most cases) or total. Extent of dis-

ability is largely a medical judgement. In all provinces but Quebec, this judgement is made by a physician satisfactory to the WCB; in Quebec, the worker may choose the physician without restriction, but the CSST may seek a second opinion. On occasion, this has given rise to some controversy in Quebec.

In Quebec, Ontario, and some other provinces, the basis of compensation for partial disability has been modified through a 'deeming' system. An assessment is made of the worker's 'deemed' earning capacity; the pension award is related to the difference between that deemed earning capacity and the disabled worker's normal earnings prior to disablement. In principle, 'deemed earning capacity' may but need not be the same as the worker's actual earnings in a new job. The relevant clause of the Ontario Workers' Compensation Act reads that a partial disability award will be 90 per cent of the difference between previous net earnings and 'the net average earnings the worker is *likely to be able to earn* after the injury in suitable and available employment' (section 43, subsections 3 and 7; italics added). Quebec's formulation is: 'If the worker remains unable to carry on his employment but becomes able to carry on suitable employment, his income replacement indemnity will be reduced by the net income that he *could earn* from that employment' (Quebec, Act respecting Industrial Accidents and Occupational Diseases, 1985, Explanatory Notes; italics added). 'Deeming' awards could presumably be made even though the disabled worker may challenge the WCB's idea of 'suitable and available employment'.

In most provinces, a lump-sum benefit is provided as well as income replacement; it, too, is scaled to the worker's normal annual income.

Other benefits are paid, in addition to health care costs and income replacement for the disabled worker. In the event of death, a lump-sum death benefit is payable to the estate, and a pension is paid to dependent survivors. In most jurisdictions, the survivors' pensions are a proportion of the worker's net covered income. Sometimes, if the worker died while receiving benefits, the benefits simply continue; in a few provinces and territories, survivors' pensions are paid at a flat rate. In most provinces—not all—a surviving spouse's benefits will not be affected by remarriage.

A significant departure was a 1989 amendment to the Ontario Workers' Compensation Act, creating a so-called 'dual award' system: in addition to the standard replacement of part of lost income, victims may under certain circumstances also be compensated for *non-economic loss*. This means just what it says. It has become common in civil suits for plaintiffs to get compensation for moral, psychological, and emotional damages in addition to the direct material damages suffered. Workers' Compensation law effectively excludes workers from suing for such damages; the amendment is intended to make up for that exclusion.

WC benefits are not taxed as income; for tax purposes, they are regarded as an insurance settlement. By contrast, EI benefits and Canada/Quebec Pension Plan benefits are taxable (see Chapters Three and Five).

WC benefits are calculated strictly on insurance principles. Neither the recipients' other means, if any, nor their needs are considered. The determination of benefits takes up a lot of the attention of WC administrators and experts and often gives rise to disagreement, as one might imagine. Administrators' decisions may be appealed, ultimately, to the Board or Commission itself, and no further. Since appeals are taken against decisions made by the WCB, this has the appearance of making the WCB the final judge in its own cases. Some provinces have created structures to put some distance between the WCB and its own appeals process. In Manitoba, appeals are determined by a separate body, the Appeals Commission, which is bound, however, by the policies established by the WCB. In Quebec, appeals may go through several stages, culminating before the independent Commission d'appel en matière de lésions professionnelles, which, like Manitoba's Appeals Commission, is bound by CSST policies.

In 1985 Ontario created a special Workers' Compensation Appeals Tribunal that is more at arm's length from the WCB. This separation of the quasi-judicial from the other functions wrapped up in WC (to be discussed in greater detail later) has created certain problems. There have been instances where Appeals Tribunal decisions had far-reaching consequences for the Workers' Compensation Board, insufficiently thought through by the Tribunal, or so the Board has thought.

If an appeals process is to respect the demands of due process, it will take some time. To avoid such delays where possible, mediating machinery may be put in place. In 1988, for example, Manitoba appointed a 'Fair Practices Advocate' whose task is to iron out with WCB staff concerns over procedural fairness expressed by clients: delays, inadequate communication, and disagreements over decisions relating to benefit entitlement.

Only rarely may a WCB decision be taken on appeal to a court. (See 'Legislation' and 'Criticisms' below.)

Financing of WC: Who Is Insured? Who Pays?

If the WC program has a true insurance character, then the question 'Who pays?' comes down to the question 'Who is insured against what risk?' Superficially, the answer might seem to be obvious in the case of WC: the worker is the one who is exposed to the risk of losing income and incurring health care costs (leaving aside hospitalization insurance and medicare, not in existence when WC was initiated). But it is not wise to jump to conclusions. The question is discussed below (see 'Background to Workers' Compensation'); it will be seen that the immediate risk absorbed by the program is the risk the employer would otherwise face of lawsuits for damages at the hands of injured workers. Contributions to the WC fund are therefore made *only by employers*. It is illegal for employers to try to exact from employees a covert contribution.

Contributions are based on each worker's income, with, as usual, a minimum and a maximum. They are subject to 'industry experience rating';

that is, the rates assessed depend on the experience of injuries and illnesses in each industry. The contribution rates range from less than 1 per cent of insurable income in the industries with the safest conditions to as much as 25 per cent in Ontario in the industries shown to be most dangerous. Average rates are, however, a great deal less onerous than that—3 per cent of total insurable payroll is fairly standard. In keeping with insurance principles, there are provisions in the legislation of almost all jurisdictions, provincial, territorial, and federal, to reward individual firms in whose workplaces relatively few accidents occur and few illnesses arise, and to penalize firms with bad records.

The responsible Board or Commission sets its premium rates so as to collect from all employers enough money to meet all claims, basing its expectations on past experience. All WCBs are obliged by law to break even financially, an obligation that in recent years, however, has been honoured more in the breach than in the observance. Until the early 1990s the WCBs in almost all provinces had been operating at annual deficits for many years, borrowing short-term to make up the difference, somewhat to the concern of their respective governments. The most spectacular examples were the $2.6 billion and $1.25 billion dollar deficits incurred in Ontario in 1985 and 1990 respectively (WCB, Ontario, 1993).

In economic terms, as far as competition allows, businesses naturally include their Workers' Compensation costs in the prices of their goods. Thus, in the words of the famous slogan used in the political promotion of WC laws, 'The price of the product must bear the blood of the working man.' This should be taken literally: in the end, the costs of workers' injuries and illnesses are part of the costs of production, and those costs ought to be shared by the consumers of those products and services. If the total costs are such that a producer cannot deliver at a price consumers will pay, that producer will deserve to run out of consumers.

It should be understood that to speak of *employers* as paying all the bills is correct from an accounting standpoint, but from an economic perspective, WC premiums are a compulsory addition to payroll. They are part of the cost of hiring employees, and employers undoubtedly take this into account in bargaining over the wages and salaries they will pay. So, in theory at least, employees may be thought of as sharing the burden of WC costs, to some indeterminate extent.

WC, which covers certain health care costs, co-exists with medicare and hospital insurance, which provide health care to everybody. When an injured worker receives health care services, hospital or medical, in a hospital or from a physician, the provincial health care authority will bill the Workers' Compensation Board for the services. It may make little difference to an injured worker whether his health care is paid for by WC or by medicare, but it does make some difference for policy. WC is paid for entirely by employers. For them, it is a cost of doing business. Medicare is paid for by a mix of taxes and contributions (see Chapter Nine). The health care bills are therefore paid out of two different pots.

Administration

Workers' Compensation was created partly to speed up cash settlements and to avoid cumbersome legal procedures. Such procedures were costly to both parties, and the expenditure was socially questionable, since the proceedings sought to determine fault in matters where compensation was more urgent that the question of fault. But even if taken out of the normal courts, compensation cases still have some judicial aspects: evidence must be gathered and judgements must be made based on the evidence. Alongside this quasi-judicial work, there is an insurance business to be administered: funds must be collected from contributors, managed, and disbursed to clients. Finally, regulations must be made applying the principles of the law to specific situations. As in the case of Employment Insurance, it is generally agreed that this combination of administrative, quasi-judicial, and quasi-legislative functions is better carried out by an agency with special expertise, endowed with a good measure of autonomy, than by a regular department of government subject to the political authority of the government of the day. WC programs are therefore administered by 'Workers' Compensation Boards' or 'Commissions'.

The Board itself, the governing body of the organization, is made up of a small number of people, usually with relevant qualifications, appointed by the government, representing the significant interests involved: business, labour, health care, the public at large. In most provinces, the chair is required to give full time to the work of the WCB. Appointments to the WCB are for a stated number of years, giving the members some independence from the political preferences of the government of the day. The WCB is assured by law of adequate financing for its operations; this buttresses its independence.

Traditionally, in each province one minister, usually the minister of labour or the equivalent, speaks for the WCB when questions arise in the provincial legislature, and presents the reports of the WCB to the legislature. The current (Conservative) government of Ontario has introduced an entirely new element into this picture. In response to the serious financial and other tribulations of Ontario's Workers' Compensation system, the premier of the province has appointed a minister in charge of Workers' Compensation. It is too early to say what this unprecedented step means for the status of the Ontario Board.

In Quebec, since 1980, the responsible body has been called the Commission de la santé et de la sécurité du travail (CSST), a name that reflects the enlarged responsibilities assigned to it by the terms of the Health and Safety at Work Act of 1979 and the Act respecting Industrial Accidents and Occupational Diseases of 1985 (the latter governs the workers' compensation program). Offices are located in all the administrative regions of the province, where claims are adjudicated (with an internal appeal mechanism), certain health services are dispensed, safety inspections of workplaces are made, and so on.

With variations, the picture is similar in other provinces. To illustrate the variations, Ontario separates responsibility for compensation from responsibility for the regulation of health and safety conditions under the Occupational Health and Safety Act, originally enacted in 1978. Health and safety are entrusted to the Workplace Health and Safety Agency, which reports to the minister of labour; workplace inspections are carried out by staff of the Department of Labour, not the WCB. Similarly, Manitoba has a Workplace Health and Safety Division in the Department of Labour; the Division's costs of operation are met by a payment from the Workers' Compensation Board—that is to say, by the employers of the province of Manitoba.

For convenience, some large employing organizations, mostly provincial governments and public utilities, but also including large Crown corporations, some large cities, and government-owned or licensed operations, are empowered to run their own workers' compensation programs. They must, of course, observe all the provisions of the law.

Add-ons: Victims of Criminal Acts; 'Good Citizens'

We have seen how (Un)Employment Insurance has been used as a convenient vehicle for the administration of sickness and maternity benefits programs. In somewhat similar fashion, the Commission that is in charge of WC in Quebec also has responsibility for two programs for financial relief of victims of criminal acts, neither of which has anything to do with work injuries or illnesses, and both of which are financed out of the government's revenues, not the CSST's.

The first is the indemnification program authorized by the Act respecting the Victims of Criminal Acts. It has long been recognized as a gap in our system of justice that neither the direct victims of criminal violence nor unlucky innocent bystanders had much recourse for the financial costs to them of injuries resulting from criminal acts. (Convicted violent criminals are rarely worth suing for damages in civil courts, and, where no one has been convicted, as is often the case, there is no criminal to sue.) Now, some compensation is paid out of general taxation to people who have lost work or who have been disabled as a result of a crime, and to the survivors of persons killed as a result of a crime. It is not necessary that anyone be convicted. In similar fashion, the Manitoba Workers' Compensation Board administers that province's Criminal Injuries Compensation Act.

The second is a tax-financed program of compensation to persons who suffer financial losses through acts of 'good citizenship'—e.g., assisting in the apprehension of a criminal or rescuing a victim of a crime or other mishap. This program is authorized by the Act to Promote Good Citizenship. Statutes of this kind, generally known as 'Good Samaritan' laws, have become quite common.

Other provinces have long had similar legislation (e.g., Ontario's Compensation for Victims of Crime Act, first passed in 1971), but the laws are not necessarily administered by the Workers' Compensation authority.

BACKGROUND TO WORKERS' COMPENSATION

Workers' Compensation represents a marriage of legal reform and social insurance. Its history provides a fascinating illustration of the adaptation of industrial society to an emergent need—not that everybody considers the adaptation to have been wholly satisfactory.

In the years of developing industrialism, when machinery was new, primitive, and relatively unfamiliar, work accidents were exceedingly frequent. In liberal countries, including Great Britain and its Dominions and the United States, an injured worker had the ordinary common-law right to sue his employer for damages. For a long time, employers were able to resort to legal defences that had grown out of centuries of judicial treatment of ordinary damage suits. In the British common-law tradition, the guiding precedents had emerged over the centuries from the so-called law of master and servant, a reflection of the fact that in pre-industrial society the relationship of master to servant was more characteristic than that of employer to employee. When sued, employers could argue that the worker had contributed to the injury by his/her own negligence, or that some other worker had been at fault and was therefore the one who ought to be sued; or they could argue that when the worker accepted the job, and the wages, he/she accepted all the risks that went with it.

These defences were often successful, partly because they may sometimes have had some basis in fact, but more because they put an exceedingly difficult burden of proof on the plaintiff worker. Besides, the typical worker was less able to afford a costly court case than the typical employer. On the one hand, this worked to the manifest disadvantage of workers. On the other hand, workers occasionally did win their cases, and employers were obliged to pay heavy damage awards. Such large, unpredictable expenditures were upsetting to business firms. And if the employer could not pay, as sometimes happened, and had to go out of business, that was of no comfort to the injured worker, nor to fellow workers who would lose their jobs. A major preoccupation of many early workers' associations and trade unions was, in fact, the administration of funds to which members would contribute regular premiums, and which would pay modest benefits to members who were disabled by injury. However, this proved to be a heavy burden for such organizations to bear. So both sides, employers and employees, had a latent interest in systematic collective protection against the risk of injury, regardless of who was at fault. Furthermore, the conviction grew that society at large was also the loser, on account of the costs, delays, and hardships that ensued from the process of litigation.

In response to this problem, modern workers' compensation developed in three steps.

(1) The first step came before the recognition of any common interest in a collective remedy. Workers' organizations and advocates of reform decried the unfairness of the conventional judicial treatment of work accidents. The old-fashioned judicial concern over who was at fault had become irrelevant in the context of the industrial factory. It created an intolerable bias against the complaining worker, because fault, being almost inevitably widely distributed, was often impossible to locate. And in any case, as far as the injured worker's broken bones and lost earning capacity were concerned, it hardly mattered who was to blame. So the first step was legislation to change the courts' handling of industrial injury cases. 'Employers' liability' laws were passed in many places to nullify some or all of the employers' legal defences, mentioned above. Damage suits arising from work injuries still went to the normal civil courts, but the question before the court was no longer 'Was the employer at fault?' but 'Did the injury arise out of the work?' If the answer was 'Yes', the worker was compensated.

Wherever this change took place, it naturally increased the workers' chances of success in damage suits. There remained the inequality in the ability of the worker and the employer to afford a court case, especially if it were drawn out for a long time. It was difficult for the worker to get witnesses, for the only witnesses were fellow employees, who might have some misgivings about testifying against their employer. And the suing worker, win or lose, could not expect the employer to be overly enthused about taking him or her back as an employee. Still, employers' liability laws did increase the likelihood that the employer would be hit with a costly settlement for damages. The risk to the employer had increased, radically.

Confronted with such a risk, many employers did the rational thing: they insured themselves with private insurance companies against damages arising out of employers' liability, thus making their annual costs regular and predictable. This answers our earlier question: Who is insured against what risk? Some workers also insured themselves privately against the loss of income through work accidents.

(2) The second step in the development of workers' compensation policy dealt with the consequences that ensued when a substantial number of employers failed to insure themselves voluntarily. (a) Employers who refused to insure gained a short-run edge, albeit a risky one, over their competitors who did undertake the expense of insurance. The latter naturally resented the advantage thus enjoyed by their less conscientious rivals. (b) The employees of uninsured employers were exposed to great risk, for

even if they won a lawsuit under the new, more liberal employers' liability rules, their employers might not have had the money available to pay.

The second step, then, was to enact laws *obliging* employers to insure themselves against liability for damages arising from work accidents. The insurance policies, and their administration by the insurance companies, would presumably have to meet standards specified in the legislation. This second step is, in fact, as far as many states in the United States have gone: compulsory WC insurance with private insurance carriers.

(3) The third step acknowledges certain realities of the business world. The insurance business is profit-seeking and competitive. Insurance companies are motivated to contest disputable claims and to look for ways to avoid the riskiest, least profitable client groups. To avoid abuse, many jurisdictions—including, to repeat, all provinces and territories in Canada—have preferred a single, publicly administered, non-profit, universal, compulsory social insurance plan whereby health care costs are paid, and income replacement benefits are awarded, according to pre-arranged schedules, with no contest whatsoever regarding the questions of either fault or liability. Workers' Compensation is, in effect, no-fault insurance. Employers give up the resort to legal defences against paying compensation. In return, workers accept the agreed-upon compensation schedules and give up the right to sue for damages for health care costs and loss of earnings.

For the insured costs, the worker does not claim against his/her employer but against the workers' compensation fund. Logically, since employers had formerly paid the damages when liability was proved, and then had paid the premiums to insure themselves against such damages, *employers pay the whole cost of Workers' Compensation.*

As an ideal, the prevention of accidents and of conditions harmful to health is more appealing than compensation for injuries and illnesses after they occur. As workers' compensation was progressing through the development just described, people disagreed as to the best way to motivate employers to prevent accidents. Some participants in the debate, including some from the side of labour, notably the American Federation of Labor itself, felt that vigorous enforcement of employers' private liability would be a more potent motivator than the compulsory collective insurance approach. They argued that the public insurance system diffuses the cost of compensation for accidents across entire industries; less scrupulous employers might skimp on prevention, knowing that the costs of compensation to any of their workers would be borne as much by their competitors as by themselves.

Some observers feel that the virtual surrender of lawsuit by workers has proved too great a price to pay for the greater certainty of payment and convenience of administration obtained through compulsory public workers' compensation. But opinions vary on this issue, even among those close to each other on the ideological spectrum. Within the covers of *The*

'*Benevolent*' *State*, all of whose authors share a critical left-wing perspective, one article, recounting the early history, refers approvingly to 'labour's proposal to remove compensation claims from the courts' while another criticizes workers' compensation for having 'eliminated the workers' right to sue the employer directly in court' (Moscovitch and Drover, 1987, p. 22; Vigod, 1987, p. 180). The extent of the exclusion of work injury cases from the jurisdiction of the ordinary courts has recently come under question, thanks to jurisprudence under the Canadian Charter of Rights and Freedoms. (See 'Criticisms' below.)

LEGISLATION

In Canada, workers' compensation is one field over which there has never been any dispute as to jurisdiction: it belongs to the provinces. Why? As we have already seen, welfare is generally considered to be a provincial concern. And the regulation of employment is considered to fall under the heading of 'property and civil rights', clearly assigned to the provinces by Section 92 of the Constitution Act. Employment fits under the 'property' heading because employment is a form of contract; you have an enforceable contract with an employer as soon as you go to work and the employer agrees to remunerate you, whether or not you have a formal contract in writing.

Around the turn of the twentieth century, several provinces (Quebec being among the first) enacted employers' liability legislation, by which, as we have seen, employers were held responsible for the consequences of work accidents even if not at fault. Following the publication of the historic Meredith Report (Ontario, 1914), Ontario, the most highly industrialized province, was the first to enact a Workmen's Compensation Act, creating a workers' compensation program, in 1914. (An earlier Ontario Act, although called the 'Workmen's Compensation for Injuries Act', was really an employers' liability act.) Almost all the other provinces followed suit within a few years, but Quebec did not pass its 'Workmen's Compensation Act' until 1931, almost the last to do so.

Quebec's current workers' compensation legislation is embodied in the 1985 Act respecting Industrial Accidents and Occupational Diseases (Acte sur les accidents industriels et les maladies professionnelles). The framework for the regulatory aspect of work safety is set forth in the 1979 Act respecting Health and Safety at Work (Acte sur la santé et la sécurité au travail). As the name implies, the 1979 Act covers many aspects of work safety other than the running of the workers' compensation program. The 1985 Act replaced the former 'Commission des accidents au travail' (Workmen's Compensation Commission) with a new 'Commission de la santé et de la sécurité du travail' (CSST). The former Commission had some authority to establish safety regulations, conduct inspections of work premises, and so on, but the CSST has a wider, more explicit mandate in those areas. In the other provinces, the responsible agencies are known as Workers' Compensation Boards. As we have seen, the Quebec pattern of conferring responsibility for both workers' compensation and health and safety regulation in the hands of one agency is not the norm.

On the question of constitutional jurisdiction over workers' compensation, students should not be confused by the fact that there is a federal Workers' Compensation Act and a federal body administering it. That Act covers certain industries whose work involves so much mobility that it might be difficult to say in which province a given employee works. For the same reason, there are a federal Minimum Wage Act and a federal Labour Code regulating industrial relations.

CRITICISMS

Benefit Levels

As with other benefit-providing programs, WC is often accused of giving inadequate benefits to recipients. Spokespeople for recipients claim that the WCBs try to keep down the benefits they disburse. Since all but a couple of WCBs have been in serious financial trouble, it is true that claims administration has become more vigilant, but the effort has concentrated on earlier return to work, not on the scale of benefits. Maximum benefits are based on the worker's total rate of prior earnings, up to the maximum that is insured; partial benefits are calculated as a percentage of earnings lost, or, as noted above, a percentage of earning capacity *deemed* to have been lost. The determination of actual benefits to be paid allows room for a certain amount of discretion.

To a degree, the former adversarial relationship between workers and employers has been replaced by an adversarial relationship between workers and the WCB, but realistically the WCB does not have quite the urgent motivation to limit costs that the employers have.

Unfunded Liabilities

Many of the substantial awards made to victims of accidents and illnesses are long-term, even lifetime indemnities. Each Board/Commission is therefore responsible for heavy annual liabilities extending indefinitely into the future; every year, a WCB has to meet liabilities incurred in the past. Only in a few provinces (Alberta, Saskatchewan, and Manitoba) do the WCBs have, at present, financial reserves sufficient to assure beyond question that they will be able to meet those liabilities: that is, they could meet those obligations without drawing on future revenues. The other provinces all have 'unfunded liabilities', i.e., they must commit large shares of their annual revenues, well into the future, to meet obligations from the past. By law, a private-sector insurance company would not be allowed to incur such unfunded liabilities. Up to a point, this is not too alarming, because the financing of WC is a legal obligation for employers, much like taxes—

WCB revenues are not going to dry up, and WCBs are unlikely to be allowed to go broke. But it does mean that present contributors are paying currently for large costs incurred in the past. All provinces are, of course, entirely aware of the problem and are aiming to eliminate or at least reduce their unfunded liabilities.

Delays

Other critics point to extended delay in the processing of claims and appeals. In fact, the time from initial claim to final settlement has lengthened in many jurisdictions. This contravenes the objective of WC to avoid lengthy procedures.

In Quebec, in the early 1990s, the average length of time between the commencement of a claim and the *initiation* of a prescribed rehabilitation program was about 600 days. As the chair of the CSST pointed out, many workers thereby lost their legal right to return to their previous employment, having exceeded the time limit for such guaranteed return. A focused effort over a couple of years has reduced this delay to an average of 90 days (Institute of Public Administration of Canada, 1996). Dealing with this issue is delicate, because a constructive solution, with no losers, requires the good will of all parties: the injured worker, the physician first involved in the diagnosis (the treating physician may be either chosen by the worker without restriction, as in Quebec, or chosen from a recommended roster), the WCB's rehabilitation experts, and the employer.

Time Lapse Before Return to Work; Rehabilitation

The increases in benefit expenditures experienced by all WCBs is largely attributed to a widespread increase in the duration of benefits received by beneficiaries ('time on claim'). This increase may simply reflect a growing recognition of the seriousness of the disabilities suffered by workers. Another interpretation is that WC programs have focused too narrowly on the issue of compensation at the expense of what ought to be a primary objective: enabling the worker to return to work as soon as possible.

Some jurisdictions, Quebec and Ontario included, have legislated a right of return to work for the worker, obliging the employer to accept the worker back after an absence from work, and at the former job, if possible. There is, of course, a time limit on the obligation, but the limit can be made high enough to cover most cases.

Moral Hazard

Employers are sensitive to the level of benefits paid, since it will be reflected in the levels of contributions they will be called on to pay. Moreover, they argue, generous benefits naturally dampen an injured worker's incentive to return to the job. This is an example of 'moral hazard', the risk that the parties might manipulate the rules of the program to satisfy their own private interest, at the expense of the others who contribute to the fund. Such critics claim that, as a result, workers often continue receiving benefits, with the implied complicity of the doctors, after they are quite able to return to work. This complaint has been voiced particularly strongly in Quebec where, as noted above, the worker has unrestricted choice of physician for the treatment of an occupational injury or illness, which includes the initial judgement of ability to return to work. (Quebec's CSST has the right to seek a second opinion from a doctor of its own choice if doubtful about the initial medical opinion.) Employers' spokespeople are likely to find the Boards too lenient in the determination of claims, especially the duration of claims; labour spokespeople are likely to find the opposite.

As corroboration, these critics point to the deficits experienced until 1993 or 1994 by the Ontario and Quebec and most other Workers' Compensation authorities, Saskatchewan and Alberta being the exceptions. Quebec's accumulated deficit, by 1993, had reached $2.2 billion; in the ten years from 1984 to 1993, the Ontario WCB ran successive annual operating deficits totalling $9.5 billion (Ontario, WCB, 1993, p. 36). Quebec's CSST brought its finances under control by means of an exceedingly vigorous administrative effort, which focused on simplifying claims and appeal procedures and sharply reducing the delays mentioned above. In Ontario's case, far-reaching legislative changes have been promised by the (Progressive Conservative) government in power in 1997.

As with EI at the federal level, the finances of workers' compensation are in separate accounts from those of the provincial governments, but outcomes like these represent a large claim upon the future financial resources of the provinces concerned. All the indebted WCBs in the provinces are presently engaged in programs of redressment: another instance of the pervasiveness of the deficit- and debt-reduction imperative that dominates social policy at present.

Illness Benefits

Work-related illness has always been troublesome for workers and, in a different sense, for administrators of WC programs. The usual legislative structure would seem to be satisfactory: the Act includes schedules of illnesses known to be related to working conditions in listed industries; where a worker contracts an illness associated with his/her industry or occupation, he/she is eligible for benefits. Illnesses not included on the schedule for the worker's occupation must be investigated case by case.

In real life it is often not that simple. A worker may well have worked in a number of different places, never long enough in any one to establish a presumption that the illness is associated with work. Illnesses usually take time to develop, and the whole range of a person's experience, work and other, may contribute in one degree or another to state of health. From the worker's point of view, that makes it difficult for a sick worker to prove that a particular condition was linked to work, and difficult therefore to substantiate a claim for WC benefits. From the point of view of the WC system, the insurance program is designed to meet certain needs and is provided with resources necessary to meet those needs, and not other needs. WC is not designed to be a generalized sickness indemnity program. To repeat: we have no general sickness indemnity program anywhere in Canada.

Prevention

The harshest criticism is less concerned with benefits than with the contribution WC makes or fails to make to the prevention of accidents and the maintenance of healthy working conditions. (This aspect is discussed under the heading 'Alternative Policy #2: Occupational Health' in Chapter Nine). Workers' compensation does not deal with the basic problem of occupational injury and illness itself. Rather, it provides health care for workers who are injured and for some who become ill at work; it mitigates the financial consequences to the income earner of such injury and illness; it provides a measure of financial motivation for employers to make workplaces safe and healthy. But to the extent that the problems of occupational health grow out of structural factors in industrial society, workers' compensation cannot reach the roots of the issue.

WC, the Courts, and the Charter of Rights and Freedoms

Reference has been made to some concern over the extent to which liability for the financial consequences of work accidents and injuries has been taken out of the judicial process. Workers are denied access to the courts to sue for a considerable range of damages they may suffer. This aspect of WC has troubled critics both liberal and radical since workers' compensation was first devised. Against this criticism must be weighed the difficulties formerly faced by injured workers through judicial handling of their claims in the civil courts. It is not possible at one and the same time to avoid the inconveniences of the judicial process and to enjoy its protections.

Over the years, cases have arisen where the classification of certain accidents and *all of their consequences* as wholly work-related has been disputed. As a general rule, the determination made by the appropriate provincial WCB that a given case was work-related has been accepted as final; that is, the courts have entirely refrained from considering a case where the WCB has exercised jurisdiction. But a change has occurred as a result of the Canadian Charter of Rights and Freedoms (see above, 'Constitutional Restrictions on the Powers of Governments'). Section 15 of the Charter, the 'equality rights' clause, guarantees all Canadians equality 'before' and 'under' the law. A case in Newfoundland involving the accidental death of a worker went to the Supreme Court of Canada for a determination as to whether the denial of access to procedures outside the workers' compensation system amounted to a denial of equality before the law. The Supreme Court decision left the issue in such doubt that several provinces later collaborated in putting it before the Supreme Court in a 'reference' case (that is, an exceptional proceeding, described in Chapter One, whereby the Supreme Court may resolve a disputed question brought before it *without* an actual case to be decided). The Court then upheld the constitutionality of the legislation that enshrined the trade-off of secure, no-fault insurance for loss of the right to sue (*Piercey v. General Bakeries and the Queen in Right of Newfoundland* [1986], 31 Dominion Law Reports, 4th series, p. 373; [1989], 56 DLR, 4th series, p. 765).

In an Alberta case, a party other than the injured worker's employer was involved in an accident (a collision of vehicles) in which the worker suffered serious, permanent disability. The issue was whether the worker could sue the other party for damages beyond the compensation award. The Alberta Court of first instance agreed that the claimant could sue, but the Court of Appeal reversed that decision. After some time, a private settlement was reached out of court, but many interests were at stake, including the constitutionality of the clauses in all Workers' Compensation Acts that denied the right to sue an employer and that empowered the WCBs to make judgements as to further suits. Thus, the case proceeded to the Supreme Court of Canada. The Court produced a cautious ruling, saying that the plaintiff (the accident victim) had indeed been discriminated against by the manner in which the Workers' Compensation Act had been interpreted, but that the Act was intrinsically constitutionally sound in excluding lawsuits as a general rule (*Budge v. Public Trustee for Alberta and Workmen's Compensation Board* [1991], 77 DLR, 4th series, p. 361).

OCCUPATIONAL HEALTH AND SAFETY

The maintenance of healthy and safe conditions in workplaces would reduce the injuries and illnesses that arise out of work, and would accordingly reduce expenditures on workers' compensation. We have seen that the provinces have given to certain bodies the task of regulating health and safety at work; in Quebec, this function is assigned to the same body, the CSST, that runs the workers' compensation program. Chapter Nine will address the topic of occupational health and safety as a field of policy that can contribute significantly to the well-being of the population.

PROBLEMS FOR DISCUSSION

1. In general, employers occupy a strong power position *vis-à-vis* workers. With regard to WC in particular, as the sole financiers, employers may be inclined to feel that it is 'their' program. Workers may feel poorly placed to exert pressure, if not represented by alert unions. Recently, some representatives of employers, including the Conseil du patronat du Québec, have proposed that workers share in meeting the costs of WC; it cannot be

claimed, therefore, that employers badly want to pay the whole bill just in order to be able to dominate WC. And judging from the tone of their complaints, employers do not feel that they always do get their own way.

2. The impact of the imposition of a high contribution rate on employers in a particular industry—15 per cent, 20 per cent, or 25 per cent—provides food for thought. A 20 per cent rate means that $200 is added to every $1,000 the employer pays in wages, up to the maximum covered income, i.e., up to an added $8,000 a year per employee, if the maximum is $40,000 (it is actually much *higher* than that in all but two provinces). In such a case, for every five workers whose incomes are at or below the maximum, an employer in such an industry is in effect paying the salary of a sixth in WC premiums. If work in the particular industry is so unsafe that that much money is required to cover the costs, then that is what the rate must be. More attention to safety would bring the rate down.

3. The attention that has been given in recent years to work dangers and, conversely, to the promotion of work safety has revived the issue of the right of workers to sue for some aspects of damages arising from work conditions other than loss of income. *Consumers* have been more successful in suing over unsafe products than *workers* have been over unsafe working conditions. In the court cases referred to above, based on the Canadian Charter of Rights and Freedoms, the workers who had been injured were seeking additional redress, not from their own employers but from third parties involved; that is a different matter from claiming that the employer had not maintained safe working conditions.

4. While the rules vary somewhat from province to province, in most instances the worker has some degree of choice regarding the doctor who makes the first judgement of the extent of disability. In Quebec, as we have seen, the worker has unrestricted choice of physician, while in other provinces, the worker can choose from among a panel of approved physicians, and the Board usually has the right to seek a second opinion. WC administrators often express concern that this gives undue influence in decisions to doctors who (a) are not specialized in occupational health, (b) may be inclined, especially if they happen to be the workers' regular doctors, to tell their patients what they want to hear about how substantial a compensation claim they are entitled to, and (c) bear no responsibility for the ensuing costs. On the other side, it is argued (a) that the physicians' professional ethics would deter them from exaggerating the seriousness of any patient's condition, and (b) that, given the WCB's interest in limiting compensation awards, it would be unfair to entrust assessments of disability totally to doctors who are too close to the Boards.

Over the past twenty or thirty years, there has been a fairly steady lengthening of the periods of time during which disabled workers have received benefits. This could mean that disabling conditions are becoming more serious; that benefit periods were inappropriately short in the past; or that benefit periods are inappropriately lengthy now. All WCBs have addressed the matter of the duration of benefit periods in the last few years, partly by enhancing their rehabilitation programs.

5. As with health care generally, there has been in workers' compensation experience an increasing recognition of disabilities other than physical or neurological, disability as a result of work-related stress being the most common example. This has posed difficulties in defining disability and in assessing its seriousness.

STUDY QUESTIONS

The student should answer or discuss the following questions by referring to authoritative sources, to items in the bibliography, or to this text.

1. How big a program is WC in any given province? How much is paid out in benefits in a year? How many people receive benefits? How much is collected in contributions? Is your province's program in a deficit, surplus, or balanced situation?
2. Does your province's WCB have extensive 'unfunded liabilities'? Whether the answer is yes or no, what does this mean for workers and employers?
3. Who pays for WC? What is the rationale for the costs being met thus?
4. Do most recipients collect benefits for short periods—say, two or three weeks—or for longer periods of time, say, two months or more? Is there a discernible trend?
5. In your province, how many deaths occur among covered employees in a typical year? How much do death benefits cost the WCB? What benefits would be received by the spouse of a worker killed in a work accident, with two dependent children?
6. What body runs WC? Of what background are the members of the governing body? Do they represent special interests?
7. Explain the links between the financing, administration, and benefits of WC.
8. What is the difference between 'employers' liability' legislation and 'workers' compensation' legislation?
9. In the legislation, how forceful is the presumption of liability when a worker contracts a disease recognized to be associated with working conditions in his/her occupation?

BIBLIOGRAPHY

Bradet, Denis, Bernard Cliche, Martin Racine, and France Thibault (1993). *La loi sur les accidents du travail et les maladies professionnelles*. Cowansville, Quebec: Y. Martin. Informative, somewhat specialized and technical.

Canada. Department of Labour. *Workmen's Compensation in Canada*. Revised annually. Facts on WC programs across the provinces. Obviously useful for anyone contemplating comparisons.

Canada. Human Resources Development Canada (1994). *Occupational Injuries: Their Costs in Canada 1989–1993.*

Canada. Human Resources Development Canada (1993). *Inventory of Income Security Programs in Canada.* Chapter 6, 'Provincial Workers' Compensation Programs'. This *Inventory*, issued at irregular intervals, is an invaluable review of all income programs, provincial and federal, in Canada.

Canada. Statistics Canada (1995). *Work Injuries 1992–94.* Catalogue No. 72–208. An annual series.

CCH Canadian. *Canadian Workmen's Compensation.* Toronto: CCH. Issued at regular intervals. The most exhaustive updated compendium of Canadian laws and programs.

Guest, Dennis (1997). *The Emergence of Social Security in Canada.* 3rd edn. Vancouver: UBC Press. See especially Chapter Four.

Institute of Public Administration of Canada (1996). 'Quebec Workers' Compensation Board', *Management* 7/2.

Jennissen, Theresa (1981). 'The Development of Workmen's Compensation in Ontario', *Canadian Journal of Social Work Education* (Fall).

Manitoba. Workers' Compensation Board (1994). *Five-Year Operating Plan.*

Moscovitch, Allan, and Jim Albert, eds (1987). *The 'Benevolent' State: The Growth of Welfare in Canada.* Toronto: Garamond.

Moscovitch, Allan, and Glenn Drover (1987). 'Social Expenditures and the Welfare State: The Canadian Experience in Historical Perspective', in Allan Moscovitch and Jim Albert, eds, *The 'Benevolent' State: The Growth of Welfare in Canada* (cited above).

Nash, Michael. *Canadian Occupational Health and Safety Law Handbook.* Toronto: CCH Canadian. Intended for lawyers, but informative for all readers concerning relevant law in all provinces. Reissued at intervals.

Ontario (1914). *Final Report on Laws Relating to the Liability of Employers to Make Compensation to Their Employees for Injuries Received in the Course of Their Employment which Are in Force in Other Countries* (the Meredith Report). Sessional Paper No. 53, Legislature of Ontario. The seminal document on which Ontario based its pioneering Workmen's Compensation Act, closely followed by all the other provinces.

Ontario. Occupational Health and Safety Act. *Revised Statutes of Ontario, 1990*, Ch. O-1.

Ontario. Task Force on Vocational Rehabilitation (1987). *An Injury to One Is an Injury to All: Towards Dignity and Independence for the Injured Worker.* Toronto: The Task Force.

Ontario. Workers' Compensation Board (1993). *Annual Report.*

Québec. Commission de la santé et de la sécurité du travail (CSST). *Annual Report.* In other provinces, Workers' Compensation Board, *Annual Report.*

Quebec. Minister for Social Development (1978). *Health and Safety at Work.* A so-called White Paper, explaining the Government's intentions before the presentation of the new Act respecting Health and Safety at Work. Contains valuable information about trends in work-related deaths, injuries, and illnesses in Quebec. Translated from the French, *La Santé et la Sécurité au Travail.*

Quebec. An Act respecting Health and Safety at Work. *Statutes of Quebec, 1979.* In other provinces, Workers' Compensation Act.

Reasons, C.E., L.L. Ross, and C. Paterson (1981). *Assault on the Worker.* Toronto: Butterworths. An attack on both Canadian industries and governments (mostly of the Western provinces) on the score of work safety.

Thomason, Terry, François Vaillancourt, Terrance J. Bogo, and Andrew Stritch (1995). *Chronic Stress: Workers' Compensation in the 1990s.* Toronto: C.D. Howe Institute. The 'stress' in the title is the financial stress felt by all workers' compensation programs in Canada. The authors analyse the factors that have contributed to the deficits of WCBs, and measures of reparation that have been instituted.

Vigod, B.L. (1987). 'History According to the Boucher Report: Some Reflections on the State and Social Welfare in Quebec Before the Quiet Revolution', in Moscovitch and Albert, eds, *The 'Benevolent' State: The Growth of Welfare in Canada* (cited above).

Weiler, Paul (1989). *Reshaping Workers' Compensation for Ontario.* Toronto: Ontario Ministry of Labour.

CHAPTER 5

Incomes for Older People

THE NATURE AND DIMENSIONS OF THE PROBLEM

Social changes that are in themselves benign, or at worst harmless, often turn out to have regrettable consequences when they run together. A combination of such social changes has created a serious problem in meeting the material needs of older people.

One benign change has been a long-term improvement in overall health, resulting in a great extension of length of life over the last few generations. This alone has meant an enormous increase in the *number* of older people in the population. A second change has been the long-term decline in the birth rate. Those two factors—more old, fewer young—working together have inevitably increased the *proportion* of our population who are over any given age. The proportion of the elderly is almost certain to continue to increase for many decades. It will undergo a sudden boost when the 'baby boomers', born between about 1950 and 1965, begin reaching the age of sixty-five, around the year 2015. The consequent projected increases in both the numbers and the proportion of the population over ages seventy-five, eighty, and eighty-five are particularly striking.

A third factor has been the acceptance of a conventional age of retirement from paid work. The notion of retirement had and has little meaning in an agrarian society: an older person simply continues working to the extent that he or she and the family consider appropriate. Only when society has rationalized work as industrial society has (so many hours per day, so many days per week, incomes earned in accordance with work), does it make sense to think of a point when one's age renders one unable to produce sufficiently to remain in the work force. If more and more people come to live longer and longer beyond the age up to which they are commonly considered able to earn their living, we have a many-sided problem of income support for older persons.

Also important are other considerations: a humane value according to which an older person is felt to deserve a rest, and, at a certain stage in the maturation of industrialism, a concern that jobs be vacated in favour of young workers. Although this concern is widely felt, its economic validity is debatable.

Society may once have relied on the family to support the elderly, but the 'family' is not what it used to be. It now includes more older and far fewer younger members than before. The typical family today embraces more generations than ever before; the four-generation family is no longer exceptional, though this may change as many families postpone the birth of the first child. The contemporary family is less well adapted for the task of supporting its more numerous old than was the pre-industrial family, partly because the urban industrial family is likely to be more scattered, more 'nuclear', and partly because in such a society the support of older persons requires money, not just a share in the family's stock of goods, a room in the house, a place at the table.

Poverty among the old has manifested itself in virtually all industrial countries (though perhaps no more so than in pre-industrial societies, if in different ways). The prevalent social ethos of the capitalist part of the industrial world calls on the individual to look after himself or herself to a large extent; thus, the individual is expected to make some provision for old age—to save, to accumulate some capital, to arrange for a pension. But up to the present time, relatively few working people have earned incomes sufficient to allow them to save enough to make adequate unassisted provision for the years after they cease to earn. History shows country after country coming to recognize the manifestations of poverty among the elderly and adapting, often slowly, to the need to support a large and increasing class of people who are effectively separated from employment.

The economic history of the past three-quarters of a century or so has naturally affected the current financial position of older people. The Great Depression of 1929–39, when unemployment was exceedingly high and incomes exceedingly low, adversely affected the work career of almost anyone who reached adulthood during the 1930s (and who would therefore be over seventy-five in 1995). Emerging from the Depression in 1939, Canada entered a long period of inflation; personal savings would have had to be managed astutely to have retained their real value over such a long period, and anyone who entered retirement with a pension that yielded a fixed income, as most pensions normally do, has suffered considerable loss in purchasing power over time.

Still, for employed people in general, *some* saving has been possible; what was needed was a service to convert part of what they could save while earning into old age incomes, with some degree of efficiency and security. It is characteristic of an economically liberal society that where a need is perceived and money is available to pay for a service that meets it, enterprise will seek to provide the service. Individual arrangements with financial institutions that convert lifetime savings into post-retirement incomes have therefore been in existence for a long time. So have employer-based pensions, which began simply as current expenditures of companies but developed into separately funded pension plans. Partly as an outcome of collective bargaining, partly as a way of attracting and holding employees, more and more employers have made pension plans part of the compensation packages offered to employees (see 'Private Pensions' below). Alternatively, employee groups have created pension funds for themselves. Insurance companies and other financial firms have increasingly applied their expertise in the management of money to managing pension funds.

Spurred by escalating economic growth after World War II, the private pension business has grown to the point where pension funds have long been one of the largest reservoirs of capital in industrial nations (Drucker, 1976). Private pensions are regarded as a major income source for older Canadians. Yet, as we shall see, there is a limit to the extent to which a population can be covered by employment-based pension plans, and, unavoidably, the amounts eventually paid to pensioners by private plans vary greatly, some being very small indeed.

Finally, among factors contributing to inadequacy of income among older people, the relationship between age and gender weighs heavily. Women considerably outlive men; in all age cohorts over the age of about fifty-seven, women outnumber men, and the disproportion increases sharply with increasing age. As is well known, women are likely to have earned less money in their lives than men; women are much less likely to have private pensions or, for that matter, Canada or Quebec Pension Plan pensions in their own right; such pension funds as they have accumulated are likely to be smaller than men's, and, since women live more years, the *annual* pensions payable to them are further diminished (unless the terms of their pension plans specify otherwise). The private pensions of many older women were arranged subject to that rule. It is now illegal in all provinces except Quebec for private pension plans to pay different pensions to recipients on the basis of gender. Like the public pension plans, they use what are called 'unisex' pension computations.

Most women marry, and until fairly recently, they tended not to have lifetime work incomes and pension arrangements of their own; they have been the dependants of their breadwinner husbands. Because women live longer and because women typically marry men older than themselves, most women become widows, and most widows do not remarry. For their incomes in their later years, they have depended heavily on arrangements made for them by their late spouses. For various reasons, such arrangements leave many widowed women meagerly provided for. This pattern is gradually being modified as more women have working careers, incomes, and pensions of their own, but it will prevail substantially for many years to come. Thus, in statistical terms, the problem of poverty among the old is predominantly a problem of poverty among older women. There are many more older women than older men, and there is a much higher proportion of poor older women than of poor older men. This is one of the most striking instances of what has come to be called the 'feminization of poverty', others being the predominance of women among lone parents and the high proportion of women among low-paid workers.

PUBLIC TRANSFER PROGRAMS FOR OLDER PEOPLE

In Canada, as early as the turn of the twentieth century, older people were observed to represent a large proportion of welfare (or 'public charities') cases. This in itself was implicit evidence that the poverty of older people was a systematic social problem, not essentially a problem of individual hardship, let alone individual failure. Despite some recognition that this was the case, Canada long relied for relief on charity mechanisms, private and public. These relieved only the destitute aged.

The burden of relief of the aged poor in the various provinces was such that in the early 1900s serious proposals began to be made to involve the federal government in income support for older Canadians. Action from this direction was slow in coming. Part of the reason for the delay in federal intervention was, of course, the constitutional issue: all forms of 'charity', public as well as private, were the responsibility of the provinces. Even if a federal government in power were willing to undertake substantial expenditures to assist the aged, the terms of federal intervention in a provincial field would have had to be negotiated. But a large part of the cause of delay was that successive federal governments were not all willing to act.

The first step eventually taken, while modest, was an important innovation in federal-provincial relations as well as in the Canadian social welfare structure. All provinces were providing some kind of income support for needy older people, whether directly through specific programs for the elderly or through more general public assistance programs, or indirectly through public support of religious or philanthropic bodies (see, e.g., Splane, 1965). The variety of modalities of assistance complicated the project of designing a federal program acceptable to all provinces. In 1927, the provincial and federal governments having finally attained the necessary level of agreement, the Parliament of Canada enacted the Old Age Pensions Act, incorporating a simple principle: the federal government undertook to pay half (later 75 per cent) of all public income assistance made through provincial government programs to needy older persons. Thus the *costs* were to be shared, but the *responsibility*—both political and administrative—remained with the provinces.

Politically speaking, the cost-sharing terms had to be agreeable to just about all the provinces, though in principle each province was free to refuse to take part. Because the federal government was authorized by the Old Age Pensions Act only to spend its own money (which it is allowed to do under the prevailing interpretation of the 'spending power'), not to execute a program in a field outside its jurisdiction, it was generally accepted that the Act was within the constitutional powers of the federal Parliament. No constitutional amendment was considered necessary, and no party challenged the validity of the Act in the courts.

The Old Age Pensions Act was a milestone as the first instance of federal-provincial cost-sharing in the field of welfare. It was also a milestone in the development of the Canadian federal governmental system, the first really significant use of the 'spending power' to pursue a *national policy objective* in a field of *provincial jurisdiction*. We shall see how important the cost-sharing device has been when we examine public assistance in Canada, and again when we look at health services programs. The federal-provincial financial interplay in the field of social policy has become one of the most prominent items in the national agenda.

With this sharing of costs by the federal government, all public programs of support for older people remained in the hands of provincial welfare departments and, in some provinces, local government welfare authorities, until the 1950s. All such programs were needs-tested assistance programs, aimed at relieving poor people; none was directed toward the older population as a class.

Old Age Security (OAS)

Toward the end of World War II and in its immediate aftermath, many proposals were put forward, some under official federal government auspices, for programs of 'social reconstruction'. Among them were proposals for systematic attention, on a national scale, to the problem of incomes of elderly people. For a variety of reasons, including perhaps a certain complacency induced by the unexpectedly long-lasting post-war prosperity, as well as the vagaries of federal-provincial relations, most of the concrete reconstruction proposals fell by the wayside for many years.

The Family Allowances Act was passed in 1944, a year before the end of the war. The next major federal initiative in the field of social welfare was the creation of the Old Age Security program (OAS) in 1951. OAS was not a cost-sharing program; it was strictly a federal program, and therefore a novelty in the field of social welfare. As with unemployment insurance, before Parliament could enact the Old Age Security Act, it was necessary to secure an amendment to the BNA Act empowering Parliament to legislate with regard to old age pensions.

Old Age Security was designed as a *universal* income payment, payable to virtually all Canadian residents over a stated age: seventy at first, later sixty-five. It is an *income transfer* program, paid for entirely out of the resources of the federal government. At first it was financed by a special earmarked slice of certain taxes, ostensibly to allow the building up of a fund that would help secure the program. In time, this manoeuvre was dropped, so OAS is now paid for out of normal government revenues.

The amounts paid have increased substantially over the years, and for some years have been indexed automatically at intervals to provide some measure of protection against inflation (an attempt in the late 1980s to bypass the indexation as a deficit-reducing measure created such a furore that the government dropped the idea).

In recent years, OAS has lost entirely its character as a universal program. Beginning in 1988, older Canadians with annual incomes exceeding an amount something over $50,000 (which was quite high up the income scale for all individuals in Canada, let alone for all individuals over sixty-five), indexed each year according to a measure of the cost of living, have had their OAS benefits 'clawed back' at a special rate, in addition to the income tax paid on these benefits. The amount clawed back depended on one's income other than OAS: the higher the income from other sources, the less the OAS benefits. The rate of reduction is 15 per cent of other income beyond the reduction point. Older individuals sufficiently high on the income scale—something well over $80,000 per year—found all of their OAS benefits taken back in this way.

The clawback procedure was clearly inconvenient and even irksome. After a few years, the federal government began to do the clawing back in advance: now, the Old Age Security branch of Human Resources Development Canada calculates each person's net OAS entitlement, reduced if and as called for based on the *previous year's* income, and sends monthly cheques in the appropriate amount; this might be regarded as a 'clawforward' rather than a 'clawback'. OAS benefits are thereby income-tested at the upper-income ranges, and a certain proportion of older Canadians have, since the first introduction of the clawback, received *no* Old Age Security at all. Thus, the program is no longer a universal trans-fer. As we shall see, at roughly the same time, the other universal federal transfer program, Family Allowances, was superseded by an income-tested Child Tax Benefit structured so as to yield no benefits to upper-income families. By the mid-1990s, the principle of universality in income transfer programs was dead and buried.

The program is intended for residents of Canada, so there is a residence requirement: a certain number of consecutive years, or a larger number of accumulated non-consecutive adult years, before one's sixty-fifth birthday. Thus a person who immigrates into Canada as an adult, or a Canadian who resides outside the country for a large part of his or her adult life, may suf-fer a reduction of OAS pension. To mitigate the loss to adult immigrants, Canada has signed reciprocal agreements with a large number of countries, whereby adults who emigrate from one country to another may claim age-related benefits from both countries. With sufficient Canadian residence, a person will continue to receive OAS benefits anywhere in the world; oth-erwise, to continue receiving OAS after age sixty-five, a person must main-tain some minimal degree of 'residence' in Canada.

Since the government knows the birth date of just about every citizen, a person approaching the age of sixty-five will be notified that his/her OAS eligibility is imminent. The process of establishing one's entitlement has been made very simple.

Guaranteed Income Supplement (GIS)

In 1966, in conjunction with the creation of the public contributory pen-sion programs (the Canada and Quebec Pension Plans, discussed in the next section), the federal Parliament amended the Old Age Security Act to create the Guaranteed Income Supplement. As the name implies, this legislation authorizes the government to provide a supplement to OAS in order to guarantee to all older Canadians a certain minimum total income from all sources.

The income level that is guaranteed is the sum of the universal OAS payment and the maximum GIS payment. With the exception of those whose OAS is reduced on account of the residence requirement, *every Canadian resident over sixty-five should have an income at least equal to OAS plus maximum GIS.* The maximum GIS is paid to an OAS recipient with no other income at all. If an OAS recipient has a small personal income (e.g., from a job or a pension), his/her GIS will be reduced according to a scale that amounts to a reduction of a dollar of supplement for every two dollars of personal income. It follows that once one's personal income is twice the current maximum GIS, one's GIS disappears. Like OAS, the maximum GIS is indexed, so the actual dollar amounts change at regular intervals.

For married couples, both over sixty-five, the GIS of each individual is based on their combined income. On the premise that two can live more cheaply than one, the GIS paid to each member of a married couple is somewhat less than the GIS of a single individual with half their total income. The previous discussion of the OAS clawback has no bearing on GIS entitlements; the income levels at which the clawback takes effect are too high to affect recipients of GIS benefits.

At first, the maximum GIS was *smaller* than OAS. In time, the government decided that since GIS was going to those older people with the lowest incomes, GIS should be increased more than OAS, which goes to *all* older people. Now the maximum GIS is much larger than OAS.

Since GIS depends on the recipient's income, it is an example of a *selective, income-tested* transfer payment: the more income a person has from *all* sources other than OAS itself, the smaller the GIS. A person must apply for GIS, and must reveal his/her current income.

The combination of OAS and GIS is one of the few examples in the world of a functioning 'guaranteed annual income' program. To repeat, except for the relatively few who fail to meet the eligibility requirements, all Canadians over sixty-five are assured an income equal to OAS plus the maximum GIS. That guarantee falls short of the conventionally accepted low-income-defining income level except for couples in rural communities (see Appendix A, 'Poverty Lines', pages 277–9). In most instances, it will even fall short of provincial public assistance benefits. The majority of older Canadians, however, *do* have additional income of some kind, enough to raise them above 'low income' status.

Spouse's Allowance, Widowed Spouse's Allowance

With time, a hardship-creating anomaly was observed in the working of OAS/GIS. In some cases, the income-earning husband of an older couple would reach age sixty-five, would begin receiving OAS, and, if his income were small, would receive GIS as well. His wife, being under sixty-five, would receive nothing. Thus, if the wife was not employed, the income of the couple would be almost entirely the husband's OAS+GIS, not intended to support two people. Yet, when the wife reached age sixty-five, the couple's income would jump to OAS+GIS for married couples. The same would apply if the wife were older, but the case described is the more typical. To correct this, in 1975 the Spouse's Allowance program was created. Where one spouse is over sixty-five and the other spouse is between sixty and sixty-five, the Spouse's Allowance puts them in much the same position as if both were over sixty-five. The details of the calculations are different in the two cases, but the resulting incomes are nearly the same. Once the younger spouse reaches sixty-five, Spouse's Allowance terminates, as both will then be entitled to OAS+GIS at the married rate. A special clause protects the benefits of a spouse who has been receiving Spouse's Allowance and who becomes widowed before reaching sixty-five.

Provincial Top-ups
Even with OAS+GIS, the elderly make up a disproportionately large part of Canada's poor. OAS+GIS combined yield an income low enough to qualify an elderly couple for public assistance in most if not all provinces. To avoid the needs test required by public assistance, most provinces have created additional income supplements of their own for the elderly. These include the Saskatchewan Family Income Plan; in Nova Scotia, Seniors' Special Assistance (which, in addition to an income supplement, provides a housing subsidy for both homeowners and tenants); the Alberta Assured Income Plan; and in British Columbia, a special program for seniors under the Guaranteed Available Income for Need Act, known as GAIN. The largest maximum supplements are those payable in Alberta and Ontario (Human Resources Development Canada, 1993; National Council of Welfare, 1996). These supplements raise the incomes of recipients to roughly the income levels of public assistance recipients. It is interesting that the GAIN for seniors program in British Columbia is declining in scope with the passing years, even though B.C. has a large elderly population; one can infer that the more basic income programs—GIS and CPP—are gradually bringing more people up to the levels of income guaranteed under GAIN.

The (Proposed) Seniors Benefit (1996)
As of 1996, the package comprising OAS, GIS for individuals and for married couples, Spouse's Allowance, and Widowed Spouse's Allowance, had become somewhat complicated. As explained above, all were now income-tested; the income tests for OAS, the 'clawback' or 'clawforward', took effect at fairly high income levels while the income tests for the GIS component took effect at more modest levels. GIS is not taxable, OAS is.

In 1996, the federal government announced its intention to introduce a new program, the Seniors Benefit, that would combine all these programs into one, and reduce their reach. For people with little or no income, the maximum benefit was to increase modestly, but in the structure of the program there was to be little change from the familiar OAS/GIS.

The first major difference in the proposed Seniors Benefit was in the reduction of benefits for those with moderate or better incomes: the income level at which the reduction of benefits would begin—$25,921— was to be much lower than previously. Also, the rate of reduction was to increase to 20 per cent of outside income above that level, compared to the prevailing 15 per cent. Accordingly, the income level at which benefits were reduced to zero was much lower. The second major difference was to affect couples. Hitherto, the GIS entitlement for married couples had indeed been based on the income of the couple, but for couples whose joint incomes were high enough to take them beyond the GIS range, OAS entitlements have been based on their *individual* incomes. The Seniors Benefit would be calculated in terms of the *combined* incomes of couples. And, more restrictive still, the proposal would subject couples' benefits to the same reduction point—$25,921—as was to apply to individuals.

The proposed Seniors Benefit was accompanied by certain tax changes. All benefits were to be free of taxation, even those received by middle-income individuals and couples, a sensible idea in that benefits were already to be sharply reduced as income rose. On the other hand, an existing income tax credit for persons over sixty-five, and a tax exemption of the first $1,000 of income from a pension plan, were both slated to be abolished.

These changes would enhance somewhat the benefits of couples with private incomes of up to about $30,000, would sharply reduce benefits for couples with incomes above that level, and would entirely eliminate benefits for couples with incomes over about $75,000. The OAS system hitherto in effect might have given full (but taxable) OAS benefits to couples with total incomes of $80,000 before benefits began to be reduced, and might have paid diminishing benefits to couples up to incomes substantially exceeding $100,000—not that there are many of these. (I say 'might have given' because OAS benefits have been based strictly on *individual* incomes: how much OAS a *couple* received has depended on how much income *each* of them had.) The proposed Seniors Benefit has been projected to cost the government much less than OAS/GIS.

As for unattached individuals, OAS currently has been undiminished (but taxable) up to a personal income of about $50,000, and thereafter has gradually diminished to zero at a little over $80,000. The Seniors Benefit would begin to diminish at the pivotal figure of $25,921, and would be eliminated at something over $50,000. The Seniors Benefit was designed to confer a far larger share of its total benefits to people with low or modest incomes, i.e., to be more 'redistributive downwards', than the former batch of programs.

The Seniors Benefit was to be phased in over a five-year period. People over the age of sixty in 1995 (sixty-five by the year 2000) were to have the option of receiving benefits under either the more familiar older scheme or the new one. Since the proposed Seniors Benefit is advantageous to that majority of older people who are in the lower income ranges, the new plan would be their rational choice. Those with incomes above the new reduction point of $25,921 would presumably choose to continue receiving OAS. After 2001, only the Seniors Benefit was to be in effect (Canada, Department of Finance, 1996).

The proposed Seniors Benefit would retain one important characteristic of OAS/GIS: it would constitute a genuine guaranteed annual income program, but at a somewhat higher level of guarantee.

The Seniors Benefit was announced as policy by the Liberal government in the 1996 federal budget statement. The necessary authorizing legislation was not presented to Parliament in 1996, as had been expected, but was set aside in the face of the 1997 federal election. The re-elected Liberal government will certainly revive the scheme in some form. At the time of publication, it is not known what modifications the government may propose. The Seniors Benefit has been well received by low-income older people and those who expect to have low incomes. It is popular also with those who are in favour of reductions in government expenditures. It has been opposed by groups that represent people in the lower middle-income range, who would lose substantially. Not only would they lose all or part of current OAS/GIS benefits, on which they may well have been relying; they may feel, as well, that the lifetime savings they have put into pensions and other preparations for retirement have been devalued, because in terms of income they will end up little further ahead for their pains (Slater, 1997; Swankey, 1997).

PUBLIC CONTRIBUTORY PENSIONS: THE CANADA AND QUEBEC PENSION PLANS

The concept of retirement necessarily implies some form of provision for material support in the years following retirement. One way in which a society may address this problem is illustrated by the Old Age Security package of programs discussed above: collect taxes every year from the taxpaying part of the population, and distribute part of the money collected to old people (some of whom, of course, are still substantial taxpayers), as we used to do; or distribute it in such a way that the less well-off older people keep it while the better-off give all or most of it back. The other way is to create mechanisms that assist people to save effectively from their earnings while they are earning, and to use the accumulated funds to provide incomes after they have ceased earning.

For a long time, the business of administering pension funds in Canada was left almost entirely to the private sector. The Canadian Post Office used to operate an annuities program, discontinued several years ago, whereby individuals made voluntary deposits whenever they could, and, beginning at a certain age, received an annuity based on the total of their deposits and the interest their deposits had earned over the years. Other than that, and the pension programs provided for public servants (federal, provincial, municipal, educational, military), the administration of pensions funded through current savings was entirely private until the 1960s.

It is appropriate here to address a possible semantic difficulty. Public servants have fairly good pension plans, to which their employers (governments) and, usually, they themselves make contributions. The funds are now usually administered by fund managers independent of the governments. Public service pension plans are considered 'private' pensions, as distinguished from the 'public' pensions, the Canada and Quebec Pension Plans, which are available to *all* Canadians as a matter of public policy.

The private pension system extends its coverage almost exclusively to income earners, though there is nothing to prevent non-earners who have the money from contributing to pensions for themselves, or from having someone contribute for them. For reasons touched on in the next section, the private pension system cannot reach all income earners, and has other limitations. There was therefore a great deal of discussion for many years about the desirability of a *public* vehicle whereby Canadians could save and earn the right to an old age pension bearing some relation to lifetime income.

Exceedingly tortuous political, economic, administrative, and financial discussions and manoeuvres led up to the twin legislation of 1966: the Canada Pension Plan Act and the Quebec Pension Plan Act—preceded, as usual, by the BNA Act amendment required to empower the federal Parliament to pass such legislation. That there were two Acts, and that there are two programs, is a reminder that as welfare programs, linked to employment, pensions are within provincial jurisdiction; the government of Canada could not therefore move into the pensions field except with the consent of the provinces. In this case, the provinces' consent was given only with the proviso that any province could create its own contributory pension program if it so wished; and Quebec so wished.

The Quebec and Canada Pension Plans are similar except for the levels of 'survivors' benefits', and for the management of the accumulated funds. A Canadian who moves in or out of Quebec carries all the pension entitlement he/she has accumulated under both plans.

Contributions

Basically, all employed persons in Canada over eighteen make *compulsory* contributions throughout their working lives to the pension plan. They contribute and acquire pension rights, no matter how many times they change jobs and no matter how many interruptions, short or long, they have in their work careers. These contributions are a certain percentage of 'yearly pensionable earnings', that is, earned income between a minimum and a maximum. Employers match their employees' contributions exactly. Self-employed persons contribute both the employee's and the employer's share on their own behalf.

Benefits: Retirement, Disability, Death, Survivors' Benefits

At age sixty-five, *whether retired from employment or not*, the person becomes entitled to a monthly pension that will be based on the contributions made and the number of years of contribution. In that sense, the Canada and Quebec Pension Plans are not, strictly speaking, 'retirement' pension programs, but almost all recipients have in fact retired. A pensioner may elect to retire from work and receive a reduced pension before the age of sixty-five, or to delay retirement and receive an increased pension at an age greater than sixty-five. Both options are subject to limits. The early retirement option is quite popular: one of about six recipients of old age pensions is under sixty-five.

While the exact calculation is complicated, at present the outcome is a pension that will give the pensioner who had a fairly normal working life about one-quarter of the purchasing power, at current price levels, of his/her lifetime *average* pensionable earnings, that is, the average of annual earnings up to the maximum pensionable earnings of each earning year. As with UI and WC, earnings above the maximum in any year are not taken into account in the calculation of benefits. Pensions are adjusted at regular intervals to reduce the impact of inflation.

Because of the experience of the two plans, serious consideration is now being given to changing all of the quantities in the equation: the rate of contributions (which has been increased already, and is scheduled to be increased again), the maximum pensionable earnings, and the income replacement ratio.

In addition to old age pensions, the plans provide *survivors' benefits* in the event that an income earner or pensioner dies leaving dependants, *death benefits* payable on the death of a pensioner or contributor (basically to help pay for the considerable costs of dying), and *disability benefits* payable to a contributor who suffers a disabling injury or illness. (If the disability resulted from a work accident, the disabled person's benefits from Workers' Compensation are reduced by the amount of the pension paid by the pension plan.)

Financial Management

The Canada Pension Plan's financial management involves all nine provinces other than Quebec as well as the federal administering agency. The plan's surplus revenues are committed by law to loans to the nine provinces (including, again, municipalities, school boards, and provincial Crown Corporations) in proportion to their residents' contributions to the fund. The earnings realized from the interest on such loans, etc., augment the funds' revenues from contributions, and are available to help pay pensions.

The existence of the Canada Pension Plan provides the participating provinces with an immense fund from which they can borrow at privileged interest rates. Some observers have criticized the financial consequences for the Canada Pension Plan of the way the provinces have used this borrowing resource.

In Quebec, contributions are received by the Régie des rentes du Québec, pensions payable are paid out, and surplus revenues are handed over to the province's special funds-management agency, the Caisse des depôts et placements du Québec, which looks after many kinds of funds for various Quebec public agencies. With the surplus pension funds, the Caisse seeks to earn as much additional revenue as possible, by interest-bearing loans and, it hopes, profitable investments. Its lending and investing are subject to certain limitations: it is obliged by law to lend money first to the province of Quebec, to provincial agencies such as Hydro-Québec, and to Quebec municipalities and school boards; then it is permitted to lend to relatively secure private-sector corporations, and it is also permitted a certain amount of capital investment in the private sector. The student should be alert to the difference between *lending* to private corporations and *capital investment* in the private sector.

A Provincial Supplementary Contributory Pension Plan: The Saskatchewan Pension Plan

To meet the needs of individuals who have no private pensions, who wish to improve upon the pensions to which they contribute via the Canada Pension Plan, or who, having no earned income, are ineligible for the Canada Pension Plan (homemakers, for example), the government of Saskatchewan has created a voluntary contributory pension plan. Administered by a Board of Trustees, the Saskatchewan Pension Plan operates much like the former Post Office Annuity program mentioned above.

THE CANADA/QUEBEC PENSION PLANS AND OTHER PROGRAMS

The functioning of the pension plans is linked with the Guaranteed Income Supplement in that widespread pension plan pensions reduce the number of people entitled to maximum or relatively large GIS payments and reduce the amounts of GIS payable to individuals. That is, the more paid out to people in Canada or Quebec Pension Plan pensions, the less must be paid out in Guaranteed Income Supplement and Spouse's Allowance (or Seniors Benefit) payments. This has some significance in that *they are not paid out of the same pot.* In fact, when introduced, at the same time as the pension plans, GIS was seen as a transitional program, which would phase itself out as more and more people began to receive C/QPP pensions. This has not worked out as planned, due partly to continuing inflation, partly to the limits on C/QPP pensions, and partly to the considerable increase in GIS that was noted above. Still, over the years, more and more people have received C/QPP pensions, and the numbers who must have recourse to maximum or partial GIS, while still large, have been declining steadily.

The same applies to the income-tested provincial top-up programs, such as British Columbia's Seniors Supplement, the Alberta Assured Income Plan, Nova Scotia's Senior Citizens Financial Aid, and Ontario's GAIN: the better off older people are through C/QPP and private pensions, the less need for such top-up programs.

PRIVATE PENSIONS

Group Pensions

Something has already been said about private pensions as part of the provision made for old age incomes. Such a brief treatment as is possible here cannot do justice to the complexity of the field of employment-based pensions, to which our attention will be confined (although there is a substantial volume of business in pensions for other kinds of groups and for individuals). Some consideration of private pensions is called for, for two reasons: (1) the availability of private pensions is invariably brought into discussions of public policy concerning the incomes of the elderly, and (2) the favourable tax treatment of contributions made by individuals and employers to pension funds, and of the funds' earnings, gives a 'public' aspect to private pensions. Participants in private pension plans are implicitly subsidized by the taxpayer at large (a category which, it must be remembered, includes themselves). Like most tax concessions, the deduction of employee and employer pension contributions from taxable income is vigorously defended by its beneficiaries, namely, the millions of employees enrolled in pension plans, and firms in the pension industry, whose services are thereby made financially more attractive to employee groups.

Private pension managers have to earn enough return on the funds they handle to pay the promised pensions, to meet their own expenses, and then, in almost all cases, to turn a profit. (A small part of the Canadian pension business is in the hands of mutual-benefit or co-operative organizations, which return all 'profits' to the contributors. It would be hasty to conclude that their clients do better financially than do the clients of profit-seeking fund managers, but there are certain potentially important differences in administration.)

Fund managers must do all this subject to the legal obligation to give first priority to the security of the funds with which they are entrusted. They naturally prefer to work with client groups that make it easier for them to realize these objectives. The pension business is thus better adapted to serve workers in the kinds of employment that provide continuous, fairly stable incomes. It is inconvenient and costly to have people joining plans and then leaving them to join other plans. Small groups of employees can be and are served, but they lose the benefit of the economies of large scale. In some industries in which firms tend to be small and/or in which employees tend to move around from employer to employer, industry-wide plans have been developed. This is particularly true of industries in which machinery exists for industry-wide collective bargaining.

There are three problems with private pensions that have proved difficult to overcome: 'vesting', 'portability', and coverage.

(i) *Vesting* means the terms upon which the entitlement to pensions is 'vested' in the pensioners, i.e., really belongs to them as individuals. Typically the pension agreement will require a number of years of enrolment in the plan before the pension rights are 'vested'. All the provinces have laws that say what the maximum vesting period may be. For the vast majority of the Canadian work force, the maximum allowable vesting period has been reduced to two years by provincial and federal legislation (the latter applies to federally regulated industries). Before the end of this vesting period, only the accumulated value of the *employee's* own contributions (with interest) 'belongs' to the employee—the value of the employer's share does not—and if the employee leaves the plan through changing jobs, a sum of money is returned to the employee, not a pension payable at retirement age. If the plan is financed entirely by the employer, as some are, then an employee who leaves before vesting is entitled to nothing at all.

In the now distant past, vesting periods could be as long as twenty years. As a result, many employed people never qualified for the pensions their employers offered. Others with some seniority were reluctant to change jobs, even if they had a good opportunity, either because they were coming close to having their pensions vested and did not want to lose the promised pension, or because they did not want to start all over again in a new pension plan. The two-year rule greatly reduces the dimensions of the vesting problem for employees.

(ii) *Portability*. Linked to 'vesting' is the problem of 'portability', i.e., 'carrying' one's pension rights from employer to employer or from place to place. Full portability means the ability to carry all of what one owns in the way of pension rights from one pension plan to another, thus avoiding the above-noted danger of having to start one's vesting period all over again each time one changed jobs. There are obvious complexities, since plans do differ significantly. As one alternative, an employed person who has had a number of jobs, and who has earned pension rights at each job, can collect a concomitant number of pensions after retirement.

(iii) *Coverage*. The third problem is the limitations on the reach of employment-based pensions. They are simply not easy to manage for the benefit of people whose employment is irregular, whose income from employment fluctuates considerably, or who work in industries where the fortunes of individual firms are highly erratic. Nor can pensions be made as attractive to people who work alone, or as members of very small groups, as they can to larger groups. All of these limitations apply with full force to the self-employed. Similarly, as an option, contributing to an old age pension may not be a high priority for a low-income worker, especially a younger one, who may feel that he or she has better things to do with the money.

Giving the Canadian pension industry full credit for its achievements, it remains true that from one-third to one-half of Canadian workers, depending on how one counts, are not covered by an employment-related pension, and some proportion is probably beyond the reach of the private pension business.

These considerations, among others, gave impetus to the development of the public contributory programs, the Canada and Quebec Pension Plans.

Individual Pensions: RRSPs

To provide an attractive alternative for individuals who did not have access to organized pension plans, or who wished to add to their pensions, in the 1960s the federal government initiated so-called 'Registered Retirement Savings Plans' (RRSPs). As explained above, approved employer-based pension plans enjoy favourable tax treatment, which provides an incentive to employers to develop them and to employees to enrol in them. Contributors to pension plans therefore buy a measure of income security in old age at a bargain price, assisted by taxpayers (many of whom do not enjoy the benefit of a pension plan themselves). RRSPs give to individuals who put money aside on their own initiative the same tax advantages as are enjoyed by contributors to employer pension plans. First, they are allowed to deduct all their contributions to an RRSP (up to a certain limit) from their income before calculating their income tax. Second, any growth in value that is realized over the years from investments of the RRSP money is also tax-free. (The limit is a limit on the individual's total yearly contributions to *all* approved pension plans, including both employment-based plans and RRSPs.) Compounded over thirty or more years of working life, these tax concessions are extremely valuable even to a person in a low-income, low-tax bracket; but like most tax concessions, they are even more valuable to a high-income earner in a high tax bracket.

The National Council of Welfare's analysis of 1992 tax data showed that the average annual RRSP tax savings to taxfilers with incomes under $60,000 was $806, while the average advantage to earners of $60,000 or more was $2,695 (National Council of Welfare, 1994, p. 35). The Caledon Institute estimated that in 1993, after a huge leap in the maximum allowable deduction, the total cost to the federal government of the tax break for all pension contributions was $23 billion—far more than OAS/GIS, or C/QPP pensions, or Unemployment Insurance benefits for that year (Caledon Institute, 1994).

When the individual eventually begins to draw the pension from an employer pension and/or an RRSP, the income, like any other pension income, is taxed. The great majority of people have much lower incomes in retirement than while employed, so the rate at which they pay income tax on their pension income is usually lower than their income tax rates while they were working. Still, the government does get back some of the taxes it gave up while the individual was contributing; and since people with pension income will receive reduced OAS/GIS or Seniors Benefit payments, if any, the government gains through reduced expenditures as well.

RRSPs have become a popular savings vehicle in Canada, though only a little more than one-quarter of taxfilers use it. Every year, in the three or four months preceding the filing deadline for income tax returns, investment companies energetically promote their services, targeting people who have not yet made their permitted RRSP investments for the year.

To underscore the relative proportions of the public and private sources of pension income, the federal Department of Finance estimated that in 1996 Canadians received $22 billion from the Canada and Quebec Pension Plans, another $22 billion from Old Age Security, the Guaranteed Income Supplement, and the Spouse's Allowance, and a considerably larger $32 billion through employment-based pensions and RRSPs.

THE JURISDICTIONAL QUESTION: FEDERAL OR PROVINCIAL?

The definitions of federal and provincial powers under the Constitution Act and its amendments have figured prominently in the above description of income programs for older people. To summarize:

- Provincial programs, such as the 'top-ups' referred to above, and any modifications that provinces care to make to their own welfare programs, are fully authorized by the jurisdiction over welfare given to the provinces in Section 92 of the Constitution Act.
- Federal programs have been authorized under successive amendments to the BNA Act. Old Age Security was made possible by an amendment in 1951, adding Section 94A to the Act. The wording of this Section *allows* the federal Parliament to legislate, but not so as to 'affect the operation of any law present or future of a Provincial Legislature in relation to old age pensions.' Technically, this would allow any province, at some future time, to move into the field now occupied by OAS/GIS. This amendment cleared the way for the subsequent Guaranteed Income Supplement and the Spouse's Allowance. When the Canada Pension Plan legislation came on stream, at the same time as GIS, it was the considered judgement that Parliament had the power needed to legislate the old age pension part of it, but not the survivors' and disability benefit programs. Section 94A was therefore amended to give Parliament this power, and then the Canada Pension Plan legislation was passed. Quebec took advantage of the prior claim of the

provinces to jurisdiction, and its legislature passed the Quebec Pension Plan Act at the same time.

- The *regulation* of private pensions is within the jurisdiction of the provinces, except for employer-sponsored pensions in industries regulated by the federal government. Thus it is possible, for example, to have different requirements for vesting of pensions in different provinces, although there is a good deal of *de facto* co-ordination of pension legislation.

TROUBLESOME ISSUES AND NEXT STEPS

(1) Canada's history of old age income provision neatly crystallizes the issue of universality versus selectivity. OAS used to be *universal*. It retained, briefly, a vestige of universality after the 'clawback' was introduced, in that it was sent to all older Canadians before being recuperated wholly or partly from those in the upper income strata. That vestige has vanished now that the clawback is applied *before* OAS is distributed. The new Seniors Benefit is a straightforward *income-tested* program that makes no claim whatsoever to universality.

The existing OAS/GIS system still assures for older Canadians a genuine 'guaranteed annual income'—an irreducible level of final income from all sources, on the basis of which they can plan. Benefits still go to many who are not in need. OAS/GIS represents the obligation of the so-called 'active' population toward the majority of the elderly (excluding a minority with incomes well above the average). The proposed Seniors Benefit would preserve that guarantee, but would restrict income supplements to persons in lower income ranges than OAS. As years pass, fewer and fewer older Canadians reach the age of sixty-five with zero personal income. This beneficent outcome can be attributed to the very wide reach of the Canada and Quebec Pension Plans and the spread, less wide but still considerable, of private pensions, including RRSPs. OAS/GIS has contributed to the laudable decline of poverty among the elderly, but poverty has declined also because many older people already have adequate incomes from other sources, including C/QPP.

As explained above, the proposed Seniors Benefit wraps up into one package the formerly separate OAS, GIS, and Spouse's Allowance programs. GIS and the Spouse's Allowance were *selective*, as are all income-tested measures. All the money has gone to relatively low-income (but not necessarily poor) people, as would all of the Seniors Benefit, but recipients must identify themselves as low income. Thus, however humanely the Seniors Benefit is administered, the members of society will be split into two groups: those who give somewhat less and get something, and those who give somewhat more and get nothing.

The principle of universality has been abandoned in the field of income security. (The other truly universal income program, federal Family

Allowances, is defunct, as is a parallel program in Quebec.) What values are at stake in terminating universality?

- Is the termination simply a common-sense way of avoiding the waste of directing tax money toward people who clearly are not in need?
- Alternatively, is it an unfair change in the rules that will penalize people who have devoted whatever effort it took to provide moderately well for themselves in old age, have meanwhile supported the OAS program through their taxes all through their working lives, and now find part of what they had counted on taken away from them?
- Was the 100 per cent withdrawal of benefits at upper-income levels merely the sharp edge of the wedge, to be followed by ever-deeper incursions? The proposed Seniors Benefit bears out this suspicion, by bringing down the income limit above which the 100 per cent withdrawal takes effect. (Actually, the indexing of the benefit-reduction threshold of OAS at the rate of inflation minus 3 per cent means that in real income terms the limit *has* been lowered every year in which inflation occurred, i.e., every year. The proposed Seniors Benefit would sidestep this danger by fully indexing all factors, including the benefit reduction points.)

More general issues are raised. Universal programs are said to enhance social solidarity, by enlisting the support of all Canadians behind programs to which almost all contribute and from which all will benefit. Such programs do not split the population into two classes, givers and takers, both aware of the division. That being said, given the uncertainties of life, few people can predict exactly what their incomes will be when they are older. An income-tested OAS/GIS or Seniors Benefit guarantees an income floor for everyone. In that light, how serious is the threat to social solidarity, with respect to Canada's provision for its older residents?

(2) By almost any measure, Canada has achieved considerable success in combatting poverty among the elderly, even if the success is far from complete. Measuring poverty by the Statistics Canada low income cut-off lines (explained in Appendix A, pages 277–9), the poverty rate among Canadians over sixty-five in 1980 was 33.6 per cent, compared to 15.3 per cent for all persons. By 1992, the rate for the elderly had dropped to 18.6 per cent, compared to 16.1 per cent for all persons. On that measure, the elderly had gained remarkably, while the population as a whole had lost some ground. Moreover, since the elderly are guaranteed an income at least equal to the equivalent of OAS plus GIS, there is a limit to the depth of an older person's poverty—the gap between an older person's income and the poverty line—and the provincial top-ups fill in a good part of that gap.

According to the National Council of Welfare, a body not given to excessive praise, 'Poverty rates for seniors have fallen sharply in every province in line with the national trend . . . , and the size of some of the reductions is almost breathtaking.' The province with the highest poverty rate among the elderly, Quebec, is one of the few provinces that has no top-up program (National Council of Welfare, 1994, pp. 7, 10, 72).

However, this somewhat rosy picture of the income situation of Canada's elderly has its darker spots, notably with respect to older women. Although the picture will change over time, today's elderly women often have small lifetime earnings, and in many cases, none at all. They are unlikely to have much income from employer pensions or Canada/Quebec Pension Plan pensions of their own. They are probably widowed, in which case they may have Canada/Quebec Pension Plan survivors' pensions, but only a few will receive survivors' benefits from their spouses' private pensions. They have relied heavily on OAS/GIS. The National Council of Welfare found in 1992 that the poverty rate among all women over sixty-five was 23.3 per cent, compared with 12.4 per cent for men. Even that was a notable improvement: in 1980 an estimated 38.4 per cent of all Canadian women over sixty-five occupied low-income status, by Statistics Canada standards. This improvement is likely to continue, as more women will have worked for pay, and they will have their own public and private pensions (National Council of Welfare, 1994, p. 72).

(3) The arguably precarious financial condition of the Canada and Quebec Pension Plans is attracting anxious attention. Having their respective governments behind them, the plans are unlikely to 'go broke', but that does not make their financial problems less serious.

Basically, what is happening is simple and has in fact been clearly foreseen since the plans were first created. Beginning in 1967, rates of contribution from workers and employers were modest. Even so, surpluses were accumulated in the first couple of decades of the plans, while the number of eligible pensioners was small and the pensions earned were small as well. Now, the much-discussed increases in the numbers of elderly Canadians and in their longevity have taken their toll. Also, following long careers in the work force, most of today's pensioners have earned fairly substantial pensions, indexed to inflation. Meanwhile, contribution rates have remained the same until quite recently.

The current annual payout of pensions by the plans has grown to the point where it exceeds the current revenues from contributions plus fund earnings. The accumulated surplus reached a peak in 1992–3 and has been declining ever since, which inevitably reduces the annual interest revenues as well. The pension plans have therefore been operating more on a pay-as-you-go basis than on a funded basis; the premiums paid by current contributors have been totally absorbed in paying current pensions, instead of contributing to a saved-up fund whose returns will help provide for their pensions in the future. The problem can be kept in check if revenues are raised sufficiently to keep the funds solvent, i.e., able to meet foreseeable obligations by the combination of foreseeable contributions and fund earnings. Note that a private-sector pension fund is legally obliged to maintain such solvency; it would not be permitted to put itself in the situation in which the CPP and the QPP find themselves.

Some people argue that the Canada Pension Plan is not in serious danger because it has always been a pay-as-you-go scheme (Townson/CCPA, 1996). This is a questionable reading of the history of the origins of the plan, and is also tantamount to deeming it efficient and fair that future work forces be called upon to pay year by year to meet ever-heavier obligations to increasing numbers of future pensioners—obligations that have been calculated on the assumption that fund earnings would meet a substantial part of the costs.

In response to this problem, the two Pension Plan Acts have been amended to implement gradual contribution rate increases over a number of years. The rate increases have already begun. In discussion of the future of the plans, it is commonly suggested that revenues may also be increased by raising the maximum pensionable earnings, the basis against which contributions are calculated, but of course it would not be equitable to raise maximum contributions without also raising maximum pensions, which would ultimately increase expenditures. Another frequently proposed measure that would improve the funds' finances is an increase in the 'normal' retirement age for purposes of calculating retirement pensions. Potentially, this would increase the number of years of contribution, and it would reduce the number of years for which pensions would be paid to any given individual; see (5) below.

Pension plan contributions are compulsory; to the employed person, they feel much like ordinary income tax, and governments are properly sensitive to the public's tolerance of compulsory levies against income. To increase the pension plans' revenues sufficiently to keep the plans solvent, in the face of well-established demographic trends, will require a good measure of public support and political will, to use an overworked expression.

Part of the financial difficulty arises from the way the surpluses are loaned, especially in the Canada Pension Plan. Provincial governments borrow from the pension plan at interest rates a little below the 'going rate' in the money market. When provinces do pay their interest and repay their loans, as the Plan may require them to do when the money is needed, they do so with money collected through taxation, most of which is paid by the very people who are contributing to the pension plan. In this way, the plan is operating rather more on a year-by-year, pay-as-you-go basis than was originally the design. This appears to be somewhat less true of the Quebec Pension Plan, whose fund management, as we have noted, is different, allowing a small proportion of hopefully lucrative private-sector investment.

With a view to improving the financial performance of the Canada Pension Plan, the current (1997) federal government has indicated that it intends to amend the Plan to allow the investment of some proportion of its funds in the private sector.

The private-sector finance industry is inclined to blame the vicissitudes of the Canada and Quebec Pension Plans on the adulteration of their strict income security goals by political considerations, and on what the private sector generally considers to be the flabbiness of public, non-competitive administration of such undertakings. The pension industry naturally tends to support steps, such as favourable tax treatment, designed to help expand the scope of private pensions, including RRSPs. But even with such help from public policy, private pensions could not cover the field as widely as do C/QPP.

(4) A different kind of problem with the design of the public plans has been their response, or lack of response, to *the peculiar income situation of women*. Contributions are based on income from employment: low-paid job, low contributions, low pension; no job, no contributions, no pension. The low incomes most women have earned, and the years without any earnings at all in the adult lives of many women, exact a fearful price in their years of retirement.

This directly affects married women who have not been employed at all. The survivors' clause in C/QPP pays some attention to women whose husbands die, but it took a judicial decision to establish the right of a woman to a share in the pension of a husband from whom she was divorced late in life. Both pension plans now provide for the equal division of CPP or QPP pension credits in the event of divorce (unless provincial family law specifically overrides such pension-splitting, as Saskatchewan's does: a recognition of the provinces' jurisdictional primacy in matters bearing upon the family) (National Council of Welfare, 1990b, p. 32).

A more frequent problem is that of women who have worked for a while, left the labour force to raise their children, then returned to work. Originally their incomes would have been averaged over all the years from the beginning of their work lives to the end, including the years during which they earned nothing. This hardship has been amended by a special 'drop-out' provision, by which the non-working years, ascribable to child care, do not count fully in the averaging of income.

To the extent that women do earn less than men, their pensions will be less; that imbalance cannot easily be corrected by a pension plan.

(5) The issue of retirement age was raised in (3) above. It is somewhat paradoxical that as longevity has increased, the age at which people actually retire, voluntarily or involuntarily, has been dropping. The proportion of recipients of C/QPP retirement benefits who are under sixty-five was 16 per cent in 1992, and this percentage has been rising; these pensioners have opted for early retirement with reduced benefits (Human Resources Development Canada, 1993, pp. 184–6). Some of these are undoubtedly reluctant early retirements, caused by unemployment or ill health, but the trend seems too strong, in Canada and elsewhere, to be wholly explained by these factors.

Meanwhile, pension fund managers, public and private, contemplating the lengthening of life after retirement, and the consequent increasing burden upon pension funds, have proposed *increasing* the age of retirement used in the calculation of pension benefits. In the United States, the statutory age of retirement for purposes of the national Social Security program is now sixty-seven. Two questions arise: Should public policy encourage earlier or later retirement? And what measures would be legitimate to motivate people in the desired direction?

(6) The last few years have seen a large, unexpected increase in outlays on disability pensions payable under the C/QPP plans. This increase has darkened an already dark financial picture. A high proportion of disability claimants are over the age of fifty-five. They do indeed suffer from some disability, but some observers suggest that many of them have been employed, have lost their jobs, have exhausted their unemployment insurance benefits without finding new jobs, and have been driven to claim disability pensions as a substitute. Some critics hint that they are driven to do so by private disability insurers, provincial Workers' Compensation authorities, and provincial public assistance authorities seeking to lessen their own burdens. A Human Resources Development Canada report even states, 'CPP disability benefits may be effectively serving as early retirement benefits', and that much of the current money-flow problems of CPP are attributable to the misuse of disability pensions. This may be going too far. Disability pensions have been part of the C/QPP package from the beginning, and, as Sherri Torjman of the Caledon Institute has pointed out, most of the increase in C/QPP disability pensions can be attributed to a deliberate, concerted effort on the part of the plans' administrators, beginning in 1990, to improve the take-up rate of disability pensions, on the suspicion that many disabled persons were unaware of their entitlements. The CPP Advisory Board itself referred in its report on the disability pensions issue to a problem of 'incapacity to apply' on the part of many potential clients, and of the need to 'improve communication to target groups' in order to make sure that all entitled disabled people were applying for their pensions (Canada Pension Plan Advisory Board, 1996; Torjman, 1996).

(7) RRSPs became a topic of controversy in the early 1990s. The limits on tax-deductible contributions to all pension plans were raised sharply, from about $3,500 to about $14,000.

Two main issues have surfaced here. First, the forgiveness of income taxes amounts to a quite considerable cost for the federal government in terms of taxes not collected; second, only those with high incomes could afford to put aside $12,000 or $13,000 a year for retirement savings, and thus enjoy the relatively large accompanying tax saving. So raising the ceiling on deductible pension contributions was seen both as expensive to the taxpayers in general and as biased in favour of the relatively wealthy. Upper-income households do in fact make far more use of RRSPs than do lower-income households, which is to be expected, since the latter find it more difficult to save (Schellenberg, 1994, p. 80). The limits have latterly been subjected to some downward tinkering by the federal government, but the precise details of the tax concession are less important than the principles at issue: the efficiency of the RRSP device as an incentive to save, and equity across income strata.

(8) Now that the universal Old Age Security benefit has been consigned to history, our Canadian old age income package comprises:

- a *selective, income-tested transfer,* whose benefits are reduced, in accordance with income, for modest to upper-level income recipients, and annihilated for people with high incomes;
- a *contributory pension plan,* whose benefits depend largely on the person's lifetime earnings;
- *private pensions,* enhanced by favourable tax treatment of contributions;
- in several provinces, *additional income-tested benefits*; and
- *welfare (needs-tested) transfer payments* to old people whose incomes are still inadequate.

The combination of these five modes provides an opportunity to consider the best way, or the best combination of ways, to secure acceptable material living standards for our old people—bearing in mind that 'old people' means 'young people (such as most of our readers) in thirty or forty years'.

Not included in the list is income from employment. A complete picture should take note that more than 10 per cent of Canadians over the age of sixty-five continue to be active in the labour force. And as the normal life expectancy post-sixty-five gradually lengthens, and as the well-known increase in the proportion of elderly people in the population continues, there is some discussion of the usefulness of pressure put on people to retire at sixty-five. As mentioned above, some people have suggested that, as in the United States, the 'normal' age up to which one contributes to the Canada/Quebec Pension Plans should be raised to sixty-seven, to lessen the pension payout of the Canada and Quebec Pension Plans. As long as unemployment looms as a major social problem, however, any such lengthening of working years will be politically unthinkable.

(9) *Income* measures have the virtue of giving the individual members of a vulnerable population a range of choice in satisfying their needs. They are free to spend the money on what *they* want, or on what they feel they need. We have persuaded ourselves that, under some circumstances, it is more effective to provide *concrete* services to certain people rather than to give them money with which to obtain things for themselves. We thus have special programs providing housing and certain health services to older people, either free or at greatly subsidized prices. One of the incessantly recurring questions in social policy is whether and when it is better to provide assistance 'in cash' than 'in kind'.

(10) One of the conundrums in the regulation of human behaviour is illustrated by the regulation of private pensions. We do have laws obliging

private pension plans to meet certain criteria, mostly designed to secure the employees' rights to the assets accumulated in the pension funds. But while there are laws governing the private pension plans, *there is no law dictating that an employer must have a pension plan in the first place.* And it is sometimes reported that the increasing governmental regulation of pensions discourages some employers from initiating or continuing pension plans for employees. Does this say anything about the potential viability of private pensions as a pillar of our system of income security for older people?

It is sometimes suggested that laws be passed obliging all employers to institute pension plans for their employees. Not surprisingly, some of the strongest support for the proposal comes from the private pension industry. What issues are raised by such a proposal?

STUDY QUESTIONS

1. What is the current proportion of Canada's population over the age of sixty-five? What are the future projections? Where would you look for data?
2. What are the current levels of OAS and GIS payments? How does the total OAS plus maximum GIS compare with estimates of income required to maintain minimum living standards?
3. The number of people receiving maximum GIS is an indication of the number of older people with no other income. How many are there? What percentage is this of the elderly? What has been the trend over time? What does this tell you?
4. What levels of benefit are currently being provided by either the Canada or the Quebec Pension Plan? What are the current contribution rates?
5. What is the difference between the Spouse's Allowance and the OAS-GIS combination for a married couple?
6. What recourse does an older person have if the income from OAS and GIS combined still leaves him, or, more frequently, her in need?
7. Who administers OAS? GIS? CPP? QPP? If you had a question or complaint, to whom would you take it?
8. What are the limitations to which the private pension system is subject, by virtue of being (mostly) profit-seeking and competitive?
9. Is there a 'legal' age for retirement? What makes you say so?
10. Summarize the economic, demographic, social, and historical factors that have left us with a problem of incomes for the elderly.
11. In what ways does the tax system encourage private saving to provide income after retirement?

BIBLIOGRAPHY

Banting, Keith (1982). *The Welfare State and Canadian Federalism.* 2nd edn. Kingston: McGill-Queen's University Press.

———— (1985). 'Institutional Conservatism: Federalism and Pension Reform', in *Canadian Social Welfare Policy*, ed. J. Ismael. Kingston: McGill-Queen's University Press. As the titles suggest, Banting's book and his article respectively discuss the impact that our federal system of government has on welfare policy in general and pension policy in particular.

Bellemare, Diane, and Lise Poulin Simon (1982). *The Attack on Universality*. Ottawa: Canadian Council on Policy Alternatives. Note the date. The 'attack on universality' is not entirely recent.

Bryden, Kenneth (1974). *Old Age Pensions and Policy-Making in Canada*. Montreal: McGill-Queen's University Press. What kinds of organizations and individuals take part in the process of making policy in Canada in such a field as old age pensions? How do they go about it? How are decisions made? Dated as to details, but not as to the game and the players.

Caledon Institute of Social Policy (1996). *Roundtable on Canada's Aging Society and Retirement Income System, June 5, 1995*. An acute, well-informed discussion of this anxious issue.

Caledon Institute of Social Policy (1994). *Old Wine in New Bottles: Privatizing Old Age Pensions*. Ottawa: The Institute.

Canada. Canada Pension Plan Act (*Revised Statutes of Canada, 1970*, Chapter C-5) (first enacted 1965); Old Age Security Act (*Revised Statutes of Canada, 1970*, Chapter O-6) (first enacted 1951); *Statutes of Canada, 1975*, Old Age Security Act Amendment (creating Spouse's Allowance). May soon be superseded by legislation bringing into existence the Seniors Benefit.

Canada. Department of Finance (1996). *The Seniors Benefit: Securing the Future*. Ottawa: Canada Communication Group. The federal government's view (naturally positive) of the changes projected in the proposed Seniors Benefit. (Cf. Slater, cited below.)

Canada. Economic Council of Canada (1979). 16th Annual Review: *One in Three*. Ottawa: The Council. An analysis of the social and economic problems arising from the aging of the population in Canada, foreseeable long ago. ('One in three' was considered to be the proportion of Canadians who had reliable, adequate pension coverage at the time.)

Canada. Hon. M. Lalonde, Minister of Finance, and Hon. M. Bégin, Minister of National Health and Welfare (1982). *Better Pensions for Canadians*.

Canada. Human Resources Development Canada. *Annual Report, the Canada Pension Plan*.

Canada. Human Resources Development Canada (1994). *Report on the Old Age Security, Child Tax Benefit, Children's Special Allowances and Canada Pension Plan*. An annual document. The title changes when the names of programs change.

Canada. Human Resources Development Canada (1993). *Inventory of Income Security Programs in Canada*, Chapter 2. This indispensable publication lists and describes, briefly but thoroughly, income programs of the federal, provincial, and territorial governments; appears at irregular intervals.

Canada. National Advisory Council on Aging (1991). *Intergovernmental Relations and the Aging of the Population: Challenges Facing Canada*. Ottawa: The Council.

Canada. Office of the Superintendent of Financial Institutions (1994). *Canada Pension Plan: Fifteenth Actuarial Report*. 'In summary', says the actuary, 'the financial projections shown in this report indicate that the existing 25-year Schedule of contribution rates requires some revision to prevent the Account from becoming exhausted by the end of 2015' (p. 3). The 'existing 25-year Schedule', already legislated, calls for an increase in rates from the present 5.0 per cent of insurable income (shared by employee and employer) to 10.10 per cent by 2016. It is this schedule that must be revised upward to avoid exhaustion. The year 2015 is less than twenty years away.

Canada. Pension Plan Advisory Board (1996). *Report of the Committee on Disability Issues.* This report confirms that there had been an unexpected increase in disability pension claims under the Canada Pension Plan, related to (1) long-lasting unemployment among older workers with disabilities, driving them out of the work force onto disability pensions, and (2) the tactics of welfare, workers' compensation, and private health and income insurance companies, all of which seek to reduce their own expenditures by encouraging their clients to claim C/QPP disability pensions. At the same time, the report confirms that no serious change in either eligibility rules or pension amounts was warranted.

Canada. Senate of Canada. *Report of the Special Senate Committee on Aging* (1966). Dated, but still interesting.

Canada. Statistics Canada (1997). *A Portrait of Seniors in Canada.* Catalogue 89-519XPE. The most authoritative storehouse of data on Canada's elderly—numbers, gender distribution, family status, etc.

Drucker, Peter (1976). *The Unseen Revolution: How Pension Fund Socialism Came to America.* New York: Harper and Row. This book opens with the words: 'If "socialism" is defined as "ownership of the means of production by the workers" . . . then the United States is the first truly "socialist" country.' A challenging thesis, to say the least! Drucker is referring to the huge amounts of shares in U.S. corporations owned by pension funds on behalf of U.S. workers.

Dulude, Louise (1981). *Pension Reform with Women in Mind.* Ottawa: Canadian Advisory Council on the Status of Women.

Frenken, Hubert (1993). 'C/QPP Costs and Private Pensions', Statistics Canada: *Perspectives on Labour and Income* 5/3 (Autumn).

Galper, Jeffrey (1971). 'Private Pensions and Public Policy', *Social Work* (May). Galper's view of the functions of pensions in a capitalist society is sharply at variance with Drucker's (cited above).

McKie, Craig (1993). 'Population Aging: Baby Boomers into the 21st Century', *Canadian Social Trends* (Summer).

Maser, Karen (1995). 'Who's Saving for Retirement?', Statistics Canada: *Perspectives on Labour and Income* 7/4 (Winter).

National Council of Welfare (1996a). *A Guide to the Proposed Seniors Benefit.* Ottawa: The Council. As good an analysis of the proposed Seniors Benefit as one will find. Unavoidably somewhat complicated: the consequences of the change would be different for unattached individuals, one-income couples, and two-income couples; and the proposed tax changes would affect different people differently.

———— (1996b). *Pension Primer.* Ottawa: The Council. Pension issues brought up to date.

———— (1994). *A Blueprint for Social Security Reform.* Ottawa: The Council. At a time when everything in the social policy field was undergoing serious review, the NCW surveyed the whole income field in this report—welfare, unemployment insurance, senior citizens' incomes, assistance to families.

—— (1990a). *Pension Reform*. Ottawa: The Council. Favours early retirement, pension credit-splitting improvements, and better survivors' benefits, especially for children. The Council makes no suggestions, however, for increasing the revenues of the Canada and Quebec Pension Plans.

—— (1990b). *Women and Poverty Revisited*. Ottawa: The Council.

—— (1984). *Sixty-five and Older*. Ottawa: The Council. Age and income in Canada.

Novak, Mark (1993). *Aging and Society: A Canadian Perspective*. Toronto: Nelson Canada.

Québec. Ministère des Affaires Sociales. *Statistiques des Affaires Sociales: Sécurité du Revenu* (current). Includes data on Régie des rentes du Québec (the Quebec Pension Plan).

Quebec. Quebec Pension Plan Act (*Revised Statutes of Quebec,* Chapter R-9) (first enacted 1965).

Québec. Régie des rentes du Québec. *Rapport Annuel.*

Ross, David P., E. Richard Shillington, and Clarence Lochhead (1994). *The Canadian Fact Book on Poverty.* Ottawa: Canadian Council on Social Development. Chapters 8 and 9 deal with incomes for the elderly.

Schellenberg, Grant (1994). *The Road to Retirement: Demographic and Economic Changes in the 90s*. Ottawa: Centre for International Statistics and Canadian Council on Social Development.

Simeon, Richard (1972). *Federal-Provincial Diplomacy: The Making of Recent Policy in Canada.* Toronto: University of Toronto Press. Chapters 3 and 11. Another study (cf. Bryden, cited above) of policy-making in Canada in fields that spill over the federal-provincial jurisdictional boundaries. The chapters cited concern programs for older people.

Slater, David (1997). *The Pension Squeeze: The Impact of the March 1996 Federal Budget. Commentary #87.* Toronto: C.D. Howe Institute. Fairly gloomy calculations of the effect the proposed Seniors Benefit is likely to have on the old-age incomes of middle-income Canadians.

Splane, Richard (1965). *Social Welfare in Ontario 1791–1893.* Toronto: University of Toronto Press. Services for, among others, the aged, long before the Old Age Pensions Act of 1927.

Swankey, Ben (1997). 'New Legislation Will Claw Back More Pension Income', CCPA *Monitor* 4/2 (June).

Torjman, Sherri (1996). 'Hysteria poor substitute for history in public debate', Ottawa *Citizen,* 11 December. In response to many statements that implied that disabled people were somewhat improperly draining the Canada and Quebec Pension Plans by claiming disability pensions, Torjman points out that the C/QPP authorities themselves had encouraged the eligible disabled to make claims to which they were entitled by law.

Townson, Monica (1996). 'The Canada Pension Plan Is *Not* Going Broke!', CCPA *Monitor* (February). Canadian Centre for Policy Alternatives.

CHAPTER 6

Public Assistance and the Working Poor; Child and Family Benefits

PUBLIC ASSISTANCE AS CONCEPT AND PROGRAM

'Public assistance' ('aide sociale') is the technical term for the program people usually have in mind when they speak of 'welfare'. Conceptually, public assistance may appear to be quite simple: (1) A certain living standard is agreed on as the minimum that society wishes all of its members to be able to enjoy (or endure). (2) Certain persons and/or families are judged unable, on account of their circumstances, to procure enough income to afford that minimum standard. (3) Society creates an instrument whereby money is collected and is given to those persons as income sufficient to enable them to meet that minimum standard of living. The appearance of simplicity is deceiving. Each of the three elements in the above description poses complex problems.

The Living Standard To Be Maintained

First, the determination of a *minimum acceptable standard of living* is a problem, as is demonstrated by the endless controversy over benefit levels. A distinction must be drawn between a minimum acceptable *standard of living* and the *income benefits* sufficient to provide that standard of living. Welfare beneficiaries are not a homogeneous class of people, with precisely similar needs; a level of benefits just about sufficient for one family might condemn another to continuous deprivation. A family might survive for a few months on a low income that would give them a living standard that would damage their health irreparably if continued for a few years. On this assumption, many jurisdictions give lower benefits to short-term welfare recipients.

The determination of welfare benefit levels is not only difficult; it is very important, both to those who get and those who give. A few dollars more or less per month will make a great difference to a family dependent on welfare; every dollar counts. Multiplied by the number of beneficiaries, the same few dollars will make a great difference as expenditure to a government 10 per cent of whose population receive welfare benefits (not an extreme). The level of welfare benefits to which a government commits itself is fraught with consequences.

While conceptions of 'poverty lines' are relevant to the establishment of public assistance levels, poverty lines and welfare benefit levels should not be confused. (See Appendix A, 'Poverty Lines', pages 273–4.) *No government in Canada commits itself to giving welfare benefits high enough to keep all welfare recipients above any particular poverty line.*

Benefit levels are ultimately a matter for political decision, made with (but not necessarily following) expert advice. Need, obviously, must be taken into account. Public assistance benefits are based usually on the prices of goods and services considered to be more or less essential to the living standard accepted as minimal. One difficulty is that welfare benefit levels are never determined by people on welfare. Even a well-meaning welfare official may have difficulty putting himself/herself in the shoes of a family head trying to meet needs that go up and down with circumstance, out of an income that is constantly meagre.

In Quebec, the determination of benefit levels was recently modified in a way that might conceivably influence other provinces, but has not done so yet. In the past, Quebec's public welfare authorities paid respectful attention to the 'minimum family budgets' worked out by the Montreal Diet Dispensary, a private agency. The Diet Dispensary prepares, for its own purposes, family budgets to meet different levels of sufficiency, from short-term emergency to long-term adequacy. It cannot, of course, be held responsible for what Quebec's welfare authorities have in past years chosen to do with its figures. The Diet Dispensary itself does not use the expression 'poverty line'.

A few years ago the Quebec Department of Income Security disavowed the 'budget' approach of the Diet Dispensary in favour of a statistical approach that bases benefit levels on the observed spending behaviour of 'working' families whose incomes from employment fall in the lowest 10 per cent in Canada (Quebec, Minister of Manpower and Income Security, 1987, pp. 16, 20). There is a kinship between this manner of determining benefit levels and Statistics Canada's empirical approach to determining low-income lines, in that both are based on observations of actual consumption behaviour rather than on the costs of specified 'market baskets'.

Quebec's linkage of welfare benefit levels with statistically determined low-end incomes from employment is an overt recognition of the other element, apart from budgetary requirements, that has always been factored into public assistance: the lowest available incomes from employment. The assumption here is that employment is an alternative source of income for many welfare recipients, and the prescription is that employment ought always to be the preferable alternative. This is our modern echo of the nineteenth-century English Poor Law principle of 'less eligibility': that the situation of the welfare recipient ought always to be less likely to be freely chosen by an individual than the alternative.

There are, however, many welfare recipients for whom employment is not an option at all, or not in the short run; in such cases, the relation between welfare rates and employment incomes is of little relevance. As a general rule, welfare programs are a little more generous, if that is the word, toward beneficiaries who are seen as legitimately dependent on public assistance for relatively long periods than they are toward others who are not so clearly cut off from the possibility of employment. For the latter, the prevailing level of wages in the lower strata of the labour market weighs even more heavily in the determination of welfare benefits.

In fact, changes in the structure of employment and unemployment in Canada in the last two decades have had the effect of radically increasing the numbers of people, normally employable, who have become dependent on public assistance for shorter or longer periods of time. From the perspective of those with such a marginal attachment to the world of work, the comparison between a welfare income and a work income may be a highly practical consideration. In the words of one welfare recipient, quoted in the National Council of Welfare's 1987 report *Welfare in Canada: The Tangled Safety Net* (p. 31), 'I think that at the present time, if I could find a job, I would refuse it. . . . After paying transportation, food, baby-sitting, and the extras you need when working . . . I would come out about $10 ahead . . . So you stay where you are.'

Welfare benefits are small, but they are stable; they include coverage of the cost of certain 'special needs' which, when needed, loom large in a low-income budget; and while welfare benefits are not fully secure, being always subject to administrative monitoring of eligibility, they are as secure for many people as earnings from the kinds of employment available to them. Welfare departments accordingly watch carefully the relationship between incomes from work and incomes from welfare.

Complications ensue when a person has *some* earnings, but not enough to attain the agreed-upon minimal living standard. In such a case, the question of whether the person is eligible for support is not clear-cut (this is the topic of the next section). Rather, the question becomes, *how much* support the person should receive—a more delicate judgement. If a person receiving welfare does manage to meet part of his or her needs through earning a certain amount of money, a *prima facie* case exists for reducing that person's welfare benefits. But by how much? If benefits are reduced by a dollar for every dollar earned, the person is no further ahead for his or her efforts. The effect would be the same as an incremental income tax rate of 100 per cent—perhaps even worse, because it costs money to go to work, a point clearly made by the welfare recipient quoted above, and underlined in the Quebec government's frequently-quoted 1987 position paper on income security. Just as for anyone required to pay income tax at a rate of 100 per cent, the incentive to accept employment would be minimized and the temptation to conceal income would be maximized.

Various jurisdictions have various rules for the adjustment of welfare benefits to take account of modest earnings, including various rates of recovery of earned income. It is all the more awkward both to live with and to administer, since people whose employment is irregular do not earn the same amounts from one month to the next. Typically, persons and families on welfare are allowed to earn a certain minimal amount per month without affecting their benefits at all. Sometimes those classed as fit for work are allowed a higher level of such 'exempt' earnings as an additional incentive to work. This is now the case in Quebec.

Like other Canadians, welfare recipients often receive income from other sources than work, notably employment insurance (which, as we know, may be paid for several months), survivors' benefits from pension plans, and disability allowances from Workers' Compensation. The public assistance program must have a rule for the adjustment of benefits to take account of income from public and private programs of these kinds. The rule often is to reduce benefits by the total of such other income.

Eligibility to Receive Benefits

Second, the determination of *eligibility* to receive benefits is likely to be contentious, for some of the same reasons. The state undertakes the total financial support of most welfare recipients. As a rule, the state will want to be quite sure that the recipient is really unable to support himself/herself adequately, i.e., is either unfit to work or, if fit to work, has good reason for not working.

Those who seem capable of earning a living may expect their eligibility for benefits to be closely scrutinized. This is equally true in countries that adhere to different socio-economic systems than ours. The distinction between 'employable' and 'unemployable' has always been crucial in welfare administration. The records of the administration of the seventeenth-century Poor Law, which prevailed from the late seventeenth to the mid-nineteenth century in England, are haunted by the figure of the 'sturdy beggar'—the able-bodied lout who could make a living if he tried, instead of taking advantage of the charitable impulses of others. The suspicion of the 'employable' welfare recipient is at the basis of all policies that aim to make the situation of such a person palpably less desirable than that of the person who works. This, again, is the meaning of the historic principle of 'less eligibility': the situation of the recipient must be, by design, less likely to be freely chosen than that of the self-sufficient person (Irving, 1989).

In Canada, the distinction between the treatment of the unemployable and the employable manifests itself in a number of ways. In some provinces, notably Ontario, the programs for the two classes of recipient are sharply separated, to the point of being separately administered. One, Family Benefits, is administered by the province, the other, General Welfare, is administered by the municipalities and counties. Similar administrative distinctions prevail in Manitoba and Nova Scotia (Human Resources

Development Canada, 1994a, ch. 4)).The plans of the current government of Ontario to hand over the responsibility for *all* public assistance to municipal and regional authorities would not eliminate the employable/ unemployable distinctions.

The province of Quebec, which formerly had one assistance program for both 'temporary' and 'permanent' beneficiaries, overhauled its welfare legislation in the late 1980s. The most basic and most hotly controversial issue is the sharply different treatment proposed for those judged 'apte' and 'inapte au travail' (Quebec, Minister of Manpower and Income Security, 1987). For the able-to-work, an Employment Incentive Program offers financial incentives to follow a training program and/or to enter the work force, and provides financial penalties for those who choose to do neither.

Finally, in addition to possibly having some income, a recipient individual or family may have some assets—'liquid assets' such as money in the bank or savings bonds, 'real assets' such as equity in a house (through having paid off all or part of a mortgage), a car, and so forth. This, too, raises delicate questions. On the one hand, should a person be permitted to receive an income from the public while still the owner of assets, possibly more valuable than the possessions of others who are *not* eligible for welfare benefits? On the other hand, does it serve any purpose to oblige a struggling household to give up quite modest possessions that are conceivably the result of years of saving, thereby reducing its ability to cope with and recover from its situation? Public assistance programs answer these questions by allowing a person or family to retain liquid assets and real assets up to a certain value, with special exemptions for certain kinds of assets (e.g., tools and equipment used in normal employment).

Collecting and Distributing

Someone must *collect* the money from somewhere, and someone must be given the responsibility of *distributing* it. In the now fairly distant past, much of the collecting and distributing of money was handled through privately supported voluntary organizations, often religious or philanthropic in character. Under the previously mentioned Poor Law of Great Britain, landowners supported local recipients of aid, with emphasis on the 'local', through special taxation based on the value of their land. Today, virtually all income support is done through government, though there are still some private agencies that give limited financial support to small clienteles. The money for public assistance is collected through the tax system as a whole; there is no tax specially earmarked for welfare. Because the 'transfer' character of public assistance is so transparent, from a majority 'us' to a minority 'them', the levels of total public assistance budgets are vulnerable to exploitation as a political issue, as in Canada in the 1990s.

Provincial governments are held politically accountable by their electorates for public assistance programs. This creates tension between the need for central control, central responsibility, and uniform administration, and the need for flexible adaptation to local and individual circumstances, especially in the larger, more diversely populated provinces. Uniform treatment of recipients may be inappropriate in many cases; but differentiated treatment may look inequitable and prejudiced, especially in the harsh light of the political arena. This consideration tends to limit the discretion allowed local administrators of welfare programs.

In dispensing benefits, administrators must implement the agreed-upon definitions of 'minimal standard' and of eligibility. Usually, benefits are determined according to a scale in which the main variables are family size and the ages of children, but in the nature of things, individual cases vary too much to allow public assistance to be calculated as determinately as pensions or Employment Insurance benefits. To cope with the variations in the situations of recipients, administrators may be allowed wide discretion. Or the opposite approach may be taken: legislation and regulations may attempt to deal with the inevitable variations by detailed rules that try to prescribe for all the possibilities. Too much discretion is feared, because it may lead to unequal, arbitrary, or even abusive treatment of recipients; but too little discretion is feared, too, because it leads to impersonal, rigid, 'bureaucratic' treatment of individuals, and to potentially unjust treatment of cases that happen not to fit the rules (National Council of Welfare, 1987; Marsh 1943/1975, p. 61; Ontario Social Assistance Review Commission, *Transitions*, 1988, ch. 7).

Quite apart from actual decisions about the amounts of money given to recipients, the *manner* of administration is significant. What kind of training should welfare administrators have? Should they have some level of skill in human relations and in counselling? Can an individual's need for financial assistance be separated from the need for help in managing and possibly improving his/her life situation? Or is that side of things none of the business of public assistance administration? Should income provision therefore be separated entirely from any kind of counselling? In the history of public assistance in Canada, as elsewhere, this was long a heated issue; an important part of the emerging profession of social work argued that most families on welfare, invariably faced with serious disadvantages, needed more than just money (Struthers, 1985).

In most places today, income support is separated from counselling. If the money comes compulsorily accompanied with advice, the recipient might get the message that unless he/she at least pretends to take the advice, the money will stop coming. Training and counselling intended to improve employability are an exception. In many jurisdictions, a person judged capable of working may come under pressure to train for a job and to take other kinds of work-related counselling. This has been more and more the case in the 1990s, with the proliferation of 'workfare' or workfare-like programs across the country (Caledon Institute, 1996).

Policing the Recipients

Should recipients be closely watched, lest they cheat the system—for example, by concealing resources, by avoiding opportunities to work, or by taking money on the side from some source or other?

For many welfare recipients, the effort spent on close scrutiny would be wasted. Female heads of household and their dependent children account for a large proportion of welfare recipients. It would often be impossible for a mother responsible for the care of young children to manage her household if compelled to earn a living through regular work. Furthermore, it is widely agreed that it is socially beneficial for mothers, at least in theory, to be enabled to stay home with their children while the children are young (equally so, perhaps, for fathers, but in this context the fathers are usually absent). Mothers with dependent children are customarily classified, therefore, as 'unemployable', not because of personal incapacity but because of their situations. In fact, the records show that female heads of household typically leave the ranks of welfare recipients as soon as they can, once their domestic circumstances allow them to take a decent job.

Equally 'unemployable' are considerable numbers of people with disabilities, for whom opportunities for paid work have been so few that close scrutiny of their willingness to work would serve little purpose. Times change, however; more positive social attitudes and advancing technology have made many people with disabilities considerably more employable than they used to be. On the understanding that many disabled people are now capable of employment, the current (1997) government of Ontario, which seems to take advantage of every opportunity to reduce its welfare expenditures, has announced its intention to tighten the definition of disability in its public assistance legislation.

Larger and larger numbers of welfare recipients in recent years have been relatively young, able-bodied, unattached people, more men than women (National Council of Welfare, 1987, Appendix B; Quebec, Minister of Employment and Income Security, 1987, pp. 8–11; Ontario, Social Assistance Review Commission, 1988, pp. 29–43). They are 'employable' if they could be matched with available employment. The fear is constantly expressed that some welfare recipients, especially the young and able-bodied, will not feel much incentive to work, and the conclusion is drawn that they must be closely watched. Welfare recipients are fairly ordinary human beings caught in difficult situations. As little as the amount of money may be that they receive, it is at least relatively secure. Living on public assistance might be understandably more attractive to some of them than the unpleasant, unreliable, poorly paid work that is their real alternative. At the best of times, work of any regularity may be difficult to find in low-end occupations. When overall unemployment rates are as high as they have been in recent years, especially in certain regions of Canada, one would expect employment to be even slacker in the lower strata of the labour market.

Since welfare levels are far from generous, some who live on public assistance have ample reason to try to add to their incomes, through work and otherwise. Any income received is supposed to be acknowledged, and beyond a certain point it will result in a reduction of benefits. Humanly speaking, this creates a certain incentive to conceal such earnings, and this in turn incites the further watchfulness of welfare administrators. The watchfulness sometimes becomes highly intrusive, as in the case of the 'man-in-the-house' rules discussed below. There is a risk that the lifestyle of the whole class of welfare recipients could be adversely affected by suspicions that can be valid for only a few.

So how closely, if at all, should welfare recipients be watched? An overzealous administrator might easily push scrutiny to the point where not only the dignity and self-respect but even the civil rights of beneficiaries are violated; there are many well-attested accounts of such violations. One recurring issue is the propriety of welfare officials receiving unsolicited information about recipients from third parties (labelled 'welfare snitches' by advocates for welfare recipients), or, worse, actively seeking such information. In recent years, Ontario created a telephone 'hot line' for the exclusive use of people who wished to give unsolicited information about possibly fraudulent welfare recipients. Quebec amended its Income Security Act to allow welfare investigators to question third parties about recipients; if third parties refused to give information, they could be fined. In contexts other than welfare, the gathering of information in these ways would be strongly resisted as a violation of privacy (Gow, Noël, and Villeneuve, 1995).

This brings into the picture another set of values. To give these civil libertarian values some measure of defence, public assistance systems in Canada are obliged by the terms of their subsidization by the federal government to establish mechanisms for appeals against both the decisions and the procedures of administrators. The appeal procedures that have been set up so far seem rather rudimentary in structure and desultory in performance (National Council of Welfare, 1987; Riches and Manning, 1989; Borovoy, 1988).

In Canada, two additional potential lines of defence have been added. First, the Canadian Charter of Rights and Freedoms provides some protection against unequal and/or arbitrary treatment by all governments or government agencies, including welfare departments. Alleged violations of the Charter must, however, be prosecuted by bringing cases to court, something no welfare recipient could afford without financial aid. Second, in the provinces there exists human rights legislation that may be invoked by or on behalf of welfare beneficiaries. The provincial human rights acts have the additional potential advantage, important in the context of welfare administration, that they are enforced by appointed commissions that will investigate alleged violations themselves, a quicker and less expensive route than the judicial. One recognizes, however, that welfare recipients are poor, hard-pressed, seldom well-informed about legalities, and understandably reluctant to risk a fight with the source of their livelihood. They are therefore not in the best position to take advantage of these recourses.

PUBLIC ASSISTANCE: A SELECTIVE TRANSFER PROGRAM

The administrative and other specific issues may be more clearly appreciated if the characteristics of public assistance are set forth and contrasted with those of other programs.

Financing Through Taxation

Public assistance is a *transfer* program, or, to use a more telling expression, a tax-and-transfer program. There is obviously no way to collect the necessary funds from potential recipients on a contributory basis, as with Employment Insurance. The only feasible source of funds is the general revenues of government, which are nearly all taxes. The money is then 'transferred' from the taxpayers to the recipients; recall that this process is termed a transfer because the recipient gives nothing specific in return.

Welfare recipients do pay taxes—not income taxes, but many other taxes, including sales and consumption taxes, and, if they are tenants, an implicit share of their landlords' property taxes; so the classes of 'taxpayers' and 'welfare recipients' overlap to some limited extent. The tax system has been adapted, however, to lighten the burden of such taxes upon the poor. Canada has introduced a federal sales tax credit, and nearly all the provinces have incorporated into their tax systems similar measures that rebate to low-income people, including welfare beneficiaries, all or part of consumption, property, and/or other taxes to which they have contributed during the year.

Administration

The financing of public assistance is therefore on much the same basis as that of any other regular government expenditure; it takes its place among the policies and priorities of the government, subject to the same kinds of pressures as other political issues. Public assistance is a responsibility of a minister, a member of the government of the province, the political head of a department. Each province has its own designation for its public welfare department, and these designations change from time to time. This subordination of the administration of public assistance to the political authority is consistent with its character as a tax-supported transfer program. It contrasts with the finance and administration of programs like Workers' Compensation and the pension plans, which have their own non-tax sources of finance, and which are managed by independent bodies, ostensibly protected from political influence.

There is a long tradition of local administration of public welfare, based no doubt on the feeling that each community best knew its own poor people and their needs. English Canada inherited this tradition. At the time of Confederation and after, the pattern of local administration continued; provinces had constitutional responsibility for supervising public charities, but they generally chose to leave the task to local governments. Unfortunately, the tax structure that evolved in Canada made it difficult, if not impossible, for municipalities to cope with the demands of public 'relief', as it was called, especially during the catastrophic years of the Great Depression. Besides, local administration was out of tune with the mobility of population demanded by modern conditions. Local authorities were strongly tempted to discourage indigent newcomers, and were even known to encourage the departures of their own poor for other localities, e.g., by giving them money for train fares. Gradually, *very* gradually, throughout the twentieth century, the provinces took over the major responsibility for finance, policy, and administration, first through special programs like Mothers' Allowances, which aroused little opposition, then through more general programs of support for the needy (Guest, 1997; Kitchen, 1987).

Several provinces (Ontario, Manitoba, and Nova Scotia) continue, as we have seen, to devolve substantial welfare responsibilities upon municipalities, including the responsibility for raising a certain proportion of the taxes needed to finance benefits. This permits a degree of flexibility in adaptation to local circumstances, but it also means that there are variations in the treatment of people from one town to another. The system cannot have uniformity and variability at the same time. Some consequences are described in the National Council of Welfare's publication, *Welfare in Canada: The Tangled Safety Net* (1987).

The previously mentioned 1988 report of the Social Assistance Review Commission of Ontario, *Transitions*, criticized the inequalities and incoherence it found in the municipally administered 'General Benefits' programs. Benefit schedules were subsequently harmonized. Similar steps have been taken in Manitoba and in Nova Scotia.

The Ontario report went so far as to recommend the unification of public assistance in one provincial program. The proposal was not implemented by the then Ontario government. On the contrary, as has been noted, the government in power in Ontario in 1997 announced a radical restructuring of public assistance, to take effect in the near future, giving to municipal institutions entire responsibility for the administration of public assistance. The apparently unexpected announcement did not include sufficiently clear provision for the expanded transfer of financial resources to reassure Ontario's municipal officials, so the project has been greeted with consternation. If it proceeds, the undertaking is much too vast to be implemented overnight. The unfolding of this project will be among the most interesting developments in public welfare in Canada since the launching of the Canada Assistance Plan in 1967.

In Quebec, public assistance is wholly managed by the province, except in the City of Montreal, which alone maintains a vestige of local administration. The Welfare Department of the city administers the public assistance program of the province within the limits of the city, entirely at the expense of the province. The city itself does not pay for even the administrative costs, being reimbursed for them by the province. The explanation is purely historical. Montreal had a 'Service de bien-être' in place long before the province did. Also, the employees were party to a collective bargaining agreement with the city, which they have never been inclined to give up in favour of that which is in force in the provincial public service. So Montreal has a welfare department, but no welfare program.

Public assistance is one of the most important areas of cost-sharing between the federal and provincial levels of government. The nature of the cost-sharing was fundamentally changed as of 1995–6, just as it was in the financing of health care and post-secondary education. The evolution of cost-sharing in public assistance is covered on pages 145–9 of this chapter. From the standpoint of administration of welfare programs, the mechanism in place from 1967 to 1996, the Canada Assistance Plan, and its successor, the Canada Health and Social Transfer, have obliged the provinces to meet certain agreed-upon standards in the structure and delivery of programs. The salient point is that although a large share of welfare expenditures (formerly about half, now less, but still substantial) is borne by the federal government, *the provinces remain firmly in charge of welfare programs.*

Eligibility for Benefits (Selective)

Public assistance is a *selective* transfer program. Someone is officially authorized to decide whether a particular individual or family is to receive benefits. The decision is made on the basis of need, not solely on that of some general characteristic that is thought to be related to need, such as age. The selection of some people as eligible to receive benefits implies the potential rejection of others.

A program that selects beneficiaries in this way directs money to cases of real need, which is desirable, but it involves much effort and some risk of maladministration. Partly to avoid this effort and this risk, we used to rely on so-called *universal* transfer payments, where everybody qualified for benefits as long as he/she met a relatively easily judged demographic criterion, such as age. Even with the removal of the universality of formerly universal programs, the criteria for their successor programs remain relatively uncomplicated: age and declared income in the case of the revised Old Age Security and the proposed Seniors Benefit, and the ages of dependent children and declared income in the case of the federal Child Tax

Benefit (which has replaced Family Allowances). The public assistance official is denied the ease afforded by such simple eligibility criteria.

The administrators of public assistance must work within guidelines laid down by law and by regulations issued by the minister or by the government (the law specifies who has the authority to make regulations). Administrators may do only what the law and the regulations authorize them to do. Conversely, they may be held to account to do what the law and the regulations oblige them to do. There may well be some tension between the necessity to run things according to the law and the necessity to make appropriate decisions in individual cases.

Finally, it is in the nature of public assistance that the recipient must play a role in its administration: he or she must step forward and apply for it. This can be self-diminishing in a society that puts a high value on independence and self-reliance. Recipients stigmatize themselves and are vulnerable to being stigmatized by others. By contrast, there is little stigma attached to applying for benefits like those provided by Old Age Security.

Similarly, the welfare recipient is expected to continue to collaborate in the administration of public assistance, by notifying the authorities of changes in circumstances that affect benefit levels and by providing relevant information when asked in due form (National Council of Welfare, 1987, pp. 87–8). Welfare laws provide for recovery from recipients of any excess money they may have received through failure to provide such information. The 'relevant information' is more extensive in the case of public assistance than in the case of an income-tested program. Should a few unwarranted details be asked for by any official, the dependent position of recipients makes it difficult for them, humanly speaking, to challenge the propriety of the questions.

Benefits

No public assistance authorities anywhere would claim to be providing welfare recipients with anything more than a minimally decent standard of living. Typically, a calculation is made of the cost to a family of that list of goods and services considered essential. As noted above, Quebec bases its benefit levels not on the cost of a list of needed items, but on the observed spending patterns of low-income Canadians. Adjustments are made for the number of adults and the number of children, of different ages, in the household, and usually for such factors as urban-rural variations in the cost of housing.

Other benefits to which recipients are entitled, such as OAS/GIS (or Seniors Benefit) or disability benefits, reduce the amounts forthcoming from public assistance, usually dollar for dollar. Welfare departments, like other organizations, look for ways to shift their expenses onto other bodies, possibly including other units of the same government. If a provincial government can place a potential welfare recipient on a federally financed program, such as EI, or in an associated work training program, it will prob-ably do so.

Welfare benefit scales for families with dependent children have always taken into account the amounts the families were receiving from Family Allowances or the Child Tax Benefit. By the end of the twentieth century, thanks to important agreements arrived at in 1996 and further discussions in 1997 between the federal and provincial governments, it is expected that *all*, or at least a large part, of the benefits for the children of families on welfare will come through income-tested federal Child Tax Benefits, paid to working poor and welfare families alike (Canada. Department of Finance, 1995, 1997). Public assistance benefit scales will then not be affected by the presence of children in families (see below, 'The "Tax Expenditure" Approach: The Federal Child Tax Benefit . . .').

As a rule, a limited amount of income may be earned by members of the household without affecting public assistance benefits. Income over this small amount is supposed to be reported, and some proportion of it deducted from the benefits paid (National Council of Welfare, 1987, pp. 38–43). The issue of work income on the part of welfare recipients is raised below. For the moment, suffice it to say that (1) the deduction of earnings (all or part) from benefits may logically be expected to dampen the incen-tive to work at anything less than a fairly decent job, and (2) the possibil-ity of substantial work is real for only a small proportion of welfare house-holds even at the best of times, and is still less real when unemployment is high.

In addition to regular income benefits, assistance programs make avail-able to beneficiaries *ad hoc* benefits for 'special needs', notably prescription drugs, prosthetic devices, and expenses related to schooling. The cost of such items could easily be prohibitive for a poor family, so provision for 'special needs' is a precious benefit. In fact, something of a dilemma may be created for welfare policy-makers, in that families on welfare may be getting things that low-income persons and families not receiving welfare need but cannot afford. Also, a family head faced with a realistic choice between relying on earnings instead of welfare may be deterred from choosing to rely on earnings by the prospect of losing access to such spe-cial benefits. This is a factor in the 'welfare trap', the combination of cir-cumstances that makes it difficult for people with dependants to 'get off welfare' and enter the labour market (National Council of Welfare, 1993, pp. 39–44; Caledon Institute, 1993). To circumvent this, and as a measure for the prevention of poverty, some jurisdictions will meet the 'special needs' of some poor people who are not permanent welfare recipients.

Non-Categorical Assistance: 'Needs Exceed Means'

In the past, throughout Canada, public assistance was organized to serve persons subject to specific categories of need. The basic question then was, 'What personal conditions render persons unable to support themselves and their families?' If it were agreed that a particular condition, e.g., old age, blindness, or widowhood with dependent children, made it impossible to earn a living in all but exceptional cases, then legislation would be passed to make financial assistance available to poor people in that needy *category*. The assistance would depend on the gap between the *needs* of the individual and the amount of income the individual might have from other sources, such as a job, a disability pension, or income-yielding investments (not very likely). This is what is meant by calling such assistance *categorical* and *needs-tested*. People received assistance (1) if they fell into certain categories, and (2) if they were in need. We have seen that by the 1920s the burden of support of the needy elderly had become so heavy that the provinces and the federal government negotiated a cost-sharing arrangement by which the federal government subsidized provincial programs in support of the needy elderly. Over time, other categories of need emerged; some of them—blindness, disability, needy motherhood, long-term unemployment—led to similar cost-sharing arrangements.

In time, the 'categorical' approach was seen to lead to gaps and inequities in provision and to much administrative clumsiness. It has been replaced, though not quite everywhere, by the 'general' approach to determination of need. The basic question becomes, 'In this case, do needs exceed means?', with little regard for the *cause* of the insufficiency. The spirit of the 'general' approach is captured in Section 3 of the former Quebec Social Aid Act: 'Social aid shall be granted on the basis of the deficit which exists between the needs of, and the income available to, a family or individual.' The clienteles of 'general' welfare programs include many of the same people, of course, as those of 'categorical' programs.

THE ISSUE OF INCENTIVE TO WORK; 'WORKFARE'

We have emphasized the centrality of the concept of income through employment in all issues of social policy. In public assistance, as we have seen, it is expressed in the concern that no one should receive welfare if capable of self-support. The history of categorical assistance provides the clearest illustration. Categorical programs were constructed around the recognition that a number of personal conditions made it virtually impossible to hold a job. Our society was reluctant to admit that someone might suffer from no disabling condition and still be unable to support a family, even though the reality was well known to anyone informed about Canadian social conditions (Marsh, 1943/1975, pp. 105–16). It was not until 1956 that the provinces and the federal government negotiated federal sharing of the costs of assistance to the long-term unemployed. This category of the needy had come to loom large in provincial welfare caseloads. The long-term unemployed did not suffer from any recognized disability, so the cost of their welfare benefits was not shareable with the federal government until the passage of the Unemployment Assistance Act in 1956. (Unemployment assistance must be clearly distinguished from

[Un]Employment Insurance, which dates from 1941; a person who had been employed regularly enough to qualify for unemployment insurance would certainly not have been eligible for unemployment assistance in any of the provinces [Struthers in Ismael, 1985].)

It is significant that the long-term unemployed were the last category of people to become eligible for federal cost-sharing before the federal government initiated the more generalized mechanism of the Canada Assistance Plan.

The fact that unemployment assistance was the last of the categorical programs illustrates our long-standing fear of giving public money to people who had the ability to earn their own living or who ought to be supported by the putative breadwinner. Today, the same concern over the motivation of welfare recipients shows up in various regulations intended to incite, or oblige, able-bodied welfare recipients to seek and accept employment or employment training, in the nearly universal practice of paying lower cash benefits to welfare recipients classed as 'employable', in regulations that reduce assistance payments to recipients who earn more than a certain amount, in the suspicion of recipients who have what might be income-earning equipment around the house (e.g., sewing machines or power tools), and in the notorious practices of some authorities whereby a female welfare recipient who is seen to have adult male company in her home may be penalized on the assumption that she is sharing in the male's presumed income (the 'man-in-the-house' rule, or, in its Quebec version, the 'pipe-in-the-ashtray' rule) (Gow, Noël, and Villeneuve, 1995, p. 34; Borovoy, 1988, pp. 164–8).

'Definition as "employable" in Canadian welfare programs,' say Evans and McIntyre, 'has consistently meant less eligibility: lower benefit levels, less security in the duration of the benefit, and more frequent and rigorous investigation of capacity and motivation to work, in comparison with categories of recipients believed unable to work' (in Ismael, 1985, p. 102). The Ontario Social Assistance Review Commission found much of this to be true of public assistance administration in that province in the 1980s. This bias is also reflected sharply in those clauses of Quebec's Act respecting Income Security (1988) that deal with those who are fit to work ('apte au travail').

The vulnerability of the 'fit-to-work' public assistance recipient is highlighted by the recent enthusiasm shown in many provinces for measures that virtually tie entitlement to benefits to participation in some form of work or work training: programs like 'N.B. Works', 'B.C. Works', and the above-mentioned distinction between 'support' for the disabled and 'supplementation' for the fit-to-work in Quebec, and so on (McFarland and Mullaly, 1996). Workfare is not a new idea, of course, but it has taken on new force for a variety of reasons:

- the fiscal situation of all provinces encourages them to cut expenditures in large-budget envelopes, of which welfare is one, and one without a truly powerful political constituency to resist the cuts;

- the considerable reduction in federal welfare-related transfers to the provinces, also primarily budget-driven, impels the provinces, in turn, to try to reduce *their* welfare expenditures;
- the change in the terms under which provinces receive their transfers through the Canada Health and Social Transfer allows them greater room to manoeuvre than did the Canada Assistance Plan;
- analysis of welfare caseloads shows a long-term increase in the numbers and proportions of young 'employable' recipients, apparently due to the long-lasting unemployment effects of the recessions of the early 1980s and 1990s, though the trend can be discerned earlier still. This trend is almost sure to be exacerbated by the contraction of the duration of benefits under Employment Insurance.

It is conceivable that policy regarding employable public assistance recipients may be positive as well as, or instead of, punitive. The Quebec program does offer added benefits to participants in training programs and provides additional encouragements, as did Ontario's 'Supports to Employment Program'. But inasmuch as widespread unemployment is what pushed so many employable people onto public assistance in the first place, the outcomes of such back-to-work programs obviously depend more on general economic conditions, particularly conditions in the labour market, than on their success in stirring up a motivation to work on the part of employable recipients.

Genuine, well-delivered training for reasonably remunerative work gives rise to no objection, nor do work experience opportunities aimed at introducing previously unemployed people to working life, or rehabilitating people long out of work. When economists speak of 'structural unemployment' as a troubling characteristic of the Canadian economy, they are saying, among other things, that a great many potential workers are poorly prepared for the jobs in demand; so in itself, a policy that emphasizes work-related preparation is not misplaced. Such a policy is not, however, easy to carry out.

The cost of welfare is cited as a justification for work-linked programs, but critics sometimes overlook the fact that effective training programs cost a good deal as well—the training staff, facilities, and administration required, all cost money. It is unrealistic to assume that appropriate training, rehabilitation, or employment can be found for the currently large number of 'employable unemployed' in a short time. When employers are assisted financially to employ the participants in such programs, a risk arises that the trainees become a pool of inexpensive labour for the employers, competing unfairly with other workers. Given persistent high unemployment in certain sectors of the economy, it is commonly suggested that welfare recipients could take jobs in the 'social economy', working for community organizations and the like. However, unless carefully managed, such a strategy could well add unfairly to the responsibilities of the host organizations, many of which have burdens enough as it is.

Hovering over the whole discussion of 'workfare' is the simple fact that the numbers of fit-to-employ on welfare closely track unemployment levels: the availability of jobs is overwhelmingly more important in determining the employment of the 'fit-to-work' than is anything else. Even modest work income, if reasonably stable, is a sufficient improvement over welfare to provide its own motivation for the vast majority of the 'employable unemployed'. Good training presumably improves anyone's chances of finding and keeping a job—but during a period when work opportunities are limited by economic conditions, work-for-your-welfare schemes are equally limited. If 'workfare' programs follow a well-known path, and simply try to record impressive numbers of participants ('head counts') for political public relations purposes, with little concern for the long-term results for those participants, they will do little good, even on their own terms.

They may even do harm, as a host of critics loudly argue. For also hovering over the concept of 'workfare' is a moral issue: the introduction of a new level of compulsion into the lives of people who already have few and narrow options. The compulsion can only be justified by assuming bad will—reluctance to work—on the part of the compelled, and it is this implicit (sometimes explicit) assumption that provokes a strong reaction from welfare advocates (CQDS, 1997; McFarland and Mullaly, 1996; Shragge, ed., 1997; CCPA, 1996; Sayeed, ed., 1995). Provincial authorities, on the other hand, are inclined to downplay the extent of the compulsion, insisting that recipients are left with some choice in the matter. However, the choice is likely to come at the price of reduced benefits.

THE FEDERAL-PROVINCIAL RELATIONSHIP IN PUBLIC ASSISTANCE

Jurisdiction over welfare is in the hands of the provinces, so that each province has its own welfare legislation and administration. The federal government pays a large (although now reduced) part of the costs of public assistance through fiscal transfers to the provinces. From now on, this will be done through the new Canada Health and Social Transfer (CHST), announced in the 1995 federal budget, which took effect in 1997. The Health and Social Transfer lumps into one undivided package the federal share of expenditures on welfare, health care, and post-secondary education. It may therefore be impossible in future to determine exactly how much of the transfer any province has dedicated to welfare.

For about thirty years (1967–96), federal financial support of welfare in nine of the provinces, and latterly in the Territories, was much more transparently arranged through the Canada Assistance Plan (CAP). Quebec alone chose to receive its federal funding through the alternative Established Programs Financing scheme (EPF), under the Fiscal Arrangements Act (see below), while still accepting the conditions stated in the Canada Assistance Plan. Although the Canada Assistance Plan came to an end in 1996, it has been an important part of social welfare history, and merits some attention here.

The terms of the Canada Assistance Plan, enacted in the very active federal parliamentary session of 1996–7, embodied the consensus that emerged from extensive intergovernmental negotiations over the rationalization of federal transfers to the provinces in support of assistance programs. Reference has already been made to the earlier history of federal cost-sharing in assistance to the elderly, blind persons, other disabled persons, needy mothers, and, finally, the long-term unemployed, these being the categories of needy persons who dominated the welfare caseloads of the provinces. In virtually all cases the federal government undertook to pay 50 per cent of the costs of benefits.

The Canada Assistance Plan expressed the policy of 'general' assistance, enabling provinces to enjoy federal sharing of costs of assistance that was distributed on the basis of need alone, without specifying *categories* of need (though it did not prohibit provinces from continuing to have categorical programs). One might therefore regard CAP as signalling a modest move away from a narrowly 'residual' conception of public assistance (for pathological cases and emergencies only).

Federal cost-sharing assists provinces in the financing not only of welfare but also, as we shall see in Part Two, of health care provision (and also post-secondary education). For health care programs, in 1977 the federal government imposed, after years of discussion, an alternative cost-sharing principle, called 'Established Programs Financing' (EPF). EPF transfers were based on each province's capacity to raise taxes on its own, to replace the 50-50 expenditure-based principle that had previously prevailed. For some time previous to 1977 the provinces had been free to switch to EPF for public assistance too, but, as noted above, Quebec alone had taken up that option. (See Appendix C, 'Federal Sharing in the Costs of Health Care', pages 296–300.) All the other provinces preferred CAP.

Naturally, CAP put some conditions upon the cost-sharing: benefits had to be needs-tested; no residence requirement could be imposed on people entering Canada or moving from one province or territory to another; a plausible appeals process had to be in place; and there were guidelines governing the exemption of modest amounts earned, and liquid assets owned, by recipients, the granting of allowances for the aged and the disabled, and the supplementation of low earned incomes.

Obviously the availability of 50 per cent cost-sharing by the federal government powerfully inclined provinces toward programs and services covered by CAP, and away from items not covered; in that sense, the federal legislation influenced the welfare policies of the provinces. To be sure, some provinces provided benefits *not* covered by the Canada Assistance Plan. For example, British Columbia has a program of 'Shelter Aid for

Elderly Renters' (known as SAFER), and another of reimbursement of the cost of therapeutic medications; since neither program was needs-tested, they were not eligible for CAP cost-sharing. Quebec operates a Parental Wage Assistance program (in French, Aide aux parents pour leurs revenus de travail, 'APPORT') which, because it is not needs-tested, was similarly ineligible for full federal subsidization. At the same time, some provinces have not done some things for which cost-sharing was available through CAP. CAP did not, therefore, pay *exactly* half of all provincial welfare expenditures.

All of these cost-sharing programs, including the Canada Assistance Plan itself, have been accepted as fitting under the constitutional umbrella of the federal Parliament's well-known 'spending power'. No constitutional amendments were needed to empower Parliament to pass such Acts (unlike UI, OAS, and CPP). Much intense negotiation and argument preceded each of the cost-sharing Acts, however, including CAP itself. (In the 1990s, the provinces were notified of the impending shift from CAP to the Canada Health and Social Transfer, but the initiative for the change lay with Ottawa.)

At first glance, one might ask what the provinces found to argue about when the federal government offered to pay 50 per cent of one of their largest budget items. The grounds for contention boil down to three. First, the federal government, accountable to its taxpayers and to its electorate, is reluctant to hand over large sums of money without exercising surveillance over how it is spent (and getting political credit for it). But this is easily seen by the provinces as trespassing on their jurisdiction over welfare. Second, the provinces argue that they are logically entitled to the resources needed to carry out their constitutional responsibilities; therefore, federal transfers to the provinces are merely correcting a flaw in the public finances of the federal-provincial system, and should not be manipulated as levers for federal influence over policy. Third, there have always been deeply rooted inequalities between the provinces, some of which are chronically 'have-not', while others tend to be consistently prosperous and still others have their ups and downs. Given this inequality, consensus among provinces concerning the best way to share the fiscal resources of the entire nation, channelled through Ottawa, is difficult to obtain— though the provinces agree readily enough on the value of preserving provincial control over welfare policy.

The final stage in the life of CAP, from 1990 to 1996, attested to the difficulties of what is called 'fiscal federalism', especially in a time when all governments feel the need to contract their expenditures. The terms of CAP imposed no maximum on provincial welfare outlays; the federal government had an open-ended commitment to pay half of whatever the provinces chose to spend, within the conditions of the plan. Over the years, provincial welfare expenditures increased markedly, and federal expenditures via CAP inevitably increased in parallel.

In 1990, the federal government imposed a change in the CAP rules. As part of an overall regime of restraint in government expenditures, Ottawa limited *annual increases* in CAP payments to 5 per cent in provinces calculated as enjoying above-average economic prosperity. The change was not made without the notice required by law, but it still shocked the provinces affected; at the time, those were Ontario, British Columbia, and Alberta. The federal government did not tell those provinces that they could not increase their welfare expenditures if they wished—only that it would limit the increase of its contribution.

To provide some sense of the impact of the 5 per cent limit, CAP payments in 1990–1 to British Columbia and Alberta would have been 8 per cent higher than in 1989–90 without the cap, so the *increases* in those provinces were 5 per cent instead of the expected 8 per cent—not much over half of what they had forecast. Ontario was hit even more seriously. An enormous rise in the number of recipients, due to the recession of the early 1990s, combined with a relaxation of eligibility requirements and a simultaneous increase in benefit levels, instituted by the NDP government then in power, led to an increase of about 40 per cent in Ontario's welfare expenditures in 1990–1—about $1.5 *billion*! Somewhat unexpectedly, Ontario found that it had to pay for nearly all, not just half, of the increase itself. (This so-called 'cap on CAP' was definitely not the outcome of federal-provincial negotiations. On the contrary, the provinces unsuccessfully contested the legality of the change in the CAP legislation before the Supreme Court of Canada.) The 5 per cent annual-increase limit remained in effect until the termination of CAP in 1996.

In 1995, the federal government announced the coming end of the Canada Assistance Plan. The new Canada Health and Social Transfer wrapped together the federal government's support of provincial programs in welfare, health care provision, and post-secondary education. No longer will any part of transfers to the provinces be based on the provinces' actual welfare expenditures. And total transfers will be sharply reduced. For 1996 and 1997, transitional arrangements were made to link the reduced transfers to expected CAP and EPF transfers for those years (the CAP estimates could only be approximate, since CAP transfers depended on amounts actually spent, a quantity that is not known in advance).

The details of the computation of federal transfers to the provinces in subsequent years remain to be clarified. As the federal Department of Finance tersely put it in its summary of the 1995 Budget (undoubtedly one of the most radical federal government budgets in Canada's history): 'The Minister of Finance will consult with provinces and territories in developing a permanent method of allocating payments among provinces under CHST for 1997–98 onward' (Canada, Department of Finance, 1995).

To round out the picture of federal-provincial financial relations, ever since Confederation the federal government has made 'equalization' payments to the provinces, obviously intending to enable the poorer provinces to provide services of tolerable quality compared to those of the wealthier provinces. Equalization payments have always been recognized as an important political and financial necessity; the Constitution Act of 1982 gave them constitutional recognition by the inclusion of clause 36(2), which reads, 'Parliament and the government of Canada are committed to the principle of making equalization payments to ensure that provincial governments have sufficient revenues to provide reasonably comparable levels of public services at reasonably comparable levels of taxation.' Equalization payments come with no strings attached: the provinces that receive them may use them for any public purpose.

PROGRAMS FOR THE WORKING POOR; CHILD AND FAMILY BENEFITS

Canada's Emerging Awareness of the Working Poor

The 'working poor' are at the centre of many current discussions in social policy. The central assumption of much social policy long was that only people unable to work needed help. Poverty among employed people therefore poses a critical challenge for policy. At the same time, it calls into play other long-standing moral assumptions, because the working poor clearly are responding to work incentives; they are playing by the standard rules, but they are not being well rewarded.

The existence of a large class of genuine 'working poor' was never really a secret, as is attested by the documentation of Terry Copp's *The Anatomy of Poverty* (1974), Greg Kealey's *Canada Investigates Industrialism* (1973), and Michael Piva's *The Condition of the Working Class in Toronto—1900–1921* (1979). Even so, the beginnings of a systematic policy response date only from the late 1960s, and the systematic policy is still struggling to evolve. The scope of the problem in the 1960s is well conveyed in Jenny Podoluk's *Incomes of Canadians*, published by Statistics Canada in 1968, and/or in the analyses made under the auspices of the Economic Council of Canada in its Fifth Annual Review, *The Challenge of Growth and Change* (1968), Chapter 6, which relied heavily on Podoluk's work. The issue of the working poor was effectively promoted in Chapter 3 of the Report of the Special Senate Committee on Poverty, *Poverty in Canada* (1971), generally known as the Croll Report, after the Chair of the Committee, Senator David Croll.

In those documents, it was shown that an astonishingly high proportion of poor people in Canada did not suffer from any of the personal or situational disabilities conventionally associated with inability to earn a living. By far most of the poor, at that date, either worked or were dependants of family heads who worked. They may not have worked full-time, year-round, nor at enviable jobs, but they would certainly have failed to qualify for public assistance under the traditional 'categorical' approach. By that time Canada had, in fact, begun to move away from the 'categorical' approach—the Canada Assistance Plan, with its implicit encouragement to 'general' needs-tested public assistance, came into being in 1966-7; but it was something of a jolt to the Canadian consciousness that, for so many employed Canadians, their needs exceeded or nearly exceeded their means, in an era of proclaimed affluence.

Once the 'working poor' are acknowledged to be very numerous, what can be done? Paradoxically, it may prove more challenging to help those who need a little than those who need a lot. As the Economic Council of Canada said (p. 114)):

> Where a low-income family is for one reason or another incapable of offering labour services, and is therefore largely dependent on government payments, the policy problem of how to aid that family is in one sense relatively simple: the major issue is the size of the income to be provided. But where there are earnings, but on an insufficient scale, or where there is unexploited potential for earnings, the policy choices are less simple. Other things being equal, it is far better to help people to help themselves. . . . But self-help takes time, and meanwhile income support in the form of government payments may be needed.

That 'policy choices are less simple' is confirmed by the ample experimentation with various approaches to the 'supplementation' of earned incomes. (In most discussions, 'supplementation' is distinguished from income 'support', which is usually taken to mean total or near-total financial sustenance of a person or family—despite the Economic Council's contrary usage in the last sentence quoted above.)

Complicated choices are called for in designing programs for those whose situations place them in the border territory between the worlds of welfare and work. The key variables to be taken into account are the duration and regularity of the recipient's employment, earning capacity over a period of time, duration of training where needed, and additional support services that may be required (e.g., for health needs or children's school needs). The crucial problem, from the point of view of enlisting the recipient in a constructive way, is to maintain the incentive to earn by providing enough additional income to reward work effort, but not on terms that incite a reduction of work effort. This is a delicate balance to strike in a sector of the labour market where jobs tend to be neither stable nor well-paid, and the reward for work effort accordingly meager. If the net improvement in total income through work plus supplementation is not enough, or is very risky, the preference of potential recipients (especially those with children) for public assistance may be pardoned, or at any rate, may be seen as rational, as in the case of the woman quoted in the National Council of Welfare publication cited above. For the target clientele of a 'supplementation' program, the incentive to work can be heightened in two ways: either positively, by promising recipients more total income and more satisfying moral rewards under 'supplementation', or negatively, by threatening to provide them with less under 'support'—the carrot or the stick. Both are in use.

Even though 'the policy choices are less simple', some modest initiatives have been undertaken in Canada that provide benefits for the working poor. Some are available not only to the poor but to the whole population, or to all but those with relatively high incomes; others relieve the low-income taxpayer, a class largely consisting of the working poor. The current move toward expanded income-tested child benefits is intended in large part to equalize the supplements provided for children in the families of welfare recipients and the working poor, so that the financial assistance a parent is given for children is not put at risk by choosing work over welfare. The word 'choosing' is used here with circumspection: to the potential recipient, the 'choice' may be between a rock and a hard place.

The Canada Assistance Plan, CHST, and the Working Poor

On the premise of the desirability of 'work over welfare', the Canada Assistance Plan explicitly allowed, on certain conditions, federal cost-sharing of supplementary payments by provinces to the working poor and other costs of work activity projects, rehabilitation, and training programs. The new social transfer mechanism, the CHST, allows provinces even greater freedom; in the federal government's own words, 'Provinces will no longer be subject to rules stipulating which expenditures are eligible for cost sharing or not' (*sic*) (Canada, Department of Finance, 1995).

In language typical of welfare legislation, the Quebec Income Security Act (Art. 1) promises to provide 'Last resort financial assistance to persons whose resources are insufficient to provide for their needs . . . taking into account the fact that the situation of persons presenting severe limitations in their capacity for employment differs from that of persons who are fit for work.'

Supplements to Low Earned Incomes

The Quebec Parental Wage Assistance program (APPORT), the 1988 successor to the work incentive program, was and is of great interest and of much potential importance. Income from work, below a stated annual rate, is supplemented out of government revenues on a sliding scale that varies according to family size and family income earned: within a narrow range, the supplement *increases* as earnings increase. And to address an obstacle that often inhibits social assistance recipients from taking jobs, the program will meet a substantial part of day care expenses.

The PWA/APPORT program is limited in its scope. The recipient must be employed with some degree of regularity; annual income from work must reach a certain minimum, as well as not exceed a maximum; the supplement is provided only in the year *after* the year in which income was shown to be low; it comes in a lump sum or in a series of instalments, depending on its size, in the year following the year of low income, not in the form of a regular monthly cheque. Its terms made it ineligible for direct cost-sharing under CAP, perhaps enhancing the political credit due the provincial government that installed it (Quebec, Minister of Manpower and Income Security, 1987, pp. 16, 33–5; Hum in Ismael, 1985; Johnson in Ismael, 1987). Under CHST, however, the transfer Quebec receives from Ottawa may be used for PWA/APPORT as for any welfare program.

At about the same time that Quebec introduced its Work Income Supplement, Ontario embarked on an experiment with a work incentive program of somewhat smaller dimensions called the WIN program, slanted much more sharply toward employable welfare recipients than toward the already employed (Johnson in Ismael, 1987). Eligibility for benefits was closely tied to eligibility for Family Benefits, Ontario's provincially-run public assistance program for persons in need of long-term support, and the Ontario benefits were time-limited, intended to smooth the transition from welfare to work.

Explicitly directed toward the working poor is the rather modest federal 'Work Income Supplement', attached to the federal Child Tax Benefit, described in the next section.

The details of the Ontario and Quebec programs and similar programs in Saskatchewan (Family Income Plan) and Manitoba (Child Related Income Support Program, known as CRISP), are of intrinsic interest, but their greater interest is the suggestion they gave of probable developments in welfare in Canada. Virtually all welfare authorities have remarked on the increase in the number of relatively young, seemingly able people who are now receiving welfare. Whether interested in saving money or in reducing dependency, welfare policy-makers are certain to continue their search for ways (1) to keep low-income workers from falling back on public assistance, and (2) to motivate able people on welfare to move toward self-support (Séguin, 1987, Johnson, 1987, and Evans and McIntyre, 1985; Human Resources Development Canada, 1994b, pp. 32–3).

The moral gains that may be claimed depend heavily on the individual situation: even if one believes that self-support, even partial, is better than total dependency, there will still be situations in which welfare is the better alternative for a person or family, and for society at large. For the finances of governments, the immediate pay-off may seem considerable; the amount of public money needed to supplement even a deficient work income is much less than the amount needed for total support. Currently, for instance, the Saskatchewan Family Income Plan (FIP) pays out approximately $5 million per year to help support about 2,000-plus families; the average pay-out of $2,500 per year is only a fraction of what an eligible family would receive through public assistance (Human Resources Development Canada, 1994a, p. 55). One must not assume that all the FIP recipients would otherwise be receiving total support from public assistance, but for those who would, the cost difference to the province is significant. And bear in mind that the existing programs of low-earnings subsidization are designedly small in scope.

One must not, of course, limit one's analysis to immediate, first-order effects—i.e., a little more money in someone's purse at apparently small cost to the public. The fact is that there are a great many low-income earners; if all were eligible for a supplement, the costs would add up. Moreover, a publicly financed wage supplement, structured as such, may easily enter into the process of wage bargaining at the lower end of the labour market; some employers, knowing that a supplement is available, might feel free to offer lower wages, with the result that the taxpayers may find themselves subsidizing low wages. The existence of a wage supplement may become an extra lever pushing welfare recipients into work and work-training programs, not always appropriately. This aspect of the issue is taken up from a different direction in the next section, on tax credits.

The 'Tax Expenditure' Approach:
The Federal Child Tax Benefit, Tax Exemptions, and Deductions

Over the past two decades, income-tested tax credits have totally replaced universal benefits in the Canadian scheme of things. At the federal level, Old Age Security, along with its income-tested supplementary programs, the Guaranteed Income Supplement and the Spouse's Allowance and Widowed Spouse's Allowance, are on the verge of giving way to the Seniors Benefit, and Family Allowances have given way to the Child Tax Benefit. In Quebec, in 1997, the program of Allocations Familiales (Family Allowances) has been replaced by an income-tested, child-related benefit. The tax credit is a technique for income redistribution intended to overcome, in an administratively unobtrusive manner, some of the perverse combined effects of the progressive income tax and exemptions of income for income tax purposes. (See Appendix B, 'The Redistribution of Income', pages 283–8.) Equally important, the tax credit makes less daunting the shift from welfare to work for welfare recipients who are able to work but who have dependent children. The tax credit approach has rapidly assumed primary importance in Canada's system of income redistribution.

Federal Family Allowances were an important part of Canadian social policy for nearly fifty years. These payments began in 1944, as a measure to shore up the capacity of families to meet the expenses of raising children, especially in view of the economic contraction expected (wrongly, as it turned out) after World War II. Also, the allowances constituted a degree of recognition of the contribution made to society by parents who raise children, at some sacrifice. Over time, important changes took place: allowances were greatly increased; they were made taxable; and within limits, provinces were allowed to influence the patterns of payment of federal Family Allowances to their residents. Only Alberta and Quebec did so, as they have done with the federal Child Tax Benefit. In 1993, Family Allowances and a Child Tax Credit that had been initiated in 1988 were replaced by the Child Tax Benefit and Work Income Supplement described below. The termination of federal Family Allowances and of Quebec's

Family Allowances program in 1997 marked the end of universal income benefits in Canada (Baker, 1995a, ch. 4).

The history of the idea of benefits for parents of young children is interesting. Industrial wages, paid by employers, cannot be expected to vary with the size of the worker's family; if they did, family heads would be at a severe competitive disadvantage in the labour market. The wage paid a worker with several dependants will not support the same standard of living, obviously, as the same wage paid a worker with few or no dependants. This may seem unfair to parents and to children, and dangerous for society if children are deprived, materially or otherwise. As well, over the typical earning career of an individual, earnings are lower in the early child-rearing years, when expenses are high, and higher in the later years. A child-related income supplement will mitigate this.

Sir William Beveridge, author of the wartime Beveridge Report that became the blueprint for Great Britain's welfare state in the years after World War II, rated family size, and the associated inability of wage earners to support several children adequately, as almost as important a cause of poverty as unemployment and illness. Leonard Marsh, in his now-celebrated *Report on Social Security for Canada* of 1943, made substantial family allowances the keystone of the income security system he proposed (Marsh, 1943/1975; Kitchen in Moscovitch and Albert, eds, 1987). Today, the Child Tax Benefit and other family benefits constitute a substantial proportion of the family incomes of low-income people, which certainly includes the working poor.

Part of the push for the enhancement of child-related benefits has come from a spate of recent criticism of the relatively small total outlays in state income support in favour of children, compared to outlays in favour of other age groups, notably the elderly, and the consequently high, and apparently increasing, number of children living in deprivation in this country. Canada's performance in this respect compares poorly with that of other industrial countries other than the United States (Baker, 1995b). The House of Commons unanimously passed a resolution in 1989 'to seek to achieve the goal of eliminating poverty among Canadian children by the year 2000.' This has mobilized activists across the country under the banner of 'Campaign 2000' (Hay, 1997).

The federal refundable Child Tax Benefit, the largest and best known of the redistributive tax credits, is, as the name implies, a benefit for persons with dependent children that is conferred through the personal income tax system. It gives most of its benefits to low-income earners.

Taxpayers with dependent children get a cash benefit, linked for income earners with a 'tax credit'—a credit in their income tax accounts with the federal government—of a certain amount per child. Income tax payable is reduced by that amount. The whole tax credit is payable to families with *family* net incomes up to a certain threshold. For families whose incomes exceed that threshold, the tax credit is gradually reduced until, at a family income somewhere above the national average, it is reduced to zero. If the taxpayer's tax *credit* is greater than the amount he or she owes in taxes, the individual receives the difference in cash. (That is why it is called 'refundable': the tax credit that is due is paid no matter how much or how little income tax the recipient family had to pay.) It is received in the form of a monthly cheque or bank account deposit.

Upscale taxpayers, therefore, realize no benefit at all from the Child Tax Benefit. The threshold and consequently the income level at which the benefit is wiped out are partially indexed to inflation; they are raised by the year-to-year rate of increase in the cost of living minus 3 per cent. Over many years, the partial indexation will greatly reduce the real value of the benefit.

Provinces have some latitude with regard to the distribution of the federal Child Tax Benefit to families within their boundaries. Quebec has chosen to shape Child Tax Benefit credits in favour of families with three or more children, as it has done with its own family benefit program. Alberta, too, has massaged the Child Tax Benefit, discriminating on the basis of the ages of children.

Benefits paid by government in this way, by means of taxes foregone, are called 'tax expenditures'; they are expenditures in the sense that they are imputable taxes that are not collected, or taxes that are reversed, like the Tax Credits. They have been a common tool in the promotion of economic development, where investors are enticed by income tax concessions into making certain kinds of investments, presumed to be in the public interest. The Child Tax Benefit is an application of the device to the field of personal income, of proportionately greatest benefit to low-income families.

Tacked on to the Child Tax Benefit is a Work Income Supplement. The exact amounts provided are less significant than the idea itself: a benefit that gives an additional reward for income-earning effort at low earnings levels. For some time the maximum was $500 *per family*. As announced in the 1997 federal budget, in 1997 the maximum supplement is slated to *increase with family size* from $605 for one child to $1,670 for four children. The supplement is paid at the rate of 8 per cent of all earnings over a very low amount ($3,750 per year): the higher the earnings, the higher the supplements, up to the maximum. In 1998, the complexity of a Tax Benefit plus a Work Income Supplement will be resolved by combining the two into a new 'Canada Child Tax Benefit' (Canada, Department of Finance, 1997; Pulkingham and Ternowetsky, 1997).

The net benefit to families on public assistance of these successive revisions of payments in favour of children is problematic, because provinces are free to modify their public assistance benefit scales, taking the Canada Child Tax Benefit into account. By agreement with the federal government, they may not do so in such a way as to reduce total income benefits to families on welfare, but since the provinces now receive much less than previously in transfer payments from Ottawa in support of welfare programs, they might well be inclined to do some judicious trimming.

As remarked above, various provinces also make use of the tax credit device to reduce the burden of taxation on lower-income people. Some credits are not refundable. This means that the credit will be extended only up to the amount of tax for which the individual is deemed liable. If the individual has no tax to pay, the credit will be zero. When the credit exceeds tax payable, the tax obligation is wiped out, but there is no cash 'refund'. An example of a non-refundable credit is the Married/Equivalent to Married Credit, applied against the income tax liability of a person with a spouse or partner who has substantially lower income or no income at all.

Income-tested tax credits have now replaced the long-standing universal income benefit programs, Old Age Security and Family Allowances; they have also substantially replaced a number of exemptions and deductions by which Canadians were once allowed to reduce their taxable incomes for income tax purposes. The one remaining *exemption* is for people over the age of sixty-five, who are unconditionally allowed to subtract several thousand dollars from their incomes before calculating their income tax. The most relevant *deduction* is the one allowed for child care expenses actually incurred (and supportable by receipts).

The difference in the effects of credits on the one hand, and exemptions/deductions on the other, is important. A credit of $500, whether refundable or not, is worth $500 to every eligible taxpayer, whether high-income or low-income. If the credit is reduced (perhaps to zero) for high-income people, as the Child Tax Benefit is, then overall it is of the greatest benefit to lower-income people. By contrast, an exemption or deduction of $1,000, for example, saves more in income tax for the high-income earner, whose incremental tax rate is high, than for the low-income earner, whose tax rate is low. 'Equal treatment' applied to unequal people may aggravate the inequalities.

This is the perverse effect, referred to above, of the combination of two well-intended measures: the 'progressive' character of personal income tax (higher rates of tax applied to additional increments of income), and a deduction from taxable income for such an expenditure as child care expenses. Two wrongs do not make a right, but when two intended goods are paired they sometimes produce an unintended result, and perhaps an evil. (See Appendix B, 'The Redistribution of Income'.)

Replacement of Quebec Family Allowances
Quebec introduced its own Family Allowances in 1967. As in its manipulation of the federal Child Tax Benefit payable to Quebec residents, Quebec skewed its own Family Allowances program to give larger benefits to each subsequent child; and for the fourth and additional children, families have received allowances more than double the allowance for a

first child. Other universal allowances paid to Quebec parents were an additional allowance for children under six, that was, for third and subsequent children, a great deal larger than the Quebec Family Allowances, and a one-time Allowance for New-born Children ('Allocation à la naissance'), that paid families $500 and $1,000 respectively for first and second children born or adopted, and $8,000 for third and subsequent children.

In a major policy change announced in early 1997, Quebec relinquished the principle of universality in income and cash benefits for families, in favour of a package of family benefits conforming more closely to the income-tested tax credit format, while still favouring three and more children in the family. Simultaneously, Quebec announced the cessation of differential benefits for public assistance recipients with dependent children, thus doing immediately what all provinces are expected to do in the wake of the creation of the federal Child Tax Benefit (see above).

Tax credits and other tax-based family benefits have become Canada's main devices for providing public financial support for childrearing. As will be noted in Chapter Ten, poverty among children in Canada has emerged in recent years as a critical national issue, and closer attention will likely be paid to the principal public instruments used to address that issue.

In 1988, the Ontario Social Assistance Review Commission, appointed to study the whole welfare system in Ontario and to make recommendations for reform, proposed a vastly extended system of income-related benefits to replace welfare benefits (now needs-tested) for the children of families in receipt of public assistance, and to replace the universal family allowances then in effect. In other words, the Review sought to make substantial support of children of *all* low and not-so-low income families a public charge. The approach is summed up in the slogan, 'Get the kids off welfare.' This proposal, seen at the time as a bold one, has clearly taken hold (Ontario Social Assistance Review Commission, 1988; Battle and Muszynski, 1995; Kesselman, 1994).

Minimum Wage Legislation

All of the programs discussed so far actually provide incomes for people. In that respect, minimum wage legislation may seem out of place here, since it merely *regulates* the terms upon which some workers are paid for their work; a minimum wage program does not in fact put money into anyone's purse or pocket. Since it applies to low-income workers, however, minimum wage legislation must be included in a review of programs that affect the working poor.

By and large, a liberal society has some confidence that most workers and employers, where each has some freedom of choice, will strike wage bargains that are at least tolerable. Many workers strengthen their bargaining positions by organizing into unions and bargaining collectively. The existence of unions affects wage bargaining even in firms where the workers are not organized into unions. But organizing a union is not easy, even among stable work groups; it is almost impossible among the low-paid workers who come under the protection of minimum wage laws.

In an imperfect world, it is difficult to say what constitutes a 'fair' wage. It *is* possible to make a stab at estimating the lowest wage that would provide a fully employed worker with some approximation to a decent standard of living, though that estimate is affected by the assumptions made about whether the earner has dependants.

There are at least a couple of situations in which, in the absence of any form of regulation of wages, workers would probably not be able to bargain for such a minimally decent wage. One is where the supply of labour of a given quality is very great compared to the number of jobs available. The worker is then in a weak position *vis-à-vis* the employer and would have to take what he/she could get. Much of the exploitation of labour occurs in situations like this. The other situation, equally realistic, is where the employer really is operating on the margin and is in danger of going out of business. This is characteristic of certain industries where competition is severe, where labour costs form a high proportion of total costs (i.e., the work is labour-intensive), and profit margins, if any, are narrow. In such cases, wages are certainly pressed downward. Enforcement of minimum wages at least puts all competing employers on an equal basis concerning wages. They *all* must incorporate the minimum wage costs in the prices they charge their customers; none can gain an edge by paying lower wages, at least, not legally. This levels out somewhat the terms of their competition with each other, but it does not help them in their competition with foreign producers, some of whom have labour costs much lower than Canada's.

The regulation of employment is a matter of provincial jurisdiction. Accordingly, all Canadian provinces have minimum wage laws. There is a federal minimum wage law, for workers in employments under federal jurisdiction, chiefly those where the work is highly mobile.

Since minimum wage laws are in effect everywhere, it is evident that they are considered to be necessary and helpful for the protection of the workers who are in the weakest positions in the labour market. For that reason, they are warmly defended by many.

Certain reservations are, however, frequently expressed. For one thing, it is possible to oblige employers to pay employees so much per hour; it is not possible to oblige them to employ any given number of workers, nor to employ them for any number of hours per week or weeks per year. Upward pressures on wage costs, such as enforcement of even a modest minimum wage, incite employers to try to get their work done with fewer employees. Interestingly, Leonard Marsh, defending minimum wage laws in his *Report*, does not shirk this latter implication: among the advantages of a fairly high minimum wage, he includes the probability that firms will be driven to more efficient operation, conceivably less labour-intensive (Marsh, 1943/1975, p. 57).

Finally, if we concede that, to an employer, some less able workers may really not be worth the minimum wage, then minimum wage enforcement works against them. Some observers contend that minimum wage laws discriminate against young, inexperienced, unskilled workers, who have little chance at being hired at a legally mandated wage but who might be hired at a lower wage (and thereby acquire better work skills). Thomas Sowell, an African-American economist, argues that minimum wage laws in the United States are anti-youth and anti-black. According to Sowell, the real 'minimum wage' for many unskilled workers is not five dollars an hour, or whatever, but zero dollars per hour for zero hours of work (Shannon and Beach, 1995).

Part of the difficulty in determining minimum wage levels is some ambiguity about what kind of worker one is contemplating: a young person living with his or her family or an adult with dependants? This question also casts a shadow on the relevance of comparisons of welfare benefits with a minimum wage income, though such comparisons are common (see, e.g., National Council of Welfare, 1993a).

Enforcement of minimum wage laws is uniquely hampered by the fact that the victim of a violation is not highly likely to complain, because of the obvious risks. Complaints are therefore often made by third parties, such as advocacy groups. In Quebec, the minimum wage law is part of an Act that regulates other work standards as well: the Work Standards Act (Acte sur les normes du travail). It is administered by the Commission des normes du travail, whose expenses are met by a levy on employers—the employers support the organization whose *raison d'être* is to police them. In other provinces, the tendency is to entrust the administration of minimum wage laws to the Department of Labour or its equivalent.

CRITICISMS AND QUESTIONS

Negative, even hostile comments about public assistance abound. Those close to the experience find that benefit levels tend to be insufficient for even a minimally acceptable standard of life. To illustrate how thin the margin of safety is, in months when the period between mailings of welfare cheques extends over an extra weekend, many recipients find themselves running short of necessities. This is confirmed by the experience of voluntary food banks across the country.

For various reasons, the administration of public assistance is often intrusive. Recipients may be subjected to inquiries that most people would consider humiliating. By counting too rigorously any earnings realized by recipients, welfare officials may be discouraging any steps toward independence, which clearly is counterproductive.

Welfare recipients are often made to feel a stigma, a feeling that 'there must be something wrong with me'. The stigmatization is not made any more bearable for recipients by the fact that it sometimes is expressed by neighbours and others who otherwise closely resemble them.

There are thornier issues at play, as in the problem of the 'employable unemployed'. On the whole, welfare programs have been liberalized to the extent that chronically unemployed able-bodied persons, with dependants, will receive assistance in most jurisdictions; but sporadic earnings complicate both their status as 'employable' and their families' status as welfare recipients.

Deprivation alters with time, like other social phenomena. Latterly, a category among the needy that has been expanding rapidly in numbers is the lone parent, almost always female. A high proportion of welfare recipients fits this description: 28 per cent in Canada in 1992, according to the federal Department of Human Resources Development (1994b). The proportion is much higher in some provinces, according to the same authority (1994a). A high proportion of *working* lone parents have total incomes that leave them poor by our most widely accepted definition. Equally striking is the proportion of families receiving welfare with heads between the ages of 20 and 40—far more than half. And while the classifications of recipients vary sufficiently from province to province to make nation-wide statistics more than a little uncertain, there appears little question that, in sharp contrast to the situation in the 1970s, nearly half or more of recipients are employable.

As the 1990s and the twentieth century pass, the overriding concern is with the maintenance of adequate public assistance benefits in the face of the financial picture confronting all governments, some more grimly than others. The federal government has changed the form (from CAP to CHST) and sharply reduced the amounts of its transfers to the provinces for social purposes. The provinces have all indicated that their resources are not sufficient to carry the burdens loaded onto them. Almost all the provinces have reduced their benefits. Most provinces are scrutinizing their criteria for eligibility for benefits, particularly the issue of the willingness of employable recipients to work, and are attaching work or work-training requirements to the receipt of benefits for those classed as fit to work. The up-front explanation for this is the deficit-and-debt issue discussed in this volume's introductory chapter. Recurring deficits and accumulated debt, however, create a golden opportunity for those who are opposed to any form of public assistance beyond minimal relief of the unquestionably indigent. Welfare is always a vulnerable target for political attack, and it is most vulnerable when the near-universal cry is to 'cut spending'. (See Alberta, Department of Family and Social Services [1995].)

Organizations like the Canadian Council on Social Development have drawn attention over the last few years to the extraordinarily high proportion of children who are living in households with poverty-level incomes. This would include *all* children in households dependent on welfare, but this is far from the whole story. Also included are children of the working poor, and children of most of the aforementioned working lone parents, especially mothers. Ironically, many such working-poor households find themselves responsible for young children early in their careers as households, when their earnings are lower than they are likely to be later. The long-term costs to society of vast numbers of deprived children are incalculable. This high incidence of poverty among the very young throws into question both the levels of public assistance for welfare families and the effectiveness of the combination of allowances and tax credits presently available to other families with children.

Some argue that, intentionally or not, the welfare system operates primarily as a powerful instrument of social control, not as an instrument to protect people from hardship. The thesis is that the large class of public assistance recipients, and others dependent on public income supplements, some of whom are potentially disruptive, is kept just sufficiently well provided for to remain reasonably quiet, and at the same time sufficiently dependent to be easily manipulated. The 'social control' thesis extends to such supplementary programs as work training programs, which are seen as relatively futile in terms of preparing people for the labour market but relatively successful in keeping them compliant with the work-oriented norms of society. The best-known statement of this thesis is an American book, *Regulating the Poor*, by Piven and Cloward (1971).

STUDY QUESTIONS

1. What is the difference between the 'categorical' and 'non-categorical' approaches to the design of public assistance programs? If a province shifted from 'categorical' to 'non-categorical' assistance, would you expect the welfare caseload to change very much?
2. The Economic Council of Canada says that when it comes to helping the working poor, 'policy choices are less simple'. Why should this be so?
3. List the public income sources available to a female head-of-household welfare recipient with two young children, in addition to her income from public assistance.
4. What advantages does the Child Tax Benefit have as a redistributive measure?
5. If you represented the interests of the federal government, what conditions would you attach to cost-sharing with the provinces in public assistance? Compare your own answer with the conditions formerly imposed by the Canada Assistance Plan Act, *Revised Statutes of Canada, 1985*, Vol. I, Chapter C-1, Article 6(2). What conditions can be imposed under the more recent Canada Health and Social Transfer mechanism?
6. Assuming they still had the same resources as before, do you think provincial governments are more likely to reduce welfare benefits now that they are free of the former Canada Assistance Plan restrictions and guidelines? Explain your answer.

7. How closely does your home province adhere to mandatory work requirements ('workfare') for welfare recipients? For *which* welfare recipients?

8. What are the pros and cons of minimum wage legislation? What is the standard minimum hourly wage in your province at the present time? Is there a different minimum wage for different workers?

9. Explain the contradiction that arises in the joint effect of a progressive income tax and a deduction from income for tax purposes.

10. Where would you look for facts and figures about public assistance in your province?

BIBLIOGRAPHY

Aitken, Gail, and Andy Mitchell (1995). 'The Relationship between Poverty and Child Health: Long-range Implications', *Canadian Review of Social Policy*, No. 35 (Spring).

Alberta. Department of Family and Social Services (1995). *Welfare Reform 1993–1995*. The responsible department points to its achievements over two-and-a-half years. Standard allowances to single parents with one child were reduced by 16 per cent, to a two-parent family with two children by 8 per cent. The total welfare caseload was reduced by 48 per cent, the caseload of 'employables' by 55 per cent. Much emphasis was placed on associated programs like 'Alberta Job Corps', 'Employment Skills Program', and 'Alberta Community Employment Program'. Total welfare expenditures declined by 42 per cent; expenditures for the disabled increased by 10 per cent; expenditures on child welfare (not income transfers) increased by over 20 per cent.

Armitage, Andrew (1991). 'Work and Welfare: A Conceptual Review of the Relationship between Work and Welfare', in Bill Kirwin, ed., *Ideology, Development and Social Welfare: Canadian Perspectives*. 3rd edn. Toronto: Canadian Scholars' Press.

Baker, Maureen (1995a). *Canadian Family Policies: Cross-National Comparisons*. Toronto: University of Toronto Press. See especially Chapter 4, comparing Canada's record with those of other nations.

———— (1995b). 'Eliminating Child Poverty: How Does Canada Compare?', *American Review of Canadian Studies* 79. In the author's view, not well.

Banting, Keith (1982). *The Welfare State and Canadian Federalism*. 2nd edn. Kingston: McGill-Queen's University Press. Does Canada's federal government structure enhance or retard the development of welfare measures, or a little of both? Public assistance is one of Banting's case examples.

Battle, Ken, and Leon Muszynski (1995). *One Way to Fight Child Poverty*. Ontario: Caledon Institute of Social Policy.

Beveridge, Sir William (1942). *Social Insurance and Allied Services* (American edn). New York: Macmillan.

Borovoy, Alan (1988). *When Rights Collide: The Case for Our Civil Liberties*. Toronto: Lester & Orpen Dennys. Intrusions upon privacy, encroachments upon freedom of decision, and fairness of procedures in welfare administration, reviewed in admirably balanced fashion by Canada's most prominent civil liberties lawyer.

Borovoy presents these issues as dilemmas, rather than as evidence of manifest ill will.

Brown, Charles, C. Gilroy, and A. Kohen (1982). 'The Effect of the Minimum Wage on Employment and Unemployment', *Journal of Economic Literature* 20.

Caledon Institute of Social Policy (1996). *Can Workfare Work? Reflections from History.* Ottawa: The Institute. Caledon's answer is, probably not.

———— (1995). *Government Fights Growing Gap between Rich and Poor.* Ottawa: The Institute. As the title indicates, this paper analyses the effect that government income programs have on income equalization in Canada. Incomes after taxes and transfers remain very unequal, but a great deal less so than *before* taxes and transfers.

———— (1995). *Colloquium on Child Poverty, January 26, 1995.* Ottawa: The Institute.

———— (1995). *One Way to Fight Child Poverty.* Ottawa: The Institute. Like the Ontario Social Assistance Review Commission's report, *Transitions* (cited below), Caledon makes the case for an expanded, income-tested child benefit, to reduce the incidence of poverty among children in Canada, which is high compared to almost all other industrialized countries.

———— (1993). *Breaking Down the Welfare Wall.* An analysis of the impact on a welfare recipient's income of moving from welfare to work. Covers in a different way some of the same ground as the National Council of Welfare's *Incentives and Disincentives.* . . . (cited below).

Callahan, Marilyn, Andrew Armitage, Michael J. Prince, and Brian Wharf (1990). 'Workfare in British Columbia: Social Development Alternatives', *Canadian Review of Social Policy* 26.

Canada. Canada Assistance Plan. *Revised Statutes of Canada, 1985*, Chapter C-1. Originally passed 1966–7.

Canada. Child Tax Benefit Act, *Statutes of Canada, 1992*, Chapter 48. The *income-tested* benefit authorized by this Act replaced the *universal* Family Allowances that had been in existence, with modifications, since 1944. Further modifications are expected.

Canada. Department of Finance (1997). *Working Together towards a National Child Benefit System.* A 'national system' will require close provincial-federal collaboration, which appears to be forthcoming.

———— (1996). *Budget Speech, Hon. Paul Martin, March 1996.*

———— (1995a). *Budget Speech, Hon. Paul Martin, February, 1995.*

———— (1995b). *Towards a National Child Benefit System.*

Canada. Economic Council of Canada (1992). *The New Face of Poverty: Income Security Needs of Canadian Families.* Ottawa: The Council.

———— (1968). Fifth Annual Review, *The Challenge of Growth and Change.*

Canada. Health and Welfare Canada. *Annual Reports*, Canada Assistance Plan Branch and Family Allowances Branch. Basic data source for the experience of these important programs, both now defunct.

Canada. Health and Welfare Canada. *Basic Facts on Social Security Programs / Précis sur les Programmes de Sécurité Sociale* (current issue). Concise descriptions of federal programs.

Canada. Health and Welfare Canada (1969). *A Chronology of Social Welfare and Related Legislation 1908–1968. Selected Federal Statutes.*

Canada. Health and Welfare Canada. *Overview: The Income Security Programs of Health and Welfare Canada* (current issue). Data on and explanations of federal government income security programs.

Canada. Health and Welfare Canada. *Report on Family Allowances, Old Age Security, Canada Pension Plan* (annual). Highly informative about all these programs. (Retitled in future, after the demise of Family Allowances.)

Canada. Human Resources Development Canada (1997). *Social Security Statistics, Canada and the Provinces, 1970–71 to 1994–95.* Updated annually.

Canada. Human Resources Development Canada (1994a). *Inventory of Income Security Programs in Canada, January 1993.* Updated from time to time. A detailed descriptive list of all federal, provincial, and territorial income programs in effect in Canada, including abundant data on recipients and on benefits; the best one-volume compilation of this material. For family benefits and public assistance, see Chapters 3 and 4.

Canada. Human Resources Development Canada (1994b). *Improving Social Security in Canada: A Discussion Paper.*

Canada. Parliamentary Task Force on Federal-Provincial Fiscal Relations (1981). *Fiscal Federalism in Canada.* Chapter VI, 'Social Assistance and Social Services: the Canada Assistance Plan'. The machinery of fiscal federalism has changed, and the Canada Assistance Plan is no more; but to understand where we are, we must know how we got there, and this report provides a lucid explanation and discussion.

Canada. Senate of Canada (1971). Special Senate Committee on Poverty, Sen. David Croll, Chair. Report. *Poverty in Canada.*

CCPA (Canadian Council for Policy Alternatives) (1996). 'Workfare a failure everywhere it's been tried'. CCPA *Monitor* (July/August).

CCSD (Canadian Council on Social Development) (1993). *Family Security in Insecure Times.* Ottawa: The Council and the National Forum on Family Security. 2 vols.

———— (1995). *Poverty among Young Families and Their Integration into Society and the Workforce: An Ontario-Quebec Comparison.* Ottawa: CCSD and Social Planning Council of Ottawa-Carleton.

(Senator) Erminie Joy Cohen, with Angela Petten (1997). *Sounding the Alarm: Poverty in Canada.* Ottawa: the author.

CQDS (Conseil québécois de développement sociale) (1997). *Le Workfare: Pour quoi faire?* Québec: CQDS. On the whole, hostile to the concept of workfare.

Copp, Terry (1974). *The Anatomy of Poverty: The Condition of the Working Class in Montreal, 1897–1929.* Toronto: McClelland and Stewart. Discusses the 'working poor', long before the 'welfare state' made them a special policy issue. See also Piva (cited below).

Cousineau, Jean-Michel (1993). *La Pauvreté et l'état: pour un nouveau partage des compétences en matière de sécurité sociale.* Montreal: Institut de recherches en politiques publiques/Institute for Research on Public Policy. Cousineau criticizes current anti-poverty policies for being unproductively concentrated on simple income distribution and too little concerned with the causes of poverty. He finds the existing federal-provincial structure partly to blame for this, and, taking interprovincial differences in resources into account, proposes a *regional* rather than strictly provincial structure.

Dyck, Rand (1976). 'The Canada Assistance Plan: The Ultimate in Co-operative Federalism', *Canadian Public Administration* 19/4 (Winter). Though the CAP is no more, its development is a good illustration of the federal-provincial relationship. (Cf. Vaillancourt, cited below.)

Evans, Patricia, Lesley A. Jacobs, Alain Noël, and Elizabeth B. Reynolds (1995). *Workfare: Does It Work? Is It Fair?*, Adil Sayeed, ed. Montreal: Institute for Research on Public Policy.

Evans, Patricia, and Eileen McIntyre (1985). 'Welfare, Work Incentives, and the Single Mother', in Jacqueline Ismael, ed., *Canadian Social Welfare Policy: Federal and Provincial Dimensions*. Kingston: McGill-Queen's University Press.

Freiler, Christa (1996). 'A National Child Benefit: Promising First Step or Final Gesture in Child Poverty Strategy?', *Canadian Review of Social Policy*, No. 38 (Fall).

Gow, James Iain, Alain Noël, et Patrick Villeneuve (1995). 'Les Contrôles à l'aide sociale: l'expérience québécoise des visites à domicile', *Canadian Public Policy/Analyse des politiques* XXI:1 (March).

Guest, Dennis (1997). *The Emergence of Social Security in Canada*. 3rd edn. Vancouver: UBC Press.

Hay, David I. (1997). 'Campaign 2000: Family and Child Poverty in Canada', in Jane Pulkingham and Gordon Ternowetsky, *Child and Family Policies: Struggles, Strategies and Options*. Halifax: Fernwood.

Hepworth, H. Philip (1985). 'Trends in Provincial Social Service Department Expenditures', in Jacqueline Ismael, ed. *Canadian Social Welfare Policy: Federal and Provincial Dimensions* (cited below).

Hepworth, H. Philip, Ron Draper, Fred R. McKinnon, Duncan Rogers, and Richard B. Splane (1987). 'Insiders Look Back: Views of the Origins of the Canada Assistance Plan', *Canadian Review of Social Policy*, No. 18 (May). As the title implies, the five co-authors were significant players in the steps that led to the creation of the Canada Assistance Plan.

Hobson, Paul A., and France St.-Hilaire (1993a). 'The Financing and Delivery of Social Policy: Fiscal Transfers for Social Assistance and Social Services', in Elisabeth B. Reynolds, ed., *Income Security in Canada: Changing Needs, Changing Means*. Montreal: Institute for Research on Public Policy.

———— (1993b). *Toward Sustainable Federalism: Reforming Federal-Provincial Fiscal Arrangements*. Montreal: Institute for Research on Public Policy. Chapter IV, 'The Federal-Provincial Dimension to Income Security', pp. 43–70.

Hum, Derek (1985). 'The Working Poor, the Canada Assistance Plan, and Provincial Responses in Income Supplementation', in Jacqueline Ismael, ed. *Canadian Social Welfare Policy: Federal and Provincial Dimensions* (cited below).

Irving, Allan (1989). '"The Master Principle of Administering Relief": Jeremy Bentham, Sir Francis Bond Head and the Establishment of the Principle of Less Eligibility in Upper Canada', *Canadian Review of Social Policy*, No. 23 (May). A rare look into early welfare history in pre-Confederation Canada.

Ismael, Jacqueline, ed. (1985). *Canadian Social Welfare Policy: Federal and Provincial Dimensions*. Kingston: McGill-Queen's University Press. Six articles in this book (listed separately in this bibliography) look at welfare programs in the provinces and at the federal government's participation in public welfare, particularly in the areas dealt with in Part One of this book.

————, ed. (1987). *The Canadian Welfare State: Evolution and Transition*. Edmonton: University of Alberta Press.

Johnson, Andrew F. (1987). 'Ideology and Income Supplementation: A Comparison of Québec's Supplément au revenu de travail and Ontario's Work Incentive Program', in Jacqueline Ismael, ed., *The Canadian Welfare State: Evolution and Transition* (cited above).

Kealey, Greg, ed. (1973). *Canada Investigates Industrialism: The Royal Commission on the Relations of Labour and Capital, 1889 (Abridged)*. Toronto: University of Toronto Press.

Kerstetter, Steve (1997). 'Fighting Child Poverty with Parental Work Income Supplements', in Jane Pulkingham and Gordon Ternowetsky, *Child and Family Policies: Struggles, Strategies and Options* (cited below).

Kesselman, Jonathan (1994). 'Policies to Combat Child Poverty: Goals and Options', in Keith Banting and Ken Battle, eds, *A New Social Vision for Canada*. Kingston and Ottawa: Queen's University School of Policy Studies and Caledon Institute of Social Policy.

Kitchen, Brigitte (1997). 'The New Child Benefit: Much Ado about Nothing', *Canadian Review of Social Policy*, No. 39 (Spring).

———— (1995). 'Children and the Case for Distributive Justice', *Child Welfare* 74 (May/June). This article appeared in a special issue of this journal devoted to child welfare in Canada.

———— (1987). 'The Introduction of Family Allowances in Canada', in Allan Moscovitch and Jim Albert, eds, *The 'Benevolent' State: The Growth of Welfare in Canada*. Toronto: Garamond. Kitchen surveys the Canadian debate about family allowances since World War I or so, reviewing the sometimes surprising positions taken by social reformers, economists, labour union leaders, Catholic clerics, social workers, and others.

Lightman, Ernie (1991). 'Support for Social Welfare in Canada and the United States', *Canadian Review of Social Policy*, No. 28 (Fall). Polls and other evidence are examined, showing a somewhat higher level of support for welfare programs in Canada than in the United States.

Linneman, Peter (1982). 'The Economic Impacts of Minimum Wage Laws: A New Look at an Old Question', *Journal of Political Economy*, Vol. 90, No. 3 (June).

Lochhead, Clarence (1997). 'Identifying Low Wage Workers and Policy Options', in Jane Pulkingham and Gordon Ternowetsky, *Child and Family Policies: Struggles, Strategies and Options* (cited below).

Maclean's (1997). 'Growing Up Poor'. 24 February.

Joan McFarland and Robert Mullaly (1996). 'NB Works: Image vs. Reality', in Jane Pulkingham and Gordon Ternowetsky, eds, *Remaking Canadian Social Policy: Social Security in the Late 1990s*. Halifax: Fernwood. The authors' research, while not based on a true sample of clients, leads them to conclude that the program is not a success. They include an interesting discussion of the shades of real meaning of 'work-fare' programs, from truly voluntary to flatly mandatory, and of the difficulty of distinguishing the one from the other in practice.

McGilly, Frank (1991). 'Ideology and Public Assistance in Canada: Reflections on the Use and Abuse of a Slippery Concept', in Bill Kirwin, ed., *Ideology, Development and Social Welfare: Canadian Perspectives*. 3rd edn. Toronto: Canadian Scholars' Press.

Marsh, Leonard (1943; republished 1975). *Report on Social Security for Canada 1943*. Toronto: University of Toronto Press.

Mellor, Earl F. (1987). 'Workers at the Minimum Wage or Less: Who they are and the jobs they hold', *Monthly Labour Review* 110 (July).

Mercier, Jacques (1987). 'Effets du salaire minimum sur l'emploi', *Relations industrielles/Industrial Relations* 40.

Mongeau, Serge (1967). *L'Evolution de l'assistance au Québec*. Montreal: Editions du jour. Of historical interest. The development in Quebec from religious and philanthropic charities to the Social Aid Act.

Morel, Michel, and André Lareau (1994). *Actes du Forum sur la fiscalité des familles*. Montreal: Bureau québécois de l'année internationale de la famille/Mouvement de caisses Desjardins. A wide-ranging series of papers and discussions on aid to families through tax measures and transfer programs.

Nakamura, Alice O., and Eric M. Diewert (1995). 'Reforming our Public Income Support Programs: A Focus on Children', in Banting and Beach, eds, *Labour Market Polarization and Social Policy Reform*. Kingston: Queen's University School of Policy Studies.

National Council of Welfare (1997). *Child Benefits: A Small Step Forward*. Ottawa: The Council.

———— (1997). *Poverty Profile 1995*. Ottawa: The Council. The poor in Canada: age, sex, region, civil status, employment status, etc.

———— (1996). *Welfare Incomes 1995*. Ottawa: The Council. A useful compilation of benefits realized by welfare recipients, including child benefits and tax credits. The Council issues an updated version every year or two.

———— (1995a). *The 1995 Budget and Block Funding*. Ottawa: The Council. An anticipation of the consequences of few-strings-attached federal block funding of provincial welfare programs.

———— (1994). *A Blueprint for Social Security Reform*. Ottawa: The Council.

———— (1993a). *Incentives and Disincentives to Work*. Ottawa: The Council. A discussion of how to remove the real material barriers in the way of welfare recipients seeking employment.

———— (1993b). *Income Assistance for Families with Children*. Ottawa: The Council. The NCW's contribution to the discussion of benefits for children in all low-income families.

———— (1993c). *Welfare and Other Income Support Programs*. Ottawa: The Council.

———— (1993d). *Working for Welfare*. Ottawa: The Council. A highly sceptical look at workfare and workfare-like schemes.

———— (1991). *The Canada Assistance Plan: No Time for Cuts*. Ottawa: The Council. A warning against threatened reductions in federal welfare outlays, a few years before the demise of CAP.

———— (1990). *Women and Poverty Revisited*. Ottawa: The Council. Women make up a disproportionately high share of Canada's (and most other countries') low-income population. Note Chapter VI, 'Single-parenthood and Poverty'.

———— (1987). *Welfare in Canada: The Tangled Safety Net*. Ottawa: The Council. A heroic attempt to summarize, explain, and compare the welfare programs in all the provinces and territories of Canada, without trying to cram in every detail, but trying to show the impact that different features of programs have on recipients. The difficulty of collecting the necessary information is reflected in the title of Appendix A: 'The Data Gap'. An invaluable tool.

Naylor, Nancy, Ruth Abbott, and Elizabeth Hewner (1994). *The Design of the Ontario Child Income Program*. Ottawa: Caledon Institute of Social Policy. Caught up in the tide of fiscal restraint of the mid-1990s, the aspiration of Ontario's NDP government to 'take the kids out of welfare' never reached the stage of a legislative proposal. To ensure public discussion of the idea, the Caledon Institute published this explanation, written by three closely involved officials of the provincial government at the time.

Ontario (1988). Social Assistance Review Commission. *Transitions*.

Phipps, Shelley (1995). 'Canadian Child Benefits: Behavioral Consequences and Income Adequacy', *Canadian Public Policy* XXI/1 (March). Briefly: Canada's modified Child Tax Benefit neither enhances the adequacy of low incomes nor incites non-earners to seek employment.

Picot, Garnett, and John Myles (1996). 'Social Transfers, Changing Family Structures, and Low Income among Children', *Canadian Public Policy* XXII/3.

Piva, Michael (1978). *The Condition of the Working Class in Toronto — 1900–1921*. Ottawa: University of Ottawa Press.

Piven, Frances Fox, and Richard Cloward (1971). *Regulating the Poor*. New York: Pantheon. Public welfare analysed as an instrument of social control. The material is American, but the idea is equally relevant to Canada.

Polanyi, Karl (1957). *The Great Transformation*. Boston: Beacon Press. How, for better and for worse, industrialism and capitalism combined to make employment, more specifically the labour market, as crucial as it is in our society. A classic.

Pulkingham, Jane, and Gordon Ternowetsky (1997). *Child and Family Policies: Struggles, Strategies and Options*. Halifax: Fernwood. See especially Part 3, 'Child and Family Poverty', and the co-editors' relentlessly critical 'Postscript' on the new Child Tax Benefit.

Québec. Ministère de la sécurité du revenu (1996). *La réforme de la sécurité du revenu, un parcours vers l'insertion, la formation et l'emploi*.

Quebec (1988). An Act respecting Income Security. Replaced the Social Aid Act.

Quebec. Minister of Manpower and Income Security (1987). *Towards an Income Security Policy*.

Quebec (1971). *Report of the Commission of Inquiry into Health and Social Services* (the Castonguay-Nepveu Report). Translated from the original French. This justly celebrated Report laid the groundwork for most of what happened during the subsequent twenty years in the development of health, welfare, and social services in Quebec, and has had considerable influence elsewhere. In these fields, it stands as the documentary monument of the 'Quiet Revolution' in Quebec.

Québec (1963). *Rapport du comité d'étude sur l'assistance publique* (the Boucher Report). Quebec's welfare expenditures exploded after the province signed on to participate in the federal Unemployment Assistance Act cost-sharing agreement. This Report examines the reasons. Though much less well known than the later Castonguay Report (cited above), the Boucher Report is also a landmark in Quebec's social history.

Québec. *Statistiques des affaires sociales, sécurité du revenu.* As the name implies, this periodical provides current statistics on income programs. Other provinces, of course, provide similar statistics.

Riches, Graham, and Lorelee Manning (1989). *Welfare Reform and the Canada Assistance Plan: The Breakdown of Public Welfare in Saskatchewan 1981–1989.* Working Paper Series, No. 4, Social Administration Research Unit, University of Regina.

Ross, David (1994) *The Changing Face of Poverty.* Ottawa: Canadian Council on Social Development.

———— (1981). *The Working Poor: Wage Earners and the Failure of Income Security Policies.* Ottawa: Canadian Council on Social Development.

Ross, David, E. Richard Shillington, and Clarence Lochhead (1994). *The Canadian Fact Book on Poverty.* Ottawa: Canadian Council on Social Development.

Sayeed, Adil, ed. (1995). *Workfare: Does It Work? Is It Fair?* Montreal: Institute for Research on Public Policy.

Scott, Katherine (1996). *The Progress of Canada's Children.* Ottawa: Canadian Council on Social Development.

Séguin, P. (1987). 'Descriptive Overview of Selected Provincial Income Supplementation and Work Incentive Initiatives', in Ismael, ed., *The Canadian Welfare State: Evolution and Transition* (cited above).

Shannon, M.T., and C.M. Beach (1995). 'Employment Effects of Minimum Wage Proposals', *Canadian Public Policy* XXI/3 (September). According to these authors, an increase in the minimum wage does raise the incomes of some workers, not necessarily all from low-income households, but such an increase eliminates some jobs as well. This is the classic dilemma concerning minimum wage legislation.

Shifrin, Leonard (1985). 'Income Security: The Rise and Stall of the Federal Role', in Jacqueline Ismael, ed., *Canadian Social Welfare Policy: Federal and Provincial Dimensions* (cited above).

Shragge, Eric, ed. (1997). *Workfare.* Toronto: Garamond. A collection of individual articles, all severely critical of provincial public assistance programs that impose work requirements upon welfare recipients.

Sividinsky, Robert (1980). 'Minimum Wages and Teenage Unemployment', *Canadian Journal of Economics* (February).

Statistics Canada (1994). *Women in the Labour Force.*

Struthers, James (1985). 'Shadows from the Thirties: The Federal Government and Unemployment Assistance, 1941–1956', in Jacqueline Ismael, ed., *Canadian Social Welfare Policy: Federal and Provincial Dimensions* (cited above).

Torjman, Sherri (1996). *Workfare: A Poor Law.* Ottawa: Caledon Institute of Social Policy.

——— (1994). 'Is CAP in Need of Assistance?', in Keith Banting and Ken Battle, eds, *A New Social Vision for Canada* (cited above). Torjman reviews various weak points in the operation of welfare programs under the Canada Assistance Plan, with special attention to the social services CAP was authorized to support. As she says, social services 'have always been the "poor cousin" of social policy in Canada' (p. 99). The subsequent change from CAP to block funding (an alternative Torjman considers in this essay) has not diluted the pertinence of her critique.

——— (1988). *Income Insecurity: The Disabled Income System in Canada.* Toronto: The G. Allan Roeher Institute. Highlights gaps and inconsistencies in support for disabled people across the country.

——— (1988). *The Reality Gap: Closing the Gap between Women's Needs and Available Programs and Services.* Ottawa: Canadian Advisory Council on the Status of Women.

Vaillancourt, Yves (1994, 1995). 'Le Régime d'assistance publique: une lecture québécoise (1960-1966)', *Canadian Review of Social Policy/Revue canadienne de politique sociale* 35, 36. A series of four articles presenting a Quebec nationalist view of the creation of the Canada Assistance Plan and the federal government's motivations.

Wellington, Alison J. (1989). 'Effects of the Minimum Wage on the Employment Status of Youths', *Journal of Human Relations* 26.

PART TWO

Health Care Services

CHAPTER 7

Hospitalization Insurance in Canada

INTRODUCTION: SICKNESS, HEALTH, AND SOCIAL POLICY

When Canadians speak of their 'health care system', they usually mean the set of arrangements whereby hospital services and medical care services are provided. Nearly all such services now come to us via public programs: hospitalization insurance and medicare.

Since legislation concerning health lies within the constitutional jurisdiction of the provinces, there are, just as in public assistance, ten different provincial programs in both hospitalization and medicare, in addition to those of the Territories. It is slightly misleading to speak of the Canadian health care system in the singular, despite the common thread of financial involvement of the federal government. Many observers expect that current reductions in the health-related transfers of funds from the federal to the provincial governments, and the consequent weakening of the federal government's leverage over the provinces, will lead to greater variations in the health care system from one part of the country to the other.

Two institutions highly valued in our society will be predictably prominent in this discussion: the hospital and the medical profession. The overwhelming bulk of the resources we presently allocate to health goes to hospitals and doctors. But for the most part, we use the services of hospitals and doctors when we are sick or injured, not to keep ourselves healthy; hence the common criticism that we have robust policies for illness and feeble policies for health. We shall raise some questions about what we are actually doing in health policy: are we spending a lot of our money and effort on the wrong things? In the health field, should we be complaining as much about *misfunding* as about *underfunding*? (Rachlis and Kushner, 1989 and 1995).

There are important health needs and health services that are not concerned with the treatment of illness; they lie outside the functions of hospitals and physicians and therefore beyond the reach of hospitalization insurance and medicare. In a subsequent chapter, we will give attention to recent developments in two such areas, occupational health and community health. Both are more concerned with the maintenance of health and the prevention of illness and accidents than with treatment. Both have developed a substantial organizational base, but neither one is identified with an institution whose social status rivals that of the hospital or of the medical profession, and neither gets resources on a scale that approaches those devoted to hospital and medical care.

Three linked themes underlie much current discussion of health issues: (1) Our conception of what health means, and of what health services ought to be, has broadened beyond the biochemical and physiological framework, with its focus on symptom/diagnosis/treatment, to include social, environmental, and cultural dimensions; as medical historian Dr Henry Sigerist has stated, 'Medicine is basically a social science.' (2) The conception of individual health has shifted to that of a *continuum* from 'well' to 'ill', rather than the dichotomy of 'well' vs. 'ill' for which most health care facilities have been designed. (3) There is, as well, concern over the imbalance between our relatively abundant supply of specialized, technologically advanced, pathology-oriented services and facilities and our less ample provision of more generalized, simpler, health-oriented services, such as home care (Courchene, 1993, pp. 51–3).

Even those most sceptical of this glorification of highly scientific medicine and highly specialized health care services must acknowledge that there are real reasons for it. Fundamentally, hospitals and doctors dominate the health field because people greatly value them. The aura surrounding doctors and hospitals no doubt owes something to the mutual reinforcement of media images and professional self-interest, but as the National Council of Welfare (1990, Foreword) ruefully states:

> The experts say we have more doctors than we need. The general public thinks we have too few. The experts think we have enough hospital beds. The public wants more. The experts have their doubts about some of the latest medical technology. The public seems completely uncritical and wants all it can get immediately.
>
> Most fundamentally, the experts believe that curative medicine is reaching its limits and that more substantial gains in health will come through preventing illness. The public still seems preoccupied with disease and clings to its faith in miracle cures.

We accord the medical profession exalted, some would say excessive, status and considerable power because, rightly or wrongly, we consider doctors to be uniquely credible in health matters. Perhaps the conventional health care system suffers from a paradox of success: when something looks as though it is working well, we may ask more from it than it can provide. The dazzling scientific and technical advances that have been made in so many special medical fields may have distracted us from potential advances along other lines.

Of the major themes underlying all Canadian social policy, discussed earlier, those most pertinent to health care policy are:

- the aging of the Canadian population;
- a demographic pattern in which men and women figure very differently;
- the federal system of government, whose current evolution is deeply influenced by developments in social policy;
- the financial crunch facing all governments in the 1990s. The proverb says that necessity is the mother of invention, but the proverb does not guarantee that we will like all of necessity's offspring. Financial necessity has lately compelled the federal government to invent a new formula to reduce its share of health care financing, and provincial governments have responded by unprecedentedly drastic changes in health care provision.

Omitted from this discussion, except indirectly, will be some important health-related social issues, such as protection of the environment, the regulation of foods and pharmaceuticals, the use of tobacco and other harmful addictive substances, and the specific problems associated with sexually transmitted diseases. The reason for these omissions is certainly not that these issues have less impact on the health of Canadians than the provision of hospital and medical care. On the contrary, the reasons for these omissions are *historical*, in that some of the most pressing of these issues have taken on such a new form in recent years that it would be difficult to say what our policies are, and *conceptual*, in that they call on disciplines not hitherto closely associated with health. Our society is still struggling to create the institutions that will deal with these issues.

The objectives of our study of health care programs are:

1. To acquire basic information, and to become acquainted with sources of further information, about hospitalization and medical care insurance—the authorizing legislation, the services provided, eligibility, involvement of service providers, finance, administration, and the intergovernmental dimension.
2. In the light of the histories of the programs and information concerning their performance, to assess their contribution to the health of the people.
3. To acquire similar knowledge and understanding of programs in occupational health and community health.
4. To lay the basis for a critical understanding of the interplay of the basic institutions in the health field and of the differences in their social standing.
5. To consider various criticisms of current health care policy.

THE HOSPITAL, HEALTH, AND SOCIETY

There are few other institutions in our society that enjoy as enthusiastic acceptance as the hospital. American sociologist Eliot Freidson ranks the hospital over the church, the legislature, and the corporation as the representative institution of our culture (scientific, technological, professional, highly organized, serving a social value of high priority) (Freidson, 1963). Ivan Illich acidly calls the hospital 'the modern cathedral'. Yet, not long ago the hospital had the standing of a necessary evil. It was only in the 1870s that Joseph Lister initiated the use of antiseptics in hospitals; prior to that, hospitals had bred as much infectious disease among patients as they cured, and were shunned by all who had the choice. As R.F. Bauhofer (*World Health*, December, 1970, p. 4) has put it:

> Once the hospital was little more than a place for the poor and the incurably sick to die in, but medical, scientific and technical advance, coupled with far-reaching social and economic change and growth, brought about a transformation. Hospitals became strongholds, isolated fortresses against illness and disease. Now this is giving place to a new concept: the hospital is emerging from isolation, reaffirming its links with the community, and developing into a center for an array of therapeutic, preventive, rehabilitative and educational medical services.

Certain social and historical processes underlie these perceived changes in what hospitals do and are expected to do.

(1) *Success creates its own problems.* We certainly no longer shun the hospital; we rush to it, so much so that it has become hard-pressed to meet the demands we put on it. Successes in health care have enormously increased the numbers of people whose state of health is somewhere between sufficiently 'well' to function independently and sufficiently 'ill' really to need hospital care. Examples are convalescents and the chronically ill. Such people as these need facilities other than those traditionally offered by hospitals. But often the hospital is given the new jobs, though not especially suited to do them, because it has done other jobs well. Partly as a result, we have lagged in developing the proper alternative facilities.

Communities and states have responded in different ways to these emerging needs. In Canada, the provinces have in fact allocated resources to the creation of new and different facilities, such as hospitals for chronic care, convalescence, and rehabilitation. Such allocation may be politically difficult to achieve, especially when people feel, however erroneously, that there is a shortage of standard hospital facilities.

(2) *The elderly constitute a special challenge to health policy.* The same demographic changes whose impact we studied with respect to old age pensions exert pressure on health care facilities, due to the special health needs of older people. Inevitably, the increase in the numbers of older people in Canada, which is due in large part to improvements in health care, adds disproportionately to the need for health-related services. As with people who are disabled or chronically ill, the needs of older people are not necessarily for the services of hospitals, but where more appropriate facilities are lacking, the hospitals bear the brunt—unfortunately, at great expense.

The costs of health care for the elderly are real, but it is easy to exaggerate their impact on health budgets. Only a minority of older people need expensive hospital care frequently.

(3) *The cost of hospital services has increased greatly.* Like most costs, hospital costs can be analysed in terms of supply and demand. In terms of supply, the equipment required for advanced health care technology is expensive to build, maintain, and use. Personnel of the required levels of expertise are expensive. On the demand side, for over a century our society has been willing to meet high levels of costs for hospital construction and operation, privately and/or publicly. To mobilize the finances needed to meet the costs of care, our society has created massive programs of insurance and prepayment, private and public. Where the expressed demand for a service is as strong as that, relative to the demand for other things, it is certain that its price will be relatively high, no matter how it is paid for.

(4) *As with any other service, the provision of hospital services to poor people has been a problem.* Even though much inequality is tolerated in our society, it has always been difficult to rationalize the double health disadvantage suffered by poor people: poorer health to begin with (because of inadequate diet, unhealthy housing, etc.), and then financial barriers to health care when they need it. In fact, thanks to public intervention over the years, the poor now have better access to hospital facilities in Canada than, perhaps, in most other countries. The remaining problems of access are no longer a matter of meeting the out-of-pocket expenses of hospitalization; by and large, these are met through public programs (although decreasing generosity in public funding in recent years has tempted hospitals to impose user charges for more and more services, at some risk to universal access). The poor may still suffer from unequal access on account of such problems as availability of information, difficulty of communication, gaps in co-ordination of services—the manifestations of 'institutional inequality'. Our proclaimed public commitment to universal access makes such inequalities difficult to excuse.

(5) Finally, there arise issues over the *control of the hospital.* These are more than mere questions of detail in administrative structures—the professional affiliation of the chief executive, the composition of the board of directors. Control over the execution of policies readily shades over into control over policies, that is, control over the nature of the services to be offered and the facilities to be acquired. In hospitals, the medical profession has always been powerful, to no one's surprise, but it has increasingly had to share power with non-medical collaborators, such as other health care providers, specialized administrators (hospital directors are now frequently non-medical managers), and financial backers (once mostly wealthy people and philanthropic organizations, now mostly governments) (Wahn, 1987). The issue of control is expressed in the creation of new governing systems for hospitals, and for other health care institutions as well. The new forms of gov-

ernance seek to reflect the new breadth given to the concept of health, while preserving the social responsibility that has characterized the health professions at their best.

A Thematic History of Hospitalization Insurance in Canada

The development of public hospitalization insurance in Canada has followed a fairly simple scenario—which is not to say that it was easily achieved. In this section we trace the process that has been followed. (The thorough account given in Malcolm Taylor's *Health Insurance and Canadian Public Policy*, 2nd edn, 1988, is unlikely ever to be improved upon.)

High Cost of Hospitalization

In modern times, hospital services are regarded as necessary in cases of accident or serious illness, and also in certain usually happier circumstances such as childbirth. Hospital services are unavoidably costly. Whether operated under private (religious, philanthropic) or public (usually municipal) auspices, the operating costs of hospitals have been subsidized either by donations of money, by the provision of work for little or no salary (as by members of religious organizations or by volunteers), by taxes, or by a combination of all three. For a long time they were subsidized also by notorious underpayment of support staff. There are a few 'proprietary' hospitals, privately operated on a 'break-even' or, in special cases, profit-seeking basis, but they are at present too few to occupy our attention. Before the advent of generalized public provision, services could thus be made available to patients at something less than their real costs, but the amounts patients were required to pay were still considerable. People who were not by any definition poor found the cost of a stay in the hospital burdensome, a long stay perhaps ruinous.

Poor people seeking hospital care used to depend entirely on some form of charity, of which a common variant was the willingness of hospitals to carry indigent patients free or at drastically reduced rates. Some hospitals felt this obligation more keenly than others; all hospitals were limited in their ability to afford such 'charity' cases. Typically, patients given such special consideration could expect, *quid pro quo*, to give something in return, such as being used for teaching rounds for medical students. And a hospital could extend its services only to those poor people who actually came to its doors; a person who stayed away, unaware of the charitable service or unwilling to accept it, could not benefit from the hospital's generosity.

Insurance for Costs of Hospital Care

In our lives, we are confronted with many risks that are sure to strike some of us, but in such a way that we cannot predict *whom*, or *when*. Many such risks inflict more or less heavy costs when they do strike. We have a standard response to such hazards: we insure ourselves against their costs. A substantial business accordingly developed in Canada, as elsewhere, in insuring individuals against the risk of heavy costs for hospital services. A large proportion of the hospitalization insurance policies were issued to organized groups, since group insurance can be offered on cheaper terms than individual insurance. Many types of groups took part, such as fraternal organizations, and the like, but in industrial society, the most accessible are groups of workers, especially those in reasonably large and stable firms where there are steady flows of income available to be tapped for insurance premiums. Much of the health insurance therefore came into force as a 'fringe' benefit bargained for by unions or offered by employers as part of their compensation packages. Favourable tax treatment of premiums, whether paid by employees or employers, encouraged the proliferation of work-based health insurance programs. Sales of hospitalization insurance as an employee benefit were quite extensive in Canada. Coverage tended, however, to be uneven, being more widespread in those parts of the country that were the most highly industrialized and urbanized.

Some private hospitalization insurance enterprises were profit-seeking; some were co-operatively organized—i.e., the policy-holders themselves bore the financial responsibility for the insuring organization, and would themselves benefit from any surpluses and suffer from any deficits; but by far the largest coverage was achieved by private organizations operated on a *non-profit* (i.e., break-even) basis, the best known being Blue Cross.

Limits of Private Insurance

There are limits to the services and to the proportion of a population that private insurance—profit-seeking, co-operative, or non-profit—can cover, with the best will in the world.

First, the insurer must collect enough revenue to meet all claims plus all the other expenses of doing business; if a profit-seeking company, the insurer must aim to make a profit as well.

Then, to stay in business, the insurers must not only be technically solvent in the eyes of public regulatory authorities, which will otherwise shut them down (to jump ahead a little, a government-sponsored health insurance plan is never in danger of being put out of business by a regulatory authority); they must also be persuasively solvent in the eyes of the clientele, who will otherwise take their business elsewhere; this is particularly true where the clientele consists of business firms and unions, which are well equipped, or should be, to judge an insurer's financial soundness.

Since the business is competitive, all insurers (including non-profit) are under pressure to get as much revenue as feasible while keeping expenses and claims under control. The insurers will be impelled to control their outlays on claims: they will avoid bad risks or, failing that, will charge bad risks higher premiums; they will contest the validity of all claims that seem questionable; they will try to limit their outlays on sales effort and administration by looking for clienteles whose policies will be easy to administer, preferably large groups with relatively stable incomes. The clienteles they *do* insure will approve of such prudent business strategies (except when their own claims are turned down), because they help keep their premiums low.

Access to private insurance is therefore unavoidably unequal. Somewhat isolated individuals with dubious health are unlikely to be insured under private schemes, or if insured will pay more for it than others. Poor people are not likely to be covered at all. Even though modest coverage can in fact be brought within the reach of many people with low incomes, they do not usually fit the pattern—they are probably not members of a preferred group.

Coverage of services by different insurers is certain to vary. This can be seen as a strength. When public hospitalization insurance was still a debatable issue, part of the argument for reliance on private auspices was to allow different clients and client groups to choose the kind of insurance coverage they wanted and to exercise some choice as to how much they could spend on it; but it means that even people with insurance might not be covered for some important kinds of expenses.

By its nature, private hospital insurance furthermore embodies features like co-insurance (the client pays a certain share of each hospital bill), deductibles (e.g., each year, the client pays the total cost of all bills up to a stated amount), and coverage limitations (people who are obviously unhealthy may not be covered). By and large, advocates of reliance on the private insurance sector have admitted these limitations but have argued that the proper solution, giving the most satisfaction for the least cost, was to allow private insurers to reach as wide a market as they could, with the cost-efficiency they could achieve and the variety they could offer in meeting the varied preferences of different clienteles, and then to provide services through public auspices for the unserved. The argument is analogous to the argument with respect to private pensions for older people. In the end, this argument has not prevailed in Canada. We have preferred, as an ideal, *universal, uniform coverage*, and this is difficult if not impossible to achieve other than through a public program. (Co-insurance and deductibles are now frequently suggested as ways to help contain the future costs of public health insurance in Canada.) (Evans, 1984, ch. 2).

Health Needs, Post-World War II

In the later stages of World War II, considerable thought and work was devoted to the possibilities of post-war economic and social reconstruction. Nobody had quite forgotten the evils of the Great Depression of the 1930s. And some of the idealism that had been mobilized to defeat the fascist powers spilled over into a resolve to build a better society when peace came. One of the aspects of Canadian life recognized as ripe for change was health care. The poor health of the Canadian people in the 1930s and

1940s was demonstrated by various widely known indicators, such as high rates of infant mortality and high incidence of tuberculosis.

The state of health of the Canadian people was documented in a 1950s study tellingly entitled the 'Canadian Sickness Survey'. The health of the population was conceded by all qualified observers to be far below the standard attainable with existing knowledge, skill, and national wealth. While medical and hospital care are only two of many factors that together determine the health of a population, almost all commentators agreed that the high cost of a trip to the hospital, and, to a lesser extent, the cost of treatment by a doctor, deterred people from seeking needed care and thereby contributed significantly to the low level of health in Canada, even among people not thought of as 'poor'.

Provincial Initiatives

Provincial jurisdiction over hospital institutions in Canada extends to the provision of hospitalization insurance. Whatever the politics of the issue, it had to be worked out in ten different political arenas.

Within each province, there were sharp disagreements over public insurance for hospital care. Aside from the merits of the case, the governments of the provinces, especially the less prosperous, are always reluctant to impose higher levels of taxes or special levies than other provinces, even for popular programs. They are afraid of losing the more mobile of their taxpayers, or turning away potential newcomers, especially businesses (which pay taxes and employ people, both of which are comforting to governments). So long as public hospitalization insurance was not in operation everywhere in Canada, some provincial governments had a reason, or an excuse, for not initiating it themselves. The fact that a few of the provinces did undertake hospitalization insurance on their own did not suffice to overcome the fears of the others concerning interprovincial tax competition.

This tax competition argument requires a close look. In a province *without* a public hospital insurance program, the citizens were paying their hospital bills somehow—out of their pockets or through private insurance. The real questions were, would they have to pay more in taxes or premiums under a public program, or less, or about the same? And would they get inferior, equal, or superior service through public insurance? Might not the citizens of a province *without* public hospitalization insurance wish to move to a nearby province *with* such insurance? Might not business managers, whose sensitivity to taxes aroused so much concern, prefer a location with a good hospital insurance program, as they might prefer a location with good roads or good schools? Might not business firms be pleased to see the cost of hospital insurance taken out of their fringe benefit packages?

Even if the political and fiscal obstacles could be overcome, there was and is considerable disparity between what the different provinces could afford with their own resources, so the creation of ten programs under unaided provincial auspices would have resulted in very uneven provision from province to province. Many government services *are* unequally provided in different provinces, but there was a certain distaste for serious inequalities in public access to hospital care. As the debates on the issue intensified, the existing inequalities buttressed the case for some kind of federal government participation.

Despite these inhibitions, by the 1950s a few provinces had taken the initiative and created public hospitalization insurance programs of their own, before a nation-wide mechanism was developed. The first province to enact and implement public hospitalization insurance was Saskatchewan, in 1946—a significant first, because Saskatchewan has never been one of the wealthier provinces. This was done during the first mandate of the government controlled by the Co-operative Commonwealth Federation (CCF), predecessor of today's New Democratic Party. This Saskatchewan government, elected in 1944, has a place in history as the first socialist government in North America. Saskatchewan built its province-wide scheme on the foundation of some pre-existing municipal and regional plans, extending their pattern to the whole province. In the Saskatchewan plan, individuals and families were required to pay 'premiums' in order to be enrolled in the program, but from the very beginning the premiums paid only a part of the total costs; by design, a large subsidy was required from the revenues of the government.

As an historic curiosity, the first provinces actually to pass legislation authorizing public programs to assist people with the costs of hospitalization were Alberta in 1935 and British Columbia in 1936; but their governments never implemented the legislation. In Alberta, the party in power, the United Farmers of Alberta, was totally wiped out in an election later in 1935, though its health insurance policy had little or nothing to do with its defeat. In B.C.'s case, possibly because of the economic situation in 1936, the legislation simply died on the shelf, despite apparent political support (Taylor, 1988, pp. 6, 14–15). These experiences show that getting legislation passed is not always the end of a struggle for change.

British Columbia created a new hospitalization program in 1948. At first it was financed by a complicated system of premiums; B.C. switched over to a special sales tax in 1954. Alberta initiated a province-wide program in 1950. The decisions taken by the three most westerly provinces to initiate hospitalization insurance programs on their own, with no financial support from the federal government, were weighty ones. Not only was the issue politically controversial, with strongly held convictions on all sides and deep vested interests at stake, but the programs committed very large sums of money indefinitely into the future. (Newfoundland already had programs in effect providing hospital and medical services to a large part of its population when it joined Confederation in 1949.)

As time went on, it was fairly clear that other provinces were willing to entertain the idea of hospitalization coverage, though they gave it varying priority. Those that were interested naturally had different preferences as to the structure of a program—services to be covered, method of financing, and responsibility for administration. All regarded the costs as a serious obstacle. There were long-drawn-out disputes, first about the proper role of the public sector (universal provider, or 'safety-net' provider for the poor?), second about the possible involvement of the federal government, and, if it were involved, about the intergovernmental division of costs and program responsibilities. The consequent delays may in themselves have helped to direct the national attention to competing political and social priorities in the 1940s and 1950s. Also, the steady, sometimes spectacular economic growth of that era enhanced the spread of private health insurance, whose very success *de facto* reinforced arguments in favour of basic reliance on the private sector (if it's working this well, why fix it?).

Federal Participation: The 'Spending Power'

In the meantime, there had been considerable agitation in favour of some intervention by the federal government. The above-noted interest in social reform at and after the close of World War II extended to the provision of health care under public auspices. The influential Beveridge Report in the United Kingdom had stipulated that a national health service was an unqualified prerequisite for a social security policy. Among Canadian proponents of such programs, the least the federal government was called on to do was to give the provinces sufficient financial support to assure an acceptable minimum standard nation-wide. Many supporters of public health care insurance went much further and called for a single nation-wide program, feeling that federal-provincial jurisdictional difficulties could be overcome.

The federal government had, in fact, had hospitalization insurance on its agenda at least since the later years of the Depression. A national health care program was one of many proposals put forward in discussions of post-war social reconstruction in the mid-1940s. The federal (Liberal) government of the day included it among the proposals it made to the provinces at an important federal-provincial conference in 1945 (Taylor, 1988, pp. 9–22, 51–4). But the enemies of compulsory public health insurance were many, passionate, and articulate, and its friends disagreed among themselves on what to do and how to do it; so the proposals languished. The Liberal Party remained in power in Ottawa without interruption from 1936 until the late 1950s; advocates of public health insurance in and out of the party regularly reminded it of its unfulfilled promise.

Through the 1950s took place the usual delicate and not-so-delicate diplomacy required to bring federal resources into play in a field that is within provincial jurisdiction, without diminishing provincial powers. The squaring of this particular circle was achieved by recourse to the doctrine of the implicit 'spending power' of the Parliament of Canada, the slender pillar that also supports federal financial involvement in provincial welfare programs, as we have seen. The subsidization of provincial hospitalization insurance programs could be seen, under Section 106 of the BNA Act, as being 'for the Public Service', and could therefore be legitimately charged to the Consolidated Revenue Fund of Canada. The final outcome was federal legislation whereby the federal government was empowered to contribute financially to any provincial hospitalization insurance program that satisfied certain performance and administrative criteria:

- a uniform minimal package of covered services;
- no charge to users;
- universal coverage;
- immediate 'portability' of eligibility from province to province; and
- explicit public accountability.

The framework for nation-wide public coverage of the costs of hospital care came into being in 1957, embodied in the Hospital Insurance and Diagnostic Services Act of that year. The similar system covering costs of medical care did not take shape until ten contentious years later. The Hospital Insurance and Diagnostic Services Act and the later Medical Care Act were ultimately merged in the Canada Health Act of 1984, which is now the authority for federal cost-sharing in both kinds of provincial programs.

The 1957 Act, it must be understood, did *not* create a program of hospitalization insurance—only provincial legislation can do that. It *did* promise federal sharing in the costs of provincial programs, on the stated conditions. All provinces that had not already done so soon enacted legislation that met the agreed-upon standards. Universal public hospitalization insurance was in effect throughout Canada by 1961.

A related part of the story was the federal policy, dating back to the late 1940s, of 'matching grants' to the provinces for the support and planning of health services—grants, that is, that matched expenditures by the provinces. The largest class of health grants was for hospital construction, in which, after nearly twenty years of depression and war, Canada had fallen woefully behind. Hospitalization insurance could not possibly have succeeded had not hospital facilities throughout the country expanded dramatically during the preceding decade. The health grants continued after 1957, but the hospital construction grants were reduced and then phased out.

The Canada Health Act indicates in general terms what costs the federal government will share, and on what terms. The federal Minister of Health (a nomenclature subject to change) is responsible for the administration of the federal cost-sharing program, but the whole federal Cabinet (the 'Governor General in Council', as the federal Cabinet is formally designated) must officially decide on any modifications to the list of services and items for which federal funds will be available. In other words, such decisions are made on the authority of the whole government. Again, one must clearly understand that Health Canada administers only the *cost-sharing*; the agencies designated by the provinces administer the actual provision of hospital services.

The federal share of the financing of hospitalization in any one province is difficult to calculate. In the first twenty years or so of the program, the federal government undertook to pay half the total costs of all the participating provinces put together; thus the total federal contribution was tied to the cost of service provision Canada-wide. But this was not the same as a commitment to pay half the costs of each province individually. The calculation of the federal contribution to each province took account of both the national per capita cost and the province's per capita cost, biasing the federal sharing in favour of provinces with per capita expenditures lower than the Canadian average. This provided an incentive for efficient management, and also was supposed to work to the relative advantage of the poorer provinces, which were expected to have low levels of such costs as wages of hospital personnel. This expectation was not fully realized; in some years, some low-income provinces had high per capita expenditures, and vice versa (Taylor, 1988, p. 236).

By the late 1970s, the more complex Established Programs Financing pattern took the place of the 50 per cent basis. From our discussion of cost-sharing in public assistance, it may be remembered that under this new system, the transfer of funds from the federal to each provincial government was not tied to actual expenditures but was inversely related to the province's estimated ability to raise its own revenue (larger transfers per capita for the poorer provinces with lower tax capacity, less for the wealthier); the transfer of funds was supplemented by the federal surrender to the province of so-called 'tax points' in certain tax fields that both levels of government exploit. This system was authorized by the Federal-Provincial Fiscal Arrangements and Federal Post-Secondary Education and Health Contributions Act. (See Appendix C, 'Federal Sharing in the Costs of Health Care', pages 298–300.)

It has already been mentioned that as of 1997, a third cost-sharing mechanism is in place, the Canada Health and Social Transfer (CHST). Undoubtedly, this transfer will be substantially less than what would have been provided under Established Programs Financing, at least for the foreseeable future. Commentators question whether the federal government will continue to be able to impose conditions upon provincial programs when the federal share of the costs of many provinces has fallen far below 50 per cent. Until the details are known, however, such questioning is purely speculative. As will be seen, under EPF federal transfers consisted of a transfer of cash and a transfer of 'tax room'—percentage points of certain taxes, dropped by the federal government and left open to the provinces to exploit. Gradually, however, the EPF transfers began to comprise more and more tax points and less and less cash. Worried provinces and commentators predicted that one day the tax room would comprise the entire federal-to-provincial transfer, and the cash component of the transfers would disappear entirely. Not only would this have left the tax-poor provinces in dire straits; it would also have weakened the federal government's influence in health policy. Therefore, in late 1996, the federal government guaranteed the provinces a minimum annual cash transfer.

Theoretically, a province could decide not to have a hospitalization insurance program at all, in which case it would simply not get a federal contribution. It would, however, be difficult for a provincial government to refuse to take advantage of a huge cost-sharing program in which all or most other provinces were taking part, and toward which its own residents were apparently contributing as federal taxpayers. Since no provincial government wants to find itself backed into this position, the provinces approach the negotiation of cost-sharing terms exceedingly carefully. This applies not only to health programs, but also to welfare programs, as we have seen, and to programs in other fields like post-secondary education. No doubt politicians and others often resort to the federal-provincial jurisdictional issue as a delaying tactic, but the issue is nonetheless real.

HOSPITALIZATION INSURANCE PROGRAMS IN THE PROVINCES

Legislation

As we have seen, all the provinces without public hospital insurance programs in place before the federal Hospital Insurance and Diagnostic Services Act came into effect (1958) initiated programs within a couple of years. In fact, three of the provinces, including Ontario, whose participation was crucial to the viability of the national scheme, were for all practical purposes merely waiting for agreement on a federal cost-sharing plan and enacted their own conforming legislation at virtually the same time.

The legislature of Quebec enacted the Hospital Insurance Act in 1961—a little later than the other provinces, but early in the historic process of modernization of Quebec society known as the Quiet Revolution.

Services

The Canada Health Act lists the services for which the federal government will share the costs. There are variations among the provincial programs, but in general, as one might expect, they cover the cost-shareable services and few services the costs of which are *not* shared. In several provinces, the services of tuberculosis sanatoria, institutions for the mentally ill, and custodial care institutions (e.g., homes for the aged) are excluded.

Some of the hospital insurance Acts of the provinces go into more detail than others. The Quebec Hospital Insurance Act, for instance, says little about services covered, nor does it define an 'insured person'. These rather critical definitions will be found in 'Regulations' that the Act empowers the Government of Quebec to make. This is a forceful reminder of the potential importance of regulations. Recall that 'Regulations' made under authority conferred by a legislative act have the force of law, without prior review by the legislature itself. The power to make regulations concerning hospitalization insurance in Quebec is not delegated to one minister, but is retained by the whole Cabinet. (The traditional statutory expression for 'the government' or Cabinet of a province is 'the Lieutenant Governor in Council'; in recent years, Quebec has used the simpler term 'the Government'.) This parallels the delegation of regulating power in the Canada Health Act to the federal Cabinet. To keep it simple, services covered in Quebec, as in most provinces, include all internal and external services normally provided in a public ward by an active treatment hospital, a convalescent hospital (not a convalescent home), or a chronic care hospital.

It is illegal for private insurers to cover the services insured under a province's hospital insurance Act. This did not put an end to private hospitalization insurance. Private insurers have moved on to cover other services, such as higher standards of comfort in hospital rooms, extra nursing attention, treatment abroad, and, very importantly, cash indemnities during periods of hospitalization and convalescence, for which there is no public provision.

Finance
The financial arrangements vary widely among the provinces. The common factor is that all provinces finance their share of the costs in whole or in part from general revenues. Ontario and British Columbia still collect 'premiums', but they cover only a fraction of expenditures. All provinces rely on the federal government's contribution.

As an example, the province's share of the financing of Quebec's hospitalization insurance comes mostly from its general revenues and in lesser part from the special health care contributions exacted from Quebec employers—at present, a little over 3.2 per cent of their payrolls between a floor and a ceiling for each employee. So that the burden does not fall totally on earned income, Quebec taxpayers with income from other sources pay an additional levy. The revenues collected are divided between the Ministry of Health (Ministère de la santé) for the support of hospitalization, and the Régie de l'assurance-maladie du Québec for the support of medicare.

The use of the word 'insurance' in this context is questionable. In each province and territory, the provision of hospital services operates on the basis of a straightforward allocation out of the government's overall budget of sufficient funds to meet the anticipated expenditures, paid for largely by taxes and by transfers from the federal government, with no application of the concept of contingent risk. In the last few years of strategic government deficit reduction, governments have slashed the amounts of money allocated to hospital services, and have left it to the hospitals to adapt their services to their clienteles to fit their budgets. This is the standard way of financing all the general services of provincial governments. Some believe that the term 'prepayment' more accurately describes the process (Taylor, 1988, p. 236).

The Quebec Hospital Insurance Act gives the minister the power to 'make contracts' with hospitals to provide insured services to insured persons 'without charge upon uniform terms and conditions'. In the contract the hospital undertakes to provide the services on those conditions and the minister undertakes to pay the hospital for services rendered. Broadly speaking, this arrangement prevails in all provinces. The patient plays no part in the transaction, except perhaps to be informed of the services that have been paid for and the amount paid for them, as is done in Quebec. Services not covered by the legislation and the regulations, and not part of the contract, are normally charged to the patient. The minister and the hospitals renew from time to time their agreement on the amounts to be remitted in return for the specified services. Agreement is not always easily reached. Especially in recent years, hospitals have claimed to be underfunded; to balance budgets, services are reduced. At the extreme, a hospital's survival may be jeopardized, as several well-publicized instances demonstrate.

Presumably, under some circumstances a minister might feel free, or even obliged, to refuse to open or to continue a contract with a given hospital. This has happened in several provinces recently. A number of hospital closings and mergers have been decreed, and some services have been transferred from one hospital to another. It is conceivable for a hospital to try to get along without such a contract, relying on its own revenues from patients' private payments, but this would be financially audacious and is exceedingly rare.

The government-to-hospital relationship in the environment of public hospitalization insurance is captured in the careful language used by the Saskatchewan Hospital Services Plan in its Annual Report of 1984–5 (p. 11; italics added):

> The Plan does not pay hospital accounts for individual beneficiaries who require care. Instead it pays *what it considers to be the reasonable net operating costs to hospitals*, that is, the total approved expenditures less revenues from sources other than the Plan. Hospitals supply the services insured by the Plan without charge to the beneficiaries. . . .
> Three primary factors determine the amount paid to each hospital. They are:
>
> • the types of services each hospital *should* provide.
> • the volume of services each *should* provide.
> • the total cost of *efficiently* providing these services.

Each of the italicized expressions gives rise to periodic negotiations between hospitals and governments.

Administration

Provinces approach the task of administration of hospital and medical care services in different ways. Some provinces put one or the other function in the hands of a quasi-independent board or commission, while others maintain responsibility within a department or government, subordinate to a minister. Some, like Ontario, make a single authority (the Ontario Health Insurance Plan, or OHIP), responsible for both hospital and medical services. Others, like Quebec, separate these responsibilities.

In Quebec, the hospitalization program is administered by a division of the Department of Health, so the above-mentioned hospital closings were authorized by the minister, while medical care insurance is administered by the Quebec Health Insurance Board (Régie de l'assurance-maladie du Québec). In all provinces, the bureaucratic nomenclature changes as successive governments reshuffle their departmental responsibilities. In any case, the relevant legislation will specify the responsible authority.

As noted above, in every province and territory, the provision of hospital services is financed mainly out of the general revenues of governments. Ongoing commitments limit the government's room to manoeuvre from year to year; within those limits the government allocates resources in accordance with its economic, social, and political priorities. In a time of fiscal constraint, hospital services are vulnerable. As we shall see in the next chapter, this contrasts with the financing of medical care insurance in provinces, including Quebec, in which medicare is paid for entirely by special earmarked contributions.

One must understand the distinction between the relationship of government to hospitals in the administration of hospital insurance, which pays for services to patients, and the relationship in the authorization and regulation of hospitals as specialized institutions, which is concerned with buildings, capital equipment, safety standards, etc. All provinces have general legislation concerned with hospitals as *organizations*, and also legislation concerned with payment for hospital *services*. In Quebec, hospitals as such are regulated under the clauses related to 'Hospital Centres' in the Health and Social Services Act; payments for insured hospital services are made under the Hospital Insurance Act. The money paid to a hospital through the Hospital Insurance Act, in return for services rendered, would not provide for building expansion or the acquisition of new equipment. And the Canada Health Act commits the federal government to contribute to the cost of hospital services in the provinces, not to the construction of facilities.

Coverage

Both 100 per cent coverage of a province's population and total portability of coverage throughout participating provinces are conditions of federal cost-sharing. Only the exceptional individual is therefore ineligible for coverage anywhere in Canada. In Ontario and British Columbia, which have retained a 'premium' mechanism, some individuals may fall through the administrative cracks through failing to pay their premiums, even though payment is difficult to evade, being compulsory. Both provinces take special measures to ensure that low-income people, especially welfare recipients, are covered.

Residents of one province temporarily in another province are covered by their home province's plan, and vice versa, as required by the federal Act. Of course, a resident of one province who receives hospital services in another may undergo some inconvenience in getting the hospital bills paid by the home province.

HOSPITAL SERVICES: A BIG BUDGET ITEM

Reading the history, one may feel that it took the federal and provincial governments a long time to iron out generally acceptable nation-wide hospital insurance, especially when one sees how highly valued it now is. Malcolm Taylor, in his definitive 1988 study *Health Insurance and Canadian Public Policy*, subtitles his account of the development of national hospital insurance 'The Case of the Reluctant Decision'. Why the reluctance? As Taylor (p. 230) explains:

> It was the largest governmental undertaking since the war and it would require federal-provincial cooperation on a scale never before known. It had been characterized by extraordinary controversy, not only on the question of whether it should happen at all, but in its timing, scope, nature, and shape.

Hospitalization insurance is one of the most costly of all governmental undertakings. Nation-wide, including the expenditures of the federal, provincial, and territorial governments, the total public bill for insured hospital services (i.e., excluding capital expenditures on buildings and equipment) runs to about $25 billion a year—more than Canada spends on national defence, for those who consider that a significant comparison. For over a decade, but with increasing urgency in recent years, governments have been looking for restraint in expenditures, obliging hospitals to provide a continuing high volume of service without expanding facilities, and/or to look for new revenues, such as user charges for some services. These efforts have generated and will continue to generate heated controversy (Swartz, 1977; Stoddart et al., 1994). In fact, cost-cutting measures have been so effective that total hospital services costs in Canada, for the accounting year ending in 1997, were less than for the preceding year, the first time this has happened since 1957.

CRITICISMS AND QUESTIONS

Canada's hospitalization and medical care insurance has become a national institution, a thread in our social fabric to which Canadians point with satisfaction when making comparisons with certain other countries, notably, of course, the United States. At the same time, so large and complex an enterprise, whatever its merits, cannot be flawless. Looking now only at the hospitalization programs, some troubling questions are frequently raised.

(1) Hospitals typically have their own governing bodies and enjoy a defined area of administrative autonomy, and they are allowed some freedom to raise revenues from voluntary donations and from charges for services not covered by public hospitalization insurance. Still, they now draw all but a small margin of their operating as well as capital revenues from public sources. Most cultures have a proverb that says 'He who pays the piper calls the tune.' Thus, hospital authorities complain that they have come increasingly under the control of central agencies of provincial governments. It is alleged that important decisions affecting hospital care are made on bureaucratic and budgetary more than on professional grounds, and without sensitivity to special local circumstances or even to the needs of patients. To a degree, this might be taken to be the predictable response of anyone who surrenders some power to someone else, and whose financial support is reduced, but the allegations are far from trivial.

(2) As noted, governments are trying to restrict expenditures in the field of hospital care, as in other fields. Because hospital costs have been large and rising, hospital care is a likely target for cost-cutting endeavours. And in point of fact, the rate of increase in hospital expenditures began to diminish drastically in the early 1990s. Much of the cost-cutting has been exercised in expenditure categories outside those covered by insurance, but any reduction in facilities or personnel without a reduction in use obviously increases the burden on the facilities and personnel that remain, so insured services have inevitably borne the brunt. Overcrowding of hospital wards, reductions in auxiliary personnel, and long delays for non-urgent operations and treatments have become common. Quality of service is difficult to maintain under such conditions. (With the recent trend toward stripping organizations to their core functions, there has been a corresponding move toward contracting out support services, like catering and laundry, and some technical services, such as some laboratory tests, to private, profit-seeking companies.)

(3) Faced with decreasingly generous public funding, hospital authorities must look to other sources of revenue. The traditional alternative source is the private donor, but many private donations fund new facilities and research, not services. Another potential source of revenue is the user of the service. For non-insured services that they want, users already pay, out of their own pockets or by means of private insurance. At present, it is illegal to charge a fee for an insured service, but the imposition of small user fees for some insured services has been proposed in some quarters.

A user fee might serve a number of purposes: it would raise some revenue, though probably not a great deal; it might deter frivolous use of hospital services, thereby reducing expenditures (which is just as useful as increasing revenues); it would constitute a token, reminding the user that the service is not really 'free', that it actually costs something. Categories of people such as welfare recipients, senior citizens receiving GIS or the full Seniors Benefit, and others to whom paying a fee would represent a real hardship, could easily be excused the user fee. Larger revenues could accrue if hospitals could, for example, collect user fees from patients opting to pay for immediate service rather than wait for non-urgent surgical procedures. However, this also implies a two-tier system, arguably implying better service for those able to pay and inferior service for the others. Unfortunately, many Canadians best able to afford user charges now travel to the United States to obtain 'better service', taking their money out of the Canadian health system altogether. Some critics ponder whether the system would benefit from allowing such patients to pay for the service they want in Canada.

Many people oppose the idea of user charges on principle, arguing that a charge of even as little as five dollars a day would discourage poor people—and not just welfare recipients, who would probably be excused from payment in any case—from using the hospital when they should. Some studies confirm that even modest user charges do indeed have their greatest impact on the use of services by low-income people (Stoddart et al., 1994; Swartz, 1977). Opponents of user charges are not appeased by the promise of special provisions for welfare recipients and low-income senior citizens because that would split the population of users into classes and would, again, create a two-tier system, one for payers, one for non-payers. It is difficult to imagine that the non-paying tier would get the same service as the paying tier. The program would inevitably lose its fundamental character as a service equally and universally available to all as a matter of right.

The notion of user charges for insured, cost-shared services was dampened by the Canada Health Act's provision authorizing the federal government to reduce its health-related transfers to any province by the amount collected in the province in the form of user fees, which left the province no further ahead. In the context of deficit-reducing budgetary policies and continuing debt burdens, coupled with diminished federal financial leverage over the provinces, governments will probably continue to flirt with the idea of user charges for hospital services, and some may go further than flirting. Since this debate will persist for some time, the student should consider his/her own stance on the matter.

(4) Any mechanism that provides and pays for a popular service, with no out-of-pocket charge to the user, is subject to abuse by both providers and users. In the case of hospital services, the service provider, knowing that the bills will be paid, may abuse the insurance program by authorizing unnecessary hospital admissions, by giving more services than are needed, by prolonging hospital stays unnecessarily (not always out of bad will: sometimes

patients may be placed and/or kept in hospital longer than is therapeutically advisable simply because they have nowhere else to go). Users may abuse the insurance in similar ways. They can really only do this with some collusion on the part of service providers, but the latter may have little incentive to refuse the users what they want, and to run the risk of bad feeling or, conceivably, unfavourable media attention. These are instances in the hospitalization field of the problem known to insurance professionals as 'moral hazard'—implicit incitement of the parties to abuse the system. This endemic weakness of insurance programs has been encountered already in the discussions of (Un)Employment Insurance and Workers' Compensation. Users cannot profit financially from abuse of hospitalization insurance—it puts no money in their pockets—but they might derive some benefits in terms of physical or emotional comfort. Public programs are conceivably more vulnerable to such abuse than private schemes, where both users and providers would more immediately feel the financial consequences—for users, higher premiums; for providers, higher costs. Roch et al. (1985) show how hospital use increased in Manitoba in the 1970s, despite the fact that this province had surpassed most jurisdictions in investing in alternative facilities such as 'care homes'.

(5) Hospitals find themselves bearing much of the brunt of the health care consequences of the aging of the Canadian population. It is well known that older people as a group make heavier demands on health care services than younger people. The health care needs of the elderly are age-specific—that is, they are not exactly the same as the needs of the younger. For one thing, older people are, by definition, more subject to chronic diseases. For another, the profile of health care needs among the elderly is influenced by the well-known fact that women live considerably longer, on average, than men. It may well be that existing health care facilities are not, therefore, fully appropriate for the needs of the elderly, but until sufficient appropriate facilities become available, services must be provided by existing facilities—notably hospitals. All hospitals in Quebec, for example, are required by an edict of the government to reserve a specified proportion of their beds for older patients. In all areas of the country, whether obliged to do so or not, hospitals do in fact accommodate disproportionate numbers of older people.

(6) From another, more fundamental, angle, many critics of health policy argue that too high a proportion of health care spending goes to the traditional health care services: hospitals and medical services. This criticism was pointedly expressed in the 1974 Green Paper issued by Health and Welfare Canada, entitled *A New Perspective on the Health of Canadians*. The argument is that a large part of the resources now allocated to hospitals, if diverted to other health functions, could do much more to improve the general level of health than hospital services can possibly do, and in the end perhaps reduce the costs of hospital and medical care. Some analysts charge that the authorities are viciously restricting hospital budgets; other analysts say we are spending far too much on the status quo services of hos-

pitals and doctors. The interested citizen may have some difficulty in reconciling such contrary criticisms, especially when some of the critics appear to favour both analyses.

Later, this book explores two rapidly unfolding fields that take health policy outside the hospital and the doctor's office, namely community health and occupational health.

(7) National concern over the status of the health care system has reached such heights that fundamental changes are envisioned. The print and broadcast media constantly raise the prospect of the impending breakdown of the system (*Maclean's*, 1994). With respect to hospital services, far-reaching changes are contemplated in what a hospital is, what it does, and, above all, how it is funded. At the same time, many quarters fiercely resist any change that would limit the access of Canadians, especially those of low income, to the services to which we have become accustomed (recall the quotation from the National Council of Welfare at the beginning of this chapter). It falls outside the scope of this book to debate various alternatives, but even an introduction to hospitalization insurance in Canada must acknowledge the threats to its existence in its present form.

STUDY QUESTIONS

1. Like all Canadian cost-shared programs, hospitalization insurance is complicated. What is the difference between the federal government's involvement and the provincial government's involvement?
2. In your home province, what are the authorizing legislation, the sources of finance, the benefits offered, the population covered, and the location of administrative authority, for the program of hospitalization insurance?
3. At the federal level, do you know the authorizing legislation, the location of administrative authority, the source of the financing, the conditions on which costs are shared, and the basis on which the federal transfers to the provinces are calculated, with respect to hospitalization insurance?
4. In terms of the Canadian Constitution, how is the involvement of two levels of government explained?
5. Do you see any conceptual difference between the use of the words 'insurance' and 'prepayment' concerning hospitalization insurance?
6. Where would you look for details concerning hospitalization insurance in your province—numbers of patients served, services provided, costs, etc.? In other provinces?

BIBLIOGRAPHY

Because many books and articles deal with both hospitalization and medical care insurance, the bibliography following Chapter Eight covers both topics.

CHAPTER 8

Medicine and Medicare

THE BACKGROUND OF THE MEDICARE ISSUE IN CANADA

Malcolm Taylor concludes his chapter on the national hospital insurance program with these words (1988, p. 238):

> The long ballet that characterized the introduction of a national program of hospital services insurance was now to be repeated over the next decade, as the nation grappled with the issue of medical care insurance.

The medicare 'ballet' indeed resembled the hospital insurance scenario in many respects, notably in that it was danced on ten different provincial stages. As with public assistance and hospitalization insurance, it must be borne in mind that there are in Canada ten provincial and two territorial medicare plans, and the provincial variations are likely to grow more significant in the twenty-first century as current trends toward greater decentralization unfold.

The Place of the Medical Profession

The doctors had played a prominent role in the hospitalization issue, but in the debate over medicare the position and the actions of the medical profession were absolutely central. Like all professionals, doctors resist the encroachment of others on their domain. They have traditionally insisted upon much latitude in defining all aspects of the doctor-patient relationship. They said in the 1960s, and some of them are saying still, that the domination of the financial aspect of the doctor-patient relationship by a third party (meaning government) will ultimately endanger the therapeutic relationship itself. Both defenders and critics of the medical profession blame observed tendencies toward assembly-line medicine on the way public medical care insurance is structured. Others are not convinced that the mechanism by which doctors are paid for their services should interfere with healing, but 'professional autonomy' is always an issue.

Professional autonomy ultimately rests on public confidence. In Canada, in the late 1980s, the doctors may have tarnished their image by certain collective actions that appeared to be unequivocally self-interested, such as the strike in defence of 'extra-billing' by some Ontario doctors (York, 1987, pp. 15–33). Even so, the public at large still evidently reposes much confidence in the medical profession—as much, surely, as it does in the politicians whom the doctors have confronted in the medicare debates. And it is interesting to observe the large measure of autonomy that the medical profession has actually retained in the public health insurance programs of modern Western nations, notably France, Great Britain, and Canada, although the profession everywhere continues to protest about centralized controls.

The High Value Put on Health and Health Care

Related to the preservation of the professional autonomy of physicians, to the method of financing of medicare, and to the public tolerance of high costs for medical services is the exceedingly high value our society places on health, more specifically on health care. 'Health' and 'health care' are not precisely the same thing. One could imagine a society in which health was prized, but in which the practitioners of the healing arts were not accorded a status bordering on the sacred, with budgets to match.

In all countries where the calculation can be made, a high proportion of combined public and private spending is seen to be allotted to health care. Canada rates quite high among industrial countries in terms of the percentage of Gross National Expenditure spent on health care (about 9 per cent). There is even a certain reluctance to suggest any limit to what ought to be spent in this direction. Ivan Illich is one critic who overcomes any such reluctance (Illich, 1975). Aneurin Bevan, the Labour Party Minister who piloted the legislation creating the National Health Service through the British House of Commons in 1946, announced that his intention was 'to universalize the Best.' The pursuit of such a paradoxical, if not oxymoronic, objective is very expensive, for each level that is 'universalized' generates a yet higher 'Best'.

This exalted valuation of health care has significant policy implications. It makes it difficult to restrain health care budgets; and it gives great inertial momentum to whatever happens to be the allocation of resources at any given moment: if it is the Lord's work, how can it be wrong? New programs related to health care must therefore not take money away from established programs—they must seek new money. This conservative force may be all the stronger at a time when governments feel compelled to make relatively less money available—precisely when innovation might be most valuable.

Private Health Insurance Coverage

The introduction of public hospital insurance nation-wide by 1961 did not diminish the interest of Canadians in private health care insurance. By the late 1960s, Canada had developed a substantial private insurance business covering the costs of medical care. In general, the story was much the same as for hospitalization insurance, including the leading place taken by non-profit organizations, notably again Blue Cross, the unevenness in coverage among classes of people, and the occasional inadequacy of the coverage afforded in particular cases.

As between hospital and medical insurance, however, there are important differences in the nature of the risk. (a) Visits to the doctor are more frequent than visits to the hospital. Some visits are fairly routine; they are not an *unpredictable* risk (though of course they still cost money). Other visits, if not routine, do not resonate with the overtones of crisis likely to accompany a visit to the hospital. (b) A doctor's bill for extensive treatment may well add up to a considerable amount, but the cost of most medical services is not likely to be high enough to upset a family's finances for years, as a big hospital bill might do. (c) People usually exercise more choice in deciding whether to go to the doctor than whether to go to the hospital. Once consulted, however, the doctor, not the patient, decides on the treatment and on the number of visits needed, presumably in the best interests of the patient.

These somewhat distinctive characteristics of medical services as a cost item, and the fairly extensive coverage of the Canadian population achieved by private health insurance, influenced the debate that rumbled through the 1950s and 1960s over what was needed in the way of a public program of medical care insurance. The questions were: Does public medical care insurance have to cover the *whole* population, or only those unable to insure themselves? Does it have to cover *all* the costs of medical services, or only the costs beyond what most people could conveniently bear ('catastrophe insurance')?

These questions may have seemed to be settled, for all practical purposes, by the provinces' political decisions of the late 1960s and early 1970s extending public medical care coverage throughout the country, and may seem, therefore, to be of merely historical interest. In the 1980s and 1990s, however, as total costs to the provinces continued to rise and as the proportion of those costs met by the federal government began to dwindle, the share of costs to be covered by the public system, and the sharing of control over practice by public authorities and the medical profession, resurrected themselves as issues. Three decades of experience with medicare have not caused all the questions to go away: quite the contrary.

The 'Medically Indigent'

The unmet needs of the poor were undoubtedly a large factor in the development of medicare in Canada. Society had acknowledged that health care was not just another consumer good, without which poor people would simply have to get along. But before medicare, Canadian society's traditional approaches to health care for the poor had been, to say the least, residual.

Many doctors, for instance, preferred not to pursue poor patients for unpaid accounts or recognized the futility of doing so, particularly in such hard times as the 1930s. 'The medical profession thus bore the brunt of providing medical care to the indigent,' as Malcolm Taylor comments, 'and their stacks of unpaid bills would be empty legacies for their heirs' (p. 4). This was still true, if on a lesser scale, in the 1950s and 1960s. Needless to say, much depended on the sensibilities of the individual physician. Also, poor persons who were reluctant to take something for which they were unable to pay would not benefit from this kind of free care.

It was also a common practice for doctors to charge fees for their services on a sliding scale, according to their assessment of the patient's capacity to pay—not an assessment they are trained to make. The more affluent patients of doctors who used a sliding scale were subsidizing the less affluent, probably unknowingly. (Some more guarded observers looked upon the 'sliding scale' as a euphemism for 'charging what the traffic will bear'.) Sliding scales and unpaid bills benefited those poor people who actually went to the doctor but could not help those whose poverty deterred them from doing so in the first place. The same could be said of free medical clinics in hospitals, also common.

The good and the bad of this pre-medicare order of things is illustrated in a sharp exchange recorded in the *Proceedings* of the Senate Committee on Poverty (which ultimately produced the Croll Report). Senator Joseph Sullivan, himself a practising physician, was stung by the complaint of a representative of the Pointe St Charles Community Clinic, located in a low-income district of Montreal, to the effect that poor people could not get medical attention (Senate of Canada, 1969–70):

> *Senator Sullivan*: Mr Chairman, I would like to comment. . . . As a practitioner for 35 or 40 years and serving in the out-patient department of a large teaching hospital for that length of time . . . I disagree completely with what the witness has said. The poor are receiving services today that they never received before. I am taking care of them, and you are talking theory, not practicality.

> *Mr Wilson* (of the Pointe St Charles Community Clinic): How many people are you not reaching? How many people do not come to out-patient departments?

> *Senator Sullivan*: I know we don't go out on the street to bring them in.

By the 1960s, various provinces had taken measures to ease the burden of medical care for welfare recipients. The costs of such services were explicitly recognized as eligible for federal cost-sharing under the Canada Assistance Plan of 1966–7. This relief was necessarily confined to beneficiaries of public assistance, who were by definition 'in need'. But many who were not welfare recipients, and who would not be considered 'poor' by any rigorous standard, found anything more than the simplest medical services prohibitively expensive. These were the 'medically indigent', sufficiently nearly poor as to be unable to afford desired medical care.

The widespread conviction that at least minimal health care was something akin to an undeclared 'right', to which there should be no financial barrier, thus encountered the reality that, despite all that had been done, a large number of people did feel themselves, in practice, denied this 'right'. It would be next to impossible to distinguish coherently between those 'able to pay' and those 'unable to pay'. Any approach based upon that distinction would have been extremely difficult to implement. And as always, doubt lingered that medical care explicitly provided for 'the poor' would be of the same quality as medical care provided for others. As the cliché has it, 'Services for the poor are poor services.'

AN OUTLINE OF THE HISTORY OF MEDICARE IN CANADA

The bare chronology of the development of medicare closely parallels that of the development of hospitalization insurance, with a time-lag of ten to fifteen years. Again the province of Saskatchewan was the first to act, in 1961, implementing its own medical care insurance program, several years before federal financial assistance was available. Again, other provinces— Alberta in 1963, Ontario in 1965, British Columbia in 1967—enacted medical care legislation on their own. And again, the federal government concluded a lengthy process of debate, partisan manoeuvring, and extremely difficult negotiations with the provinces, by presenting to Parliament, in 1966, its own legislation authorizing cost-sharing with the provinces of the coverage of the costs of medical care. The federal-provincial process had been long and tortuous enough for most tastes, but some histories of medicare suggest that in the end the federal government of the day short-circuited the consultation process, leaving some provinces with the feeling that they had been presented with a fait accompli (Taylor, 1988, p. 375).

The parallel chronology with hospitalization insurance must not be allowed to mask important differences in the evolution of medicare. The resistance of the Saskatchewan medical profession was much more vehement than it had been to that province's pioneering hospitalization insurance. It eventually took the form of a virtual cessation of normal services by physicians, convinced that their professional autonomy was compromised by some aspects of the program. One history of this episode is bluntly entitled *Doctors' Strike* (Badgley and Wolfe, 1967; see also Taylor, 1988, pp. 296–327; MacTaggart, 1973). While this was the only instance where the conflict between doctors and government was played out in such extreme fashion, it was a foretaste of the attitude of the organized medical profession: conceding the necessity of public intervention to assure all Canadians of access to services, the official medical profession has consistently resisted public control of the terms upon which services were to be offered.

An important part of the process leading up to the enactment of the federal Medical Care Act was the work of the Royal Commission on

Health Services, presided over by Justice Emmett Hall, Chief Justice of Saskatchewan. Appointed in 1961, the Commission reported in 1964. The Hall Report firmly recommended a nation-wide system of medical care insurance programs like the hospitalization insurance system—operated provincially under public auspices, with federal financial support, covering the entire population. The Report's recommendations left little room for the private insurance sector in the provision of basic coverage, and explicitly recommended against governmental programs that took the form of subsidizing individuals with low incomes in order to enable them to pay premiums. Provision of health care for low-income people was to be 'institutional', not 'residual'.

The Alberta, Ontario, and British Columbia programs that predated the federal legislation all relied heavily upon private insurance carriers for the administration of the bulk of their coverage. Ontario, in particular, had built its own program on the basis of the wide coverage the private medical care insurance business had already achieved. Continued involvement of private insurance companies was a stumbling block in view of the federal insistence upon clear public accountability for cost-shared programs. When the federal program was enacted, B.C.'s and Saskatchewan's programs were judged eligible for cost-sharing without serious amendments. The absence of Ontario, the most populous province, would have weakened, if not destroyed, the federal program, but the differences were patched up satisfactorily.

The vote in the House of Commons on the final reading of the Medical Care bill was a resounding 177-2, indicating all-party consent at the end, despite the vigour and acrimony with which the issue had been fought. And again, as with hospitalization, all the provinces fell into line before long. Quebec did not come on board until 1970, but certainly not because of any lack of interest—the Castonguay-Nepveu Commission had submitted a special report recommending a health insurance program as early as 1964. New Brunswick's program, the last, came into effect 1 January, 1971.

THE CANADIAN MEDICAL CARE ACT, 1966

The actual terms of the federal Medical Care Act resemble those of the Hospital Insurance and Diagnostic Services Act, and have been preserved in the Canada Health Act of 1984, which, as we have seen, now governs federal financial participation in both hospitalization and medical care services. To qualify for cost-sharing, a provincial program must:

- be under public administration, with clear delineation of the responsibility of the provincial government;
- make specified services uniformly accessible to all insured residents;
- cover virtually the entire resident population of the province;
- impose no residence requirement longer than three months; and

- extend coverage to all Canadians ('residents of participating provinces') in transit.

Originally, the federal government's share of the costs of insured services was 50 per cent of total costs nation-wide, each province to get one-half the national per capita cost multiplied by its enrolled population. In that way, provinces that allowed their per capita costs to rise higher than the national average would receive less than one-half their costs, those that kept their costs below the national average would get more than half. The initial medicare formula was thus somewhat different from the terms for cost-sharing of hospital insurance, though still committing the federal government to meeting half the total costs nation-wide.

Everyone concerned fully expected costs to rise somewhat, as previously unmet needs for care were met, but the rate of increase exceeded expectations. In time, as was indicated earlier in the account given of the history of hospitalization insurance, the federal government altered the basis of its funding under the Established Programs Financing (EPF) clauses of the formidably named Federal-Provincial Fiscal Arrangements and Federal Post-Secondary Education and Health Contributions Act of 1977 (now, along with federal subsidization of public assistance, rolled into the Canada Health and Social Transfer).

This complicated Act is dealt with in Appendix C on pages 298–300. Its most significant characteristic is clear: whereas previously federal contributions had been tied to the level of *total costs*, which were in effect determined by the provinces, now the federal transfers became mostly 'block grants', wholly independent of actual expenditures. The block grants were/are largely determined in accord with each province's measured *capacity to raise revenue* (i.e., its economic performance) and its population. This gives the transfers a strongly equalizing character.

The conditions attached to federal cost-sharing remained in force under EPF. But the nature of federal surveillance of provincial programs was transformed, becoming more strictly financial. The Hon. Monique Bégin, Minister of National Health and Welfare in the 1980s, tells us of her chagrin in finding that the number of federal civil servants concerned with overseeing health insurance for the federal government had diminished drastically after EPF, weakening the federal authority's ability to enforce its criteria (Bégin, 1988).

The change in the basis of federal financial support is far from being a mere technicality. The provinces must cope with possible rises in medicare, hospitalization, and certain other costs without automatic matching rises in federal transfers. It does not follow that every province was necessarily worse off under EPF than under the 50-50 regime—in fact, in the first few years, most provinces were somewhat better off. But overall, federal transfers did not rise as fast as health insurance expenditures. The Canada Health and Social Transfer, announced in the federal budget of 1995, reduces the amounts of the federal transfers, at least for the next few years, although with a guarantee of a minimum transfer of cash (as of late 1996). If provinces manage to reduce their health expenditures, as they actually did in 1996–7, their CHST grants are unaffected.

In 1984, the Canada Health Act became the statutory authority for federal cost-sharing of medical care costs, as well as hospitalization costs. The Canada Health Act made one further significant alteration in federal financial participation in medicare. At the time, the health insurance authorities of some provinces had imposed user charges for some services, and some had allowed doctors to charge patients extra fees over and above the fees paid through the public program. If widespread, this practice might have endangered the 'universal access' condition attached to federal cost-sharing. Under the Constitution, the federal government is not empowered to intervene directly in the administration of health care; what the federal government did was to reduce its transfer payments to provinces by the amounts collected by physicians through so-called 'extra billing' and by hospitals through user charges for insured services. In this way, a province's medical care system was left no better off after the extra billing. This arm-twisting stratagem seems to have been quite effective. The practice has now been prohibited in almost all provinces. The provinces in general resisted this *démarche* on the part of the federal government; the resistance of the medical profession was fierce, culminating in a virtual strike by Ontario doctors in 1986 (Taylor, 1988, pp. 444–62).

MEDICARE IN QUEBEC AND ELSEWHERE

Quebec's involvement in medicare coincided with the fullest flush of the Quiet Revolution. Through the 1960s, successive governments of Quebec had indicated interest in the creation of a public health care insurance system; it was seen, however, in the context of the province's full exercise of its constitutional autonomy. There was accordingly much ambivalence about Quebec's linkage with the federal medicare cost-sharing program— not that Quebec was in the least reluctant to claim its share of federal financial transfers, but the province was jealous of its right of control.

In 1964, while the issue was brewing throughout Canada, the provincial government obtained from the then sitting Commission of Inquiry into Health and Social Services (the Castonguay-Nepveu Commission) a report on health insurance, which provided the basis for Quebec's medicare structure. In short, the Report produced evidence to show that, by most of the conventional measures (infant mortality, life expectancy, incidence of infectious diseases), health conditions in Quebec were considerably worse than in Canada as a whole; that health standards and access to health services were exceedingly unequal, as between economic classes and as between regions; and that the proportion of people in Quebec covered by private health insurance was very low, compared to most of the rest of Canada, notably Ontario. The Report recommended a tax-supported system for the provision of medical services on a universal basis—i.e., one system for everybody (Taylor, 1988, pp. 387–8; Quebec, Commission of Inquiry, vol. I, 1964).

Legislation
The few provinces that did not have medical care insurance legislation already on their books by 1968, the year in which the federal Act came into effect, all soon did. The Health Insurance Act of the province of Quebec was passed in November, 1970. Before bringing the program itself into existence, the Quebec legislature enacted the statute that authorized the creation of the responsible administering body, the Quebec Health Insurance Board. This Act, the Quebec Health Insurance Board Act, should not be confused with the Health Insurance Act itself. Not all the provinces considered it necessary to enact the creation of the administrative body separately from the creation of the program, but all provinces must in their legislation specify the body responsible for any program.

Services
The services to be covered by the health insurance program of each province are listed with varying degrees of specificity in the respective provincial Acts. As with the hospitalization legislation, the services covered by provincial programs conform closely with the services for which cost-sharing is available through the Canada Health Act and regulations made on its authority. Provinces have a natural preference for services for which they will be subsidized, over services for which they pay the entire bill themselves.

Again using Quebec as an example, the covered services are listed in broad terms in Section 3 of the Health Insurance Act. Included are all 'medically required' services rendered by physicians, optometry services, some oral surgery, dental services provided to persons up to a certain age (to be specified in Regulations; the age has moved upward through the years), and prescribed medications and prostheses for welfare recipients and recipients of GIS. The Act gives to the provincial Cabinet the power to make Regulations specifying more precisely the 'medically required' services that are covered. The Regulations in force actually do as much to specify *excluded* services—medical examinations for employment and for life insurance coverage, services for aesthetic purposes, etc.—as to list covered services. All provinces have broadly similar lists.

No private agency is permitted to insure people anywhere in Canada for services covered by medicare. This has not prevented private organizations, notably the aforementioned Blue Cross, as well as many insurance companies, from doing considerable business insuring people against the costs of services and embellishments not covered by medicare. Coverage through such insuring organizations of all or part of the costs of dental and orthodontal treatments, eyeglasses and other aspects of optical care, and physiotherapy treatments, as well as costs of care incurred abroad, are fairly common fringe benefits offered by employers and/or included in collective bargaining agreements. These and other variants of supplementary health insurance may be considered 'private', but they have an important public aspect in view of the favourable tax treatment given both the costs of insurance (especially to employers) and expenditures on health care services by individuals.

Paying for Services

Throughout Canada, medicare operates mostly on a fee-for-service basis— that is, physicians are paid a fee for each service rendered, according to fee scales agreed on by each province's professional medical corporations and its health insurance authority, in Quebec's case, the Quebec Health Insurance Board. Doctors are not permitted to collect any additional sums from patients. Fee-for-service is the prevalent form of payment for medical services throughout the Western world, whether publicly insured or not.

Provinces have experimented with other ways of paying physicians. Some doctors work for salaries in health centres or other organizations. Nation-wide, salaries represent an increasing proportion of the total incomes of doctors, though they are still much lower than fees. In Quebec, newly certified doctors are *required* to work part-time in public health care organizations. Other doctors work for intermediate bodies (in Ontario, Health Sustaining Organizations) that enrol clients and clients' families for an annual subscription fee that entitles the client to as much or as little service as is required. This mode of remuneration is known as 'capitation', because payment is calculated per capita, not per unit of service. This pattern is common in the United Kingdom.

A physician in some provinces may opt to work entirely outside the medicare system; all patients will pay him/her out of their own pockets, although they will continue to contribute to their provincial health insurance plan. The physician will submit no bills to and receive no fees from the health insurance authority. Only a limited clientele is willing to pay for its medical care in this way, and so only a handful of physicians operate in this independent fashion.

Financing Medicare

As we have observed, the federal Medical Care Act originally committed the federal government to pay half the *total* costs of covered medical services in all participating provinces, with a twist in the formula that benefited any province whose per capita costs were below the national average, i.e., the subsidy was never exactly half the costs in any one province.

At about the same time as the preparation of the Medical Care Act, the federal government offered to restructure its fiscal transfers to the provinces on the basis of each province's fiscal capacity (i.e., estimated ability to raise taxes) rather than on the basis of actual expenditures. Quebec alone voluntarily took advantage of this offer in the late 1960s, so medicare in Quebec has received its federal contribution through EPF from the beginning.

All the other provinces stayed with the modified 50-50 formula until, in the late 1970s, the federal government gave them notice, as was provided for in both the hospital insurance and medical care insurance Acts, that it was replacing the original 50 per cent sharing by EPF. The proportion of the health care costs of any province paid for by the federal subsidy was thereafter likely to vary slightly from year to year, and could not therefore be stated precisely in advance. Beginning in 1997, the federal share will come wrapped in one bundle, the Canada Health and Social Transfer, that will include, without differentiation, transfers for health, social assistance, and post-secondary education. Apparently, it will be impossible to isolate what proportion of health care costs will have been met by the federal government.

The provinces finance their shares of medical care expenses in various ways, but the basic alternatives are special premiums, earmarked taxes, and general revenues, or some combination of these. Nation-wide, general revenues are the major source of financing: in Ontario and British Columbia, the only provinces that now collect premiums, by far most of the costs of their programs are paid out of general revenues.

Quebec finances its share of the costs of medicare by special earnings-related 'Health Fund Contributions' paid by employers, as explained in the discussion of provincial hospitalization insurance programs in Chapter Seven. To reiterate, individual taxpayers with incomes from sources other than employment make additional contributions to the Health Fund. About half the Fund goes to the Health Insurance Board, the rest to the provincial Department of Health to pay part of the costs of insured hospital services.

Administration

About half of the provinces have chosen independent board or commission administration of medical care insurance—more than in the case of hospital insurance, probably because of the more dominant position of the medical profession with respect to medical care. The doctors have wanted and have usually obtained an administrative structure overtly more independent of government.

Quebec exemplifies this point: while hospital insurance is operated by a branch of a regular department of government, the administration of health insurance is entrusted to a specially appointed commission, the Régie de l'assurance-maladie du Québec (translated somewhat curiously as the Quebec Health Insurance Board). The membership of the Board, and the assured terms of office for which the members are appointed, are set forth in Section 7 of the Health Insurance Board Act. The members must include representatives of the medical specialists and general practitioners and of health professions other than medical, plus representatives of labour, business, the hospital sector, and consumers. The Board handles the finances of the program, negotiates fee schedules with the medical profession, oversees quality of service, deals with complaints, and so on.

In Saskatchewan, the law requires a minimum of three physicians on its Saskatchewan Medical Care Insurance Commission, but some governments have shown the profession much more deference than the law requires: in 1988, out of eleven members, eight were doctors. In terms of administrative structures, Saskatchewan follows a pattern different from most, in that an 'approved health agency' may be the intermediary between the Commission and the practising physician. In these features may be seen the reflection of the history of health care insurance in Saskatchewan: the province's reliance on locally initiated municipal and regional organizations for the management of health care and other social requirements; the doctors' strike when medical care insurance was initiated in 1961.

Eligibility and Coverage

Coverage under medicare in most provinces is simple. Virtually every resident of each province is covered for that province's insured services. Some technicalities exist: in Quebec, for example, to be eligible one must obtain a health insurance card, but all that is required is proof of residence. By contrast, Ontario and British Columbia, which finance their programs partly through payment of special non-tax premiums, encounter some administrative difficulty in determining eligibility with respect to people who could not, or do not, pay their premiums. Recall that federal cost-sharing is dependent on coverage of virtually 100 per cent of the resident population.

In principle, a resident of any province is covered for services provided in any other province. Certain administrative problems accompany the handling of service to non-residents: from province to province, fee schedules vary, and billing and accounting practices vary as well. At present, interprovincial agreements exist among nine of the provinces (excluding Quebec), making it just as easy for people to get medical attention away from as in their home province. Quebec has a bilateral agreement with Ontario to cover the most frequently occurring instances of cross-province provision of services, but Quebec has been reluctant to commit itself to a coast-to-coast system of binding arrangements, on the grounds that the ensuing complications would inevitably limit its ability to modify its own health policies in future.

THE LIMITED POTENTIAL BENEFITS OF MEDICAL CARE: IS IT A 'SOCIAL GOOD'?

As noted above, society accords extraordinary status to health as a value and to the medical profession as a social institution. Yet the medical profession itself, when speaking officially, acknowledges narrow limits on its ability to affect the general health of the population. Of course a doctor will help individual patients to improve their standard of health, but the efforts of a doctor rarely affect the health of anyone but that one patient. By comparison with factors such as nutrition, hygiene, purity of food and water, road safety, work safety, fitness, and many others, medical care is a relatively modest contributor to the health of a population.

To indulge in a brief theoretical digression, this question of the contribution of medical services to overall health tests the extent to which medical care may be considered to be what economists call a *public* or *social good*. The term *social good* does not mean just something that is good for society. A pure social good is one that, by its very nature, is necessarily produced for *a whole population* if at all, a good that cannot be made available to one person to the exclusion of others, or that can do little good for one person without more or less equally benefiting others. The best examples of social goods are things like pure air, traffic safety, and public security (as provided by police forces): it is virtually impossible for you and me to enjoy them unless everyone around us enjoys them, too. The classical economist's example was the protection afforded ships by a system of lighthouses: if one fee-paying ship could be guided to safety by the lights, so could all others, without paying; it was therefore both impractical and inefficient to try to sell the service to each user individually. Some enthusiasts for free-market solutions to issues make much of the fact that lighthouse services once *were* in fact 'sold' to users; but that does not invalidate the proposition that ship safety is a public good, it only tests it. In the end, private pricing of lighthouse service was found to be futile.

The opposite of a social good is a 'private good'. My enjoyment of a chocolate bar or a microwave oven almost certainly benefits no one but myself or my immediate family circle; these are 'private goods'. In a complex society, there are only a few purely social or purely private goods, but most goods, while falling somewhere between, tend toward one class or the other.

If each individual were left free to decide for himself/herself whether to pay his/her share of the cost of a social good (for example, unpolluted air), the individual might rationally be inclined *not* to pay, since in practice it is impossible to exclude him/her from enjoying the benefit, once it was provided. He/she would cheerfully become what economists call a 'free rider'. But if many people evaded payment in that way, the cost of providing the good could not be met, *even though everyone really wanted it*. For the most part, therefore, the only way social goods can be adequately and reliably paid for is through compulsory collection devices like taxation. By contrast, private goods may be left to individual market decisions without endangering or depriving society, since one person's decision to buy or not to buy has little or no effect on the well-being of others (Barr, 1993, pp. 106–7). The identification of a good as *private* or as *social* will therefore logically help to determine how it ought to be financed: through the market by the prices paid by consumers, entirely through taxation, or by some combination of these. In the health field, measures to immunize against contagious diseases, to maintain public sanitation, and to safeguard the purity of food and water are clearly 'social goods'; services that affect only the health of the individual recipient are closer to the 'private goods' end of the continuum.

Where is medical care situated in this private-social continuum? It is not in the same class as control of epidemics or surveillance of the purity of food, which are unambiguously social goods.

(1) To the extent that universally accessible medical care does prevent the spread of illness, enhance the welfare of society by improving productivity and/or the general quality of life, or save society as a whole some discomfort or expenditure by maintaining health, it is a social good. Not to belabour the point, there is some question about how much good medical care can do along those lines (see, once more, the statement of the National Council of Welfare at the beginning of Chapter Seven).

(2) To the extent that the members of society genuinely feel revulsion at the thought that other people could not afford to go to the doctor when they needed to, and are willing to pay to avoid the revulsion, universal medical care is a mixed social and private good. It is social, because only by sharing the burden can we attain the generally desired outcome of knowing that no one will be denied services through inability to pay. It is also private, because I am ready to pay for the protection I get as an individual, and do not expect it to do much good for anyone else. Collective compassion provides its own reward, a social good. Seen in this way, medicare is partly a justifiable form of redistribution through benefits 'in kind'—medical care—as an alternative to benefits in cash. There can be little doubt that the altruistic social value of assuring access to health care for those with limited ability to pay has played an important part in the promotion of medicare in Canada, and continues to play an important part in its defence. It is rarely argued that the only or the best way to provide medical care for the wealthy is through a universal public program, which would be the argument if medical care were seen as a true 'social good'.

(3) To the extent that what Canadians are getting through medicare is simply personal insurance against one specific financial risk, entirely independently of any impact it might have upon others, then the financial protection is the good attained, more than any improvement in public health that might result. Thus viewed, medical care is closer to the 'private goods' end of the scale. Some participants in recent debates, seeing medicare squeezed between steadily rising costs and limited governmental resources, have emphasized the 'private good' (financial security) aspect. They have made proposals whose main thrust is to improve the cost-efficiency of the insurance, for example, by putting a limit on the number of services a person may get free of charge or by charging the patient a co-payment in

some circumstances. Not to debate their merits here, measures that might improve medicare's performance as an insurance policy against medical care costs, considered as a private good, would probably run counter to maintaining its performance as a mechanism for redistribution through benefits 'in kind'.

This digression is less academic than it might seem, as may be seen if we allow ourselves to get a little ahead of our story. The debate preceding the initiation of medicare *did* revolve to a great extent around whether the primary objective was to assure medical services to low-income people, or to protect the entire population against the financial hardship of medical expenses, or to spend money efficiently in the pursuit of public health. Advocates of medicare put forward all of these arguments.

After medicare had been in effect for a few years, the federal government, faced with the burden of paying half the rising costs of hospital and medical services, raised the question of whether Canadians were getting a reasonable return on the sums they were spending for medical and hospital services, compared with other possible health-related investments having much more clearly the character of 'social goods'. This question is raised in the 1974 Green Paper published by Health and Welfare Canada, *A New Perspective on the Health of Canadians*. It was partly on the basis of this concern that the federal government justified its action in altering the way it shared costs with the provinces, as we have seen in our discussion of hospitalization insurance. This has moved the provinces in turn to restructure their expenditure policies concerning their respective medicare and hospitalization programs. It is difficult to say how much of the restructuring is based on the analysis of the real benefits of medical care, and how much on the felt imperative to reduce deficits.

To separate reason from rhetoric, it is important to understand that the concept of *social goods* is an analytical tool, not a political slogan. The fact that one person considers something to be a *private* good and another person considers it a *social* good does not mean that the former is more selfish and the latter more public-spirited; it only means that they have different perceptions of the way in which the production and consumption of the good in question should be organized in order to yield, in the end, the greatest satisfaction for society. In conclusion, the concept of 'social good' is relevant to every proposal for public expenditure (Evans, 1984, ch. 3).

CRITICISMS AND QUESTIONS

(1) In Canada, medicare absorbs billions of dollars annually: more than $10 billion in recent years, close to $30 million a day. This is a great deal less than is spent on hospital services, but is still a considerable amount. Naturally, most of those billions would be spent through private channels whether medicare existed or not—perhaps even more would be spent: Canadian commentators are fond of pointing out that the United States,

without any universal programs like our hospital insurance and medicare, spends a larger proportion of GNP on health care services than does Canada (about 14 per cent to about 9 per cent). But the sheer size of total expenditures on medical services and on hospital services, by comparison with expenditures on other health-related programs, allows a reasonable person to look for confirmation that the money is being well spent.

(2) Given that medicare was intended to make it easier for Canadians to obtain medical services, it has been no surprise that demand for and expenditures on medical care have risen. Much of the increase in service demand is accounted for by the aging of the Canadian population. Even so, some observers claim that some of the cost increases are the result of overuse of the insurance program: 'Supply seldom outstrips demand when insurance creates a zero price market' (Naylor, 1986, p. 13). Only by the exercise of administrative controls that many people, including doctors, find repugnant can there be any deterrent to overuse of services by patients. Similarly, only by disciplinary measures can there be any control over a professional who abuses the program by providing excessive or unnecessarily expensive services.

In some cases, overuse of the program may be the consequence of a tacit complicity between doctor and patient, with no manifest ill will on either side. Rather than risk the unpleasantness of telling patients to go home and look after themselves, the doctor may simply repeat perfunctory services (York, 1987, p. 191). The patients are satisfied, though perhaps no healthier; it costs them nothing directly, though indirectly it will cost them something as taxpayers; and the doctor, even if professionally queasy, collects the fees for the services. Needless to say, it is difficult to prove conclusively the occurrence of abuse of this kind or to estimate its extent.

This point is another instance of the issue known to insurance professionals as 'moral hazard', the problems that arise when an insurance program creates opportunities for parties to exploit the program in their own interests, at the expense of others.

(3) The issue of extra billing, though apparently buried by the Canada Health Act, may rise from its grave as provincial treasuries become more strained (Barer, Bhatia, Stoddart, and Evans, 1994). The continuing importance of the extra-billing controversy lies in the questions which it generated: Does complete central provincial control of the levels of fees paid to physicians threaten the quality of service offered? If in fact medical care is in danger of being underfunded from present sources, where is new money to be found? Is it unreasonable to expect some of the extra money to come from patients at the time of service delivery—*some* patients, for *some* services? Or would that truly compromise the 'universal access' that has been the hallmark of medicare since its inception? Virtually every study of the impact of extra charges shows that they fall most heavily upon the poor, because the poor are deterred from seeking care. Is one possible answer to

continue 100 per cent coverage, as is now the case, but to reduce the range of services that are covered? (Stoddart, Barer and Evans, 1994).

(4) One might ask whether the fee-for-service basis, which is preserved under medicare, is still appropriate as the predominant pattern for remunerating physicians. The question has been around much longer than has medicare. George Bernard Shaw, in the preface to *The Doctor's Dilemma*, put the issue with accustomed flair: 'That any sane nation, having observed that you could provide for the supply of bread by giving bakers a pecuniary interest in baking for you, should go on to give a surgeon a pecuniary interest in cutting off your leg, is enough to make one despair of political humanity.' Is fee-for-service remuneration, as some medical spokespeople claim, a necessary support for professional autonomy? If salaried, would doctors become mere bureaucratic dispensers of service and their patients mere customers? Or would salaried physicians provide *better* service, spending more time with each patient because they would have no financial incentive to see as many patients and provide as many services as they can? (York, 1987, pp. 99–117).

The fee-for-service mode of payment makes feasible the now widespread phenomenon of privately operated walk-in clinics, where patients can see a doctor quickly, without appointments, and often without any ongoing contact between doctor and patient. Because of the easy comparison with retail trade, and because of the location of many such walk-in facilities, this kind of practice is derisively labelled 'mall medicine' by its critics. Short of close surveillance by a central authority, there is nothing to prevent a client from consulting several physicians simultaneously. Under such circumstances, one must rely on the professional ethics of practitioners to see to it that the health interests of individuals are being properly served.

Given a publicly financed system such as Canada's, discussion of the rights and wrongs of fee-for-service payment of doctors has nothing to do with the patients' ability to pay. The issue is simply whether fee-for-service payment is the best way to assure effective and economical care.

(5) A recurring problem everywhere in Canada has been that of the geographic distribution of medical services. 'Equal access' to services may be a hollow phrase for some people, if services are too far distant to be accessible at all. Year after year, doctors in outlying regions have realized larger incomes than doctors in urban centres. Yet it is still difficult to attract doctors to the more remote areas. The offer of income supplements and free facilities by some localities has had limited effect. Quebec and other provinces pay considerably higher fees for services provided by doctors in less populated regions. Even so, Quebec finds it necessary to oblige new medical graduates to practise for a few years in the less favoured regions as a condition of licensing in Quebec (in return, it is hinted, for the heavy public subsidization of their medical education); this is still resisted by the profession as an intrusion on freedom of practice.

(6) Some criticisms of medicare come from doctors. While aware that with medicare all their bills are paid, doctors nevertheless resent the rigidity and, many claim, the insufficiency of the fee structures under which they work, the regulation of their practice that has accompanied medicare, and the diminished freedom to engage in work more attuned to their personal professional interests.

(7) The role and status of the medical profession have come under some pressure as medicare has evolved. Public statements of professional spokespeople often focus on the erosion of the autonomy of the professional to practise as he/she sees fit and on the consequent threat to the quality of medical care. Critics of the profession perceive in this defence of professional autonomy a large admixture of material self-interest. Socially speaking, the issue of the role of the profession transcends the question of the relative selflessness and selfishness of doctors as individuals. The more important point is that, over a long period of time, the profession has evolved into a major social institution, upon whose self-regulated standards of practice and self-discipline society has come to rely for the exercise of exceedingly important social functions. Society may both gain and lose by a change in the role and status of the medical profession (and other professions). The net balance of gains and losses is not a trivial calculation.

(8) Financially squeezed provincial governments have explored alternatives to the dominant model of solo, fee-for-service medical practice. The next chapter looks at community health centres. These vary so greatly from location to location that it is impossible to define them with any rigour. However, they generally are organizations in which doctors collaborate systematically with other health care providers, and are paid on a basis other than fee for service, thereby short-circuiting any incentive to provide superfluous services, treatment, and consultations. Also, there is generally some kind of community and/or consumer input in the management of the organization.

The concept is most fully realized in Quebec, where 'Local Community Service Centres' (CLSCs), created by provincial legislation, now blanket the province. It cannot be claimed, however, that the population has transferred its allegiance from doctors' offices to CLSCs even for the basic, primary care the CLSCs are supposed to provide. Other provincial governments have given some encouragement to similar organizations; in Ontario, modest support has been offered 'Health Service Organizations', modelled in part on the American 'Health Maintenance Organizations'. While Canadians are reluctant to acknowledge that they have anything to learn about health care from the United States, some reports show favourable results in terms of both cost containment and improvements in health. Other reports decry the HMOs' restrictions on physicians' practices and on clients' access to service, a reflection of the organizations' eye to the bottom line.

It must be acknowledged that governments run a certain risk in pushing too hard for change. Geoffrey York, a severe critic of the prevailing system, comments: 'Ironically, the tremendous popularity of Canada's medicare system is helping to protect doctors from government regulation. The public perceives any proposed reform as a possible threat to medicare' (1987, p. 180; see also Leeson, 1995).

(9) The student of social policy may be interested in medicare's character as a *universal* service program. Its universality is defended as (a) meeting a need that may strike any member of any class of society and that many might find difficult to afford without 'insurance', and (b) enlisting all classes, including the more influential, in its support and in the surveillance of its standards, so that no class of society is obliged to accept second-rate services.

All the same, what of the proposition that medicare is in part an implicit transfer 'in kind', that is, a transfer to the less well off in the form of a specific service rather than a redistribution of money? Upper-income people contribute heavily in taxes and contributions, but analyses show that they use the services disproportionately, thus diminishing somewhat the implicit transfer in kind. And low-income people pay their part of the costs. They contribute to federal and provincial government revenues in various ways. But taken all in all, it seems likely that medicare represents a valuable transfer of resources from the wealthy to the poor (Boulet, 1978; Boulet and Henderson, 1979; Manga, 1978; Swartz, 1977). If so, would that affect one's view of the principle of universality in health care (same services to all, on the same terms)?

(10) Finally, as with all of Canada's social programs, an introduction to medicare must take into account the current fiscal difficulties of the federal government and the provinces. For whatever reasons, good or bad, governments have acted to bring their expenditures into line with their revenues. The federal transfers that support provincial health care programs have decreased. Yet, health care costs have continued to rise, though hardly at a rate that could be described, in the recurring journalistic jargon, as 'spinning out of control'. Only in the early 1990s did costs begin to recede slightly in some provinces (Health and Welfare Canada, 1994). Only in 1996–7 did total health care costs fall from one year to the next. As noted in Chapter Seven, several provinces have reduced hospitalization costs by drastic measures such as closing hospitals. Similarly drastic measures are being undertaken with respect to the costs of medical care, such as absolute limits on the total fees that a practitioner may claim in one year, and, in Quebec and British Columbia, controls limiting the numbers of new doctors allowed to practise. A province's medical care expenditures depend on the number of services performed by doctors and the fees charged for those services. The only way a province can control its medical care expenditures is by trying to control those factors. This is very different from the controls exercised in the field of hospital services, where the provincial authority can budget a total allocation for a year and leave the necessary adjustments to the hospitals.

Canada's health policy has been subjected to many critiques over the years, not the least incisive being those from public sources, such as Health and Welfare Canada's *A New Perspective on the Health of Canadians.* Nonetheless, the financial crunch of the 1990s has had a more profound effect on policy than any technical or scientific analyses. It remains to be seen whether the changes wrought by cost-cutting will serve what analysts identify as our real, long-term health needs.

STUDY QUESTIONS

1. Critics of medicare in Quebec say that the provincial part of the financing is based on a special 'regressive income tax'. Do you understand this criticism? How would you discuss it?
2. The text contrasts the financing and the administration of hospital and medical care insurance in some provinces, including Quebec. Do you understand the differences? Who administers the two programs in your province? How does the province collect the money to pay its part of the costs?
3. Explain the meaning of 'fee-for-service' medical care. What are the perceived problems with fee-for-service medicine, even when the fees are all paid by medicare? What alternatives might exist for the remuneration of doctors?
4. Where would you look for information concerning the operation of medicare in your province: patients served, services provided, costs, and so on?
5. Why is it considered appropriate in many provinces for medicare to be run by a commission? Are the reasons usually given valid?
6. Is there any such thing as 'extra billing' in your province?
7. How has the initiation of the Canada Health and Social Transfer affected hospitalization and medical care services?

HOSPITALIZATION AND MEDICARE LEGISLATION

For historic reasons, the legislation of Quebec, Saskatchewan, and Ontario is cited here. All provinces have their own hospital and medical care legislation, and there is no suggestion that the legislation of the other provinces is less important than that of these three.

Canada. Canada Health Act, 1984. Now *Revised Statutes of Canada, 1985,* Ch. C-6. Supersedes the following two historic Acts.

—————. Hospital Insurance and Diagnostic Services Act. Was *Revised Statutes of Canada, 1970,* Ch. H-8 (first enacted 1957). See especially Articles 2, 5, 7, 8.

—————. Medical Care Act. Was *Revised Statutes of Canada, 1970,* Ch. M-8 (first enacted 1966). See especially Article 4.

Ontario. Ontario Hospital Services Commission Act. *Statutes of Ontario, 1956.* Ontario Hospital Services Commission Amendment Act. Statutes of Ontario, 1957. These statutes together brought Ontario into line with the federal government's 1957 hospital insurance legislation. The Act has been amended often since 1957.

————. Medical Services Insurance Act. *Statutes of Ontario, 1965.* This was Ontario's initial Act in this field. It has been significantly amended over the intervening years. It is of special interest because of the relationship that was at first set up between the private health insurance network and the Province's health insurance administration.

Quebec. Health and Social Services Act. *Revised Statutes of Quebec,* Ch. S-4.2 (fundamentally amending the first Act, enacted 1972). Articles 81 and 132 on Hospital Centres set forth the legal and administrative framework for hospitals in Quebec. Note that this Act concerns hospitals as *organizations,* and not hospital *services.*

————. Health Insurance Act. *Revised Statutes of Quebec,* Ch. A-29 (first enacted 1970). See especially Articles 3 (services), 56 (regulations), 63-76 (contributions).

————. Hospital Insurance Act. *Revised Statutes of Quebec,* Ch. A-28 (first enacted 1961). See especially Articles 1, 2 (coverage), 4 (excluded institutions), 7 (regulations).

Saskatchewan. Hospital Services Insurance Plan Act. *Statutes of Saskatchewan, 1946,* Ch. 51. This Act is cited here in its original form, in recognition of its place in history as the first piece of legislation of its kind to be implemented in North America. The Act came into force 1 January 1947.

————. Medical Care Insurance Act. *Revised Statutes of Saskatchewan, 1978,* Ch. S-29. As with the cited Ontario statute, this was Saskatchewan's independent legislation, preceding the federal Medical Care Act. It has been amended often since 1978.

BIBLIOGRAPHY

Armstrong, Pat, and Hugh Armstrong (1996). *Wasting Away: The Undermining of Canadian Health Care.* Toronto: Oxford University Press. Chapter 2. This book is profoundly critical both of current policies of cost reduction in health care and of the prevailing policy bias in favour of medical and hospital services. This chapter puts forward and then seeks to demolish four assumptions said by the authors to underlie Canada's conventional health policies. The reader should adopt the same questioning attitude toward the critique as the authors do toward the assumptions.

Badgley, Robin F., and Samuel Wolfe (1967). *Doctors' Strike: Medical Care and Conflict in Saskatchewan.* Toronto: Macmillan.

Barer, Morris, Vanda Bhatia, Greg Stoddart, and Robert Evans (1994). *The Remarkable Tenacity of User Charges.* Toronto: The Premier's Council on Health, Well-Being and Social Justice. Ten years after the Canada Health Act was supposed to sweep away user charges in health care, a formidable team of analysts examines their persistence.

Barr, Nicholas (1993). *The Economics of the Welfare State.* Stanford, California: Stanford University Press. See pp. 81–2 and 106–7 for concise, perhaps slightly technical explanations of the concept of 'public goods' (what we have called 'social goods').

Bégin, Monique (1988). *Medicare: Canada's Right to Health.* Ottawa: Optimum. The political battles around the Canada Health Act of 1984, in which the Hon. Mme Bégin was involved as Minister of National Health and Welfare. She sees the battle as having saved universality of access to public medical care coverage.

Blishen, Bernard (1991). *Doctors in Canada.* Toronto: University of Toronto Press. A sociological view of the role and status of doctors and the medical profession.

Blømkvist, Ake (1979). *The Health Care Business.* Vancouver: Fraser Institute. Argues that outcomes would improve if market forces in health care were given freer play. The Fraser Institute consistently espouses the view that maximum market freedom produces socially optimal outcomes.

Boulet, Jac-André (1978). 'L'État au service des moins fortunés'. *Perception* (November-December). Analyses data to show that low-income groups receive the largest proportion of services from hospital insurance and medicare.

Boulet, Jac-André, and W. Henderson (1979). *Distributional and Redistributional Aspects of Government Health Insurance in Canada.* Economic Council of Canada Discussion Paper No. 146. Ottawa: The Council.

Brown, Malcolm C. (1996). 'Changes in Alberta's Medicare Financing Arrangements: Features and Problems', in Michael Stingl and Donna Wilson, eds. *Efficiency vs. Equality: Health Reform in Canada* (cited below).

Caledon Institute of Social Policy (1995). *Can We Have National Standards?* Ottawa: The Institute. The answer to the question is: not with the assurance we had before the initiation of CHST.

Canada (1980). *Canada's National-Provincial Health Program for the 1980s: A Commitment for Renewal.* Report of the Special Commissioner, Hon. Emmett M. Hall, to the Minister of National Health and Welfare. (French and English, back-to-back.) Justice Hall, Chair of the Commission whose report on health services in 1964 helped pave the way for Canada's medicare program, reported in 1980 on how well health services were functioning. (See citation of Justice Hall's 1964 Royal Commission Report, below.)

Canada. Dominion Bureau of Statistics (1953 and subsequent years). *Canadian Sickness Survey 1950–51,* etc.

Canada. Health and Welfare Canada (1974). Marc Lalonde, Minister. *A New Perspective on the Health of Canadians.* (French and English, back-to-back.) This extremely interesting analysis questions the health priorities pursued by governments over the preceding twenty-five years. Without saying so, this document built the case for Established Programs Financing as an alternative to 50-50 cost-sharing in hospital and medical care.

————— *The Health of Canadians: Report of the Canada Health Survey* (1981). After thirteen years of medicare and twenty-three years of hospitalization insurance, how were we doing?

————— (1986). Jake Epp, Minister. *Achieving Health for All: A Framework for Health Promotion.* Another ministerial statement proposing greater emphasis on health promotion and less on the provision of care.

Canada. Health and Welfare Canada, Policy and Consultation Branch (1994). *National Health Expenditures in Canada/Dépenses nationales de santé au Canada 1975–1993.* A basic source for discussion of cost trends in health care in Canada, public and private.

Canada. Parliamentary Task Force on Federal-Provincial Fiscal Relations, H. Breau, M.P., Chair (1981). *Fiscal Federalism in Canada.* (French and English, back-to-back.) Chapters II, III, and IV explain the workings of Established Programs Financing in health care, which prevailed from the late 1970s until 1996, when it was superseded by the Canada Health and Social Transfer.

Canada. Royal Commission on Health Services, Hon. Emmett Hall, Chair (1964). *Report.* Prepared the ground for the federal medicare program. See especially Chapter 10, 'The Evolution of Health Insurance in Canada', a lucid historical account. Not to be confused with Justice Hall's later, much less substantial report, *A Commitment for Renewal* (1980) (cited above).

Canada. Senate of Canada. Special Senate Committee on Poverty, *Proceedings, 1969–70,* No. 45.

Canada (1970). *Task Force Reports on the Cost of Health Services in Canada.* See especially 'Alternative Methods of Paying Physicians', pp. 229–38; 'Over-utilization of Medical Services', pp. 263–7; 'Control of Over-utilization', pp. 268–74.

Canadian Doctor. The issues of this journal throughout 1984–6 present the private practitioners' side of the extra-billing debate with vigour.

Canadian Medical Association Journal. Throughout the period of the extra-billing controversy, 1984–6, this journal published some articles about that controversy, somewhat subtler than those in the above-cited *Canadian Doctor.*

Clarke, J.N. (1996). *Health, Illness and Medicine in Canada.* 2nd edn. Toronto: Oxford University Press. Note the title: given that medicine is closely related to illness, how closely is it related to health?

Coburn, David, Carl D'Arcy, George M. Torrance, and Peter New, eds (1987). *Health and Canadian Society.* 2nd edn. Toronto: Fitzhenry and Whiteside. See especially Part II, 'Health Status and the Health-Care System', Part V, 'Health Institutions', and Part VI, 'Society, Health and Health Care'.

Courchene, Thomas J. (1995). *Redistributing Money and Power: A Guide to the Canada Health and Social Transfer.* Toronto: C.D. Howe Institute. The financial, administrative, and political implications of the latest variation in federal transfers to the provinces in support of social programs.

Crichton, Anne, and D. Hsu (1990). *Canada's Health Care System: Its Funding and Organization.* Ottawa: Canadian Hospital Association.

Crichton, Anne, Jean Lawrence, and Susan Lee (1984). *The Canadian Health Care System: Doctors and Patients Negotiate the System of Care: Case Studies.* Ottawa: Canadian Hospital Association. Some realities that are not conveyed by a simple description of medical care insurance.

Epp, Jake (1986). See 'Canada. Health and Welfare Canada (1986)' (cited above).

Evans, Robert G. (1984). *Strained Mercy: The Economics of Canadian Health Care.* Toronto: Butterworths.

Evans, Robert G., and G. Stoddart (1990). *Producing Health, Consuming Health Care.* Toronto: Canadian Institute for Advanced Research.

Field, Mark G. (1972). 'The doctor-patient relationship in the perspective of "fee-for-service" and "third-party" medicine', in E.G. Jaco, ed., *Patients, Physicians and Illnesses*. 2nd edn. New York: Free Press. Does it make any difference how the doctor is paid—fee-for-service, salary, capitation?

Freidson, Eliot, ed. (1963). *The Hospital in Modern Society*. New York: Free Press of Glencoe.

Illich, Ivan (1975). *Medical Nemesis: The Expropriation of Health*. Toronto: McClelland and Stewart.

LaMarsh, Judy (1969). *Memoirs of a Bird in a Gilded Cage*. Toronto: McClelland and Stewart. The candid political memoirs of the (Liberal) federal minister who guided the Health Insurance Act through the House of Commons in 1966 (though not through its final stage). Intriguing details about conflicts within the Cabinet, within Parliament, within the Liberal Party, etc.

Leeson, Howard (1995). 'Health Care Reform in Saskatchewan', *Management* (Institute of Public Administration of Canada) 6/4. How the NDP government of Saskatchewan went about cutting its health budget between 1992 and 1995, emphasizing a 'Wellness' model in its effort to reduce hospital costs.

Lesemann, Frédéric (1984). *Services and Circuses*. Translated from *Du Pain et des services*. Montreal: Black Rose. An analysis of the ideology underlying health care and social service reforms in Canada and Quebec.

Maclean's (1994). 'Condition Critical: How the money crunch hurts hospital care' (25 July).

———— (1995). 'A Prescription for Medicare: A 10-point plan to heal the health care system—without spending more money' (31 July).

———— (1996). 'Radical Surgery: How Ottawa's policies threaten Medicare as we know it' (2 December).

MacTaggart, Ken (1973). *The First Decade*. Ottawa: Canadian Medical Association. The Saskatchewan story, told from the standpoint of the Canadian Medical Association, but not blind to weaknesses in the doctors' position in the dispute.

McKeown, Thomas (1979). *The Role of Medicine: Dream, Mirage or Nemesis?* Princeton: Princeton University Press.

Manga, P. (1978). *The Income Distribution Effects of Medical Insurance in Ontario*. Ontario Economic Council, Occasional Paper No. 6.

———— (1987). 'Equality of Access and Inequalities in Health Status: Policy Implications of a Paradox', in Coburn et al., eds (cited above).

Meilicke, Carl, and Janet Storch, eds (1980). *Perspectives on Canadian Health and Social Services Policy: History and Emerging Trends*. Part II, 'Health Services'. Ann Arbor, Michigan: The Health Administration Press.

National Council of Welfare (1990). *Health, Health Care and Medicare*. Ottawa: The Council. A reasoned plea to the population to move discussions of health into broader fields than hospital and medical services; it even recommended that *all* policies of governments in *all* fields be assessed for their impact on health.

———— (1982). *Medicare: The Public Good and Private Practice*. Ottawa: The Council. Mainly an attack on extra billing and user fees.

National Forum on Health (1995). *Let's Talk about Our Health and Health Care*. Ottawa: National Forum on Health.

Navarro, Vincent (1976). *Medicine under Capitalism*. New York: Prodist. Medicine and its offshoot enterprises (pharmaceuticals, hospital and medical equipment, insurance) seen as means for the accumulation of capital (and power).

Naylor, C. David, ed. (1992). *Canadian Health Care and the State: A Century of Evolution*. Montreal and Kingston: McGill-Queen's University Press.

Naylor, C. David (1986). *Private Practice, Public Payment: Canadian Medicine and the Politics of Health Insurance, 1911–1966*. Kingston and Montreal: McGill-Queen's University Press. Takes the story from its beginnings to the eve of the passing of the federal Medical Care Act (1966).

Pineault, Raynald (1992). 'The Reform of the Quebec Health Care System: Potential for Innovation?', in S. Mathwin Davis, ed., *Health Care, Innovation, Impact and Challenge*. Kingston: Queen's University School of Policy Studies and School of Public Administration, 1992. Pineault comments on the reforms brought in by the Quebec Health and Social Services Act of 1991, especially the organizational aspects. He quotes the Rochon Report, to the effect that the system had become the hostage of the countless interest groups: 'groups of producers, groups of institutions, groups of community-based activists, unions, and so forth.' (p. 77).

Plain, Richard H.M. (1997). 'The Role of Health Care Reform in the Reinventing of Government in Alberta', in C. Bruce, R. Kneebone, and K. McKenzie, eds, *A Government Reinvented: A Study of Alberta's Deficit Elimination Program*. Toronto: Oxford University Press, 1997.

Quebec (1964, 1970). Commission of Inquiry on Health and Social Services, Claude Castonguay, Chair. Report. Vol. I, *Health Insurance* (1964). Vol IV, *Health* (1970), Tome I, Chapters 1 and 2, 'History of Health Services' and 'Environment and State of Health'. French and English.

Quebec. Quebec Health Insurance Board (Régie de l'assurance-maladie du Québec). *Annual Report* (current). Excellent data. Introductory material provides a good explanation of Quebec medicare. The health insurance authorities of all provinces, of course, issue similar reports.

Rachlis, Michael M., and Carol Kushner (1994). *Strong Medicine: How to Save Canada's Health Care System*. Toronto: HarperCollins.

Rachlis, Michael M., and Carol Kushner (1989). *Second Opinion*. Toronto: HarperCollins. Canadians may be getting more medical care than is good for them. (Rachlis is a doctor.)

Roch, Denis J., Robert G. Evans, and David W. Pascoe (1985). *Manitoba and Medicare: 1971 to the Present*. Winnipeg: Manitoba Health. A meticulous analysis of Manitoba experience, demonstrating the seemingly limitless capacity of people to absorb medical services—or of doctors to provide them. Over the period studied, the population of Manitoba remained stable and the number of doctors increased by nearly half, but the number of medical services billed for by doctors kept pace with the number of doctors. Robert Evans, co-author, is Canada's pre-eminent health economist.

Shillington, C.H. (1972). *The Road to Medicare in Canada*. Toronto: DEL Graphics. A journalistic history: concise, readable, and remarkably even-handed considering that Shillington was involved in the issue (as an executive of a private non-profit insurance organization).

Silver, Susan (1996). 'The Struggle for National Standards: Lessons from the Federal Role in Health Care', in Jane Pulkingham and Gordon Ternowetsky, eds, *Remaking Canadian Social Policy: Social Security in the Late 1990s*. Halifax: Fernwood. This writer gloomily concludes that under the new CHST cost-sharing mechanism, the federal government will probably not be able to enforce the administrative standards for health care embodied in the Canada Health Act.

Stingl, Michael, and Donna Wilson, eds (1996). *Efficiency vs. Equality: Health Reform in Canada*. Halifax: Fernwood.

Stoddart, Greg L., Morris L. Barer, and Robert G. Evans (1994). *User Charges, Snares and Delusions: Another Look at the Literature*. A Report to the (Ontario) Premier's Council on Health, Well-being and Social Justice. The blunt conclusion of this exceedingly thorough review of the literature on the topic is: '. . . most proposals for "patient participation in health care financing" reduce to misguided or cynical efforts to tax the ill and/or to drive up the total cost of health care while shifting some of the burdens out of government budgets.'

Swartz, Donald (1977). 'The politics of reform: conflict and accommodation in Canadian health policy', in Leo Panitch, ed., *The Canadian State: Political Economy and Political Power*. Toronto: University of Toronto Press. Swartz's analysis is explicitly Marxist; he gives an interesting alternative view of the emergence of hospitalization insurance and medicare. As he sees it, the workers and the poor have gained a little, the doctors and the associated arms of the health industry have gained a lot.

Taylor, Malcolm (1987). 'The Canadian Health Care System after Medicare', in D. Coburn et al., eds, *Health and Canadian Society* (cited above).

Taylor, Malcolm (1988). *Health Insurance and Canadian Public Policy*. 2nd edn. Toronto: University of Toronto Press. See Chapter 4 on national hospital insurance, and Chapters 5, 6, and 7 on Saskatchewan, national, and Quebec medicare, respectively. An indispensable book for an understanding of how our system of hospital and medical care insurance developed.

Wahn, Michael (1987). 'The Decline of Medical Dominance in Hospitals', in D. Coburn et al., eds, *Health and Canadian Society* (cited above).

Wolfe, B. (1986). 'Health Status and Medical Expenditure: Is There a Link?', *Social Science and Medicine* 22/10. Contrary to many current analysts (see Chapter Nine in this book), Wolfe finds that, if societal lifestyle patterns are controlled for, there *is* a positive relationship between amounts spent on medical care and general levels of health.

York, Geoffrey (1987). *The High Price of Health: A Patient's Guide to the Hazards of Medical Politics*. Toronto: Lorimer. See especially Chapter 2, 'The Strike Weapon', on the Ontario doctors' 'withdrawal of services' in 1986 over the provincial government's legislation prohibiting extra billing, and pp. 50–4 on the doctors' strike in Saskatchewan in 1962 over the introduction of medical care insurance; Chapter 4, 'How Doctors Control Their Incomes', on the fee-for-service issue; and Chapter 6, on the issue of doctors working for salaries instead of fees. York consistently adopts a critical stance toward the institutionalized medical profession (not necessarily against the work of individual doctors). He is given plenty of free ammunition by the public statements of medical spokespeople.

CHAPTER 9

Health Beyond Hospitals and Doctors

CRITIQUES OF WHAT WE HAVE NOW

Within their own frameworks, although widely cherished as a feature of Canadian life, our hospitalization and medical care insurance programs have come under some criticism. This is hardly surprising given their size, visibility, and importance. Since they occupy such a predominant place in people's perceptions of health policy, hospitals and doctors may at times be criticized for shortcomings they cannot possibly correct.

Hospitalization

Cost. Hospital care costs a great deal. A day in the hospital, even without special care, costs many times the price of a day in a good hotel, and may not be as good for your health. Some of the cost is a consequence of the high costs of modern hospital equipment and specialized personnel—what to an economist are fairly straightforward supply and demand factors. Some costs arise from the aging of the population. But some of the costs may arise from factors generated by the insurance system itself: the ease of access, the perception of the service as 'free' (while possible alternative services are not), the total separation of the service provider from any responsibility for the raising of revenue. This is a familiar process in any branch of public social policy: once a piece of machinery is created, people and groups with varying interests will try to use it to further their own ends, perhaps convinced that their ends coincide with the public interest, although they are not necessarily those for which the machinery was invented.

Under-use of alternatives that are not 'insured'. One reason for the increase in hospital expenditures is the absence, in many cases, of practicable alternatives to hospitalization. If a patient is insured for the cost of hospital care (as almost all are), but not for the cost of alternative care (as few are), the patient and/or the doctor may well prefer the hospital, whatever the therapeutic merits of the case. Often no alternative mode of care is available at all, as when the only apparent alternative is the patient's home and the home is simply not prepared to care for the patient. The patient stays in the hospital. Part of the reason for the high proportion of hospital space now occupied by older people is undoubtedly the simple fact that we need more hospital care when we are older, but it is suggested that part of it is due to the absence of appropriate alternative services. Hospital managers complain of the continual influx of psychiatric patients, who require attention even though a general hospital is not a good place for them. This, no doubt, is why, when EPF came into effect, a list of 'Extended Care Services' was added to the list for which federal cost-sharing was available, and why provincial hospital authorities are pushing to create alternative facilities.

Services versus facilities. Funds that pay for services do not create facilities. Insurance pays only for services. Services can be provided only to the extent permitted by the facilities that exist. New hospitals, and new extended care, geriatric, chronic, and convalescent facilities, represent capital expenditures; they must be paid for out of public or private funds or both. Government spending on hospital–related capital projects is financed out of the same resources as other capital expenditures of government: usually by borrowing against future revenues. It is subject to political judgement in competition with other priorities, all of which have sincere advocates.

The provision of hospital services, paid for mostly by a budget allocation out of the current revenues of government, will also be affected by other public priorities and by the revenue position of the government. With governments now in a tight budget situation, the hospitals are called upon to bear their share of the tightness: hospital services will be reduced, hospitals staffs will be cut, etc., at least until the fiscal situation improves.

Sickness versus health. With the exception of women having babies, almost everyone who goes to the hospital does so because an illness has become severe, because a serious injury has been suffered, because a condition requires intensive treatment. The hospital's understood function is to repair the body from harm that has already occurred. It would be ridiculous to say that hospital patients never learn to protect their future health, but that outcome is secondary, and is in any case confined to individual patients. There is even a body of evidence that shows that many people contract illnesses while in the hospital; for people who do not have to be there, alternatives may be not only less costly than the hospital, but healthier.

Medicare

The medical care system presents a different picture from the hospital care system. The expert services of doctors are the central issue; facilities are important, but secondary. In any case, doctors provide many of their own facilities out of their incomes from their professional fees (hence the fees doctors collect from medicare are not an accurate indicator of their personal incomes). Other facilities used by doctors may be provided under other auspices (municipal, philanthropic, proprietary) or, of course, in hospitals.

Demand for services. The demand for doctors' services is subject to many influences. Presumably the most important is the need felt by people for the treatment the doctor is trained to provide. But it is widely acknowledged that (a) people bring other agendas to their doctors than just their illnesses and injuries, and (b) even when an illness is the centre of attention of both patient and doctor, non-medical factors, with which the doctor is not explicitly trained to deal, are important. Besides, whatever the nature of the case, the contact between patient and doctor is typically initiated by the patient. The upshot of these and other considerations is that some people who are not particularly ill visit the doctor, perhaps repeatedly, and some who really are ill do not. What appears to be a demand for doctors' services may really be in large part a demand for something else. (This is a different issue entirely from that of overuse or abuse of medical care insurance, dealt with above as an instance of 'moral hazard'.)

The supply of medical services. Somewhat oversimplified, the 'supply' of medical services is the 'supply' of doctors. The supply of doctors depends on the willingness of people to undertake careers in medicine and the capacity of training institutions to train them, modified by the out-migration and in-migration of doctors. These factors are influenced by the controls exercised by the licensing bodies of the medical profession. On a world comparative basis, Canada is quite well supplied with doctors; on a Canadian comparative basis, Quebec has gone from being well below the national average, before medicare, to being now well above it. The relative abundance of doctors is no doubt related to the fact that medicare now makes a doctor's income almost unshakeably secure, and to the related fact that Canadians are free to consult physicians with very few constraints, thereby assuring doctors of busy practices, i.e., supply does elicit demand. Records suggest that when the number of doctors increases, the number of services they provide increases in proportion. In other words, the more doctors there are, the more often people will visit the doctor—which confirms Naylor's remark (above) that demand will outstrip supply in a zero-price market. Instead of being spread around, doctors' work increases. Thus, government authorities regard each new practising physician as a generator of additional costs, and some, including Quebec and British Columbia, have taken steps to restrict entrance to medical practice (Taylor, 1988, pp. 418–19, 474–8).

The investment made by Canada in the education of doctors is great, and so are the expenditures Canada devotes annually to medical services. The most generally accepted judgement of the resulting supply of doctors is that it is sufficient. Most analysts conclude, however, that the supply of services is badly distributed in geographic terms and somewhat askew in terms of certain specialties. The heavy use made of the available services is such that patients must often endure extended delays for ostensibly non-urgent treatments and procedures. (American opponents of public health insurance often support their case by depicting in catastrophic terms the long line-ups in Canadian doctors' offices and the lengthy waiting lists for surgery in Canadian hospitals. Without denying that delays do occur, few Canadians would recognize the exaggerated picture sometimes given of their health care system.) When demand presses heavily on supply, some form of rationing is inevitable, explicit or not. Canada rations health care by waiting; we have so far resisted rationing by price.

The foregoing criticisms take hospitalization insurance and medicare as the given frame of reference; what is questioned is their performance. Even within that frame of reference, questions arise concerning the extent to which the system itself is contributing to its own difficulties. Are the problems of hospitals and medical services partly a result of the terms on which services are provided? And to some extent, does apparently successful therapeutic treatment stimulate the demand for more therapeutic services, needed or not?

THE GREEN PAPER OF 1974 AND OTHER CRITICISMS

There is a quite different order of questioning, in which our programs of health care insurance are not taken as the frame of reference. The point of departure is rather the basic idea of health as a social policy objective; the provision of health care is seen as only one means toward the attainment of that objective—highly valued, but not an end in itself, and not necessarily the centrepiece of policy, as it certainly has been. The question then is not whether the health insurance system is efficient, but whether we are spending our human and material resources on the right things in order to get a satisfactory return in terms of health.

Naturally, we all feel that our governments could afford to be more generous with programs we favour, simply by curbing their extravagant spending on programs we do not support. That being said, governments are in fact sensitive to the costs of programs, for many reasons. Understandably, then, one government might be all the more sensitive about costs it met that were rising as a result of decisions made by *other* governments.

The Federal Government and the New Perspective

The Canadian federal government represented itself as being in that position with respect to hospital insurance and medicare: locked into its commitment to pay 50 per cent of the nation-wide costs of programs that were subject to *provincial* legislation and administration. The federal government itself imposed, as conditions of its cost-sharing, criteria that were bound to put a rising floor under the costs: universal access, no financial barriers to obtaining service. Still, the federal government soon began to seek a formula for cost-sharing such that the provinces would feel a million dollars' tax impact, rather than half a million's, for every million dollars they spent on health care. Even at the time that the federal medicare legislation was being prepared, the federal government proposed to the provinces a different model of cost-sharing, called (as we have seen) Established Programs Financing.

The motives of governments are a legitimate subject of speculation. The federal government's motivation may well have been to reduce its costs as well as to improve health. What is beyond speculation is that the total costs of health insurance escalated steadily, allowing a reasonable observer to ask whether the increasingly heavy spending was meeting needs in the most effective way. The effectiveness of costly health care programs and the consideration of alternative policies were the topics of the 1974 Green Paper, titled *A New Perspective on the Health of Canadians.*

Having established the overwhelming proportion of our health-related spending that currently goes to what it calls 'the Health Care Organization' (i.e., hospitals and doctors), the Green Paper looked for available indicators of the overall health of the Canadian population; it then tried to identify the factors associated with inferior health and then questioned the capacity of the Health Care Organization to contribute very much to the control of such factors. All students of social policy in the health field would do well to read the Green Paper for themselves, bearing in mind that it is a political document, intended to persuade.

The Green Paper is also interesting as an exemplar of imaginative, persuasive use of data. It is interesting as well for its observations about shortcomings of the available data in the health field; almost all our population-based health statistics classify people simply as well, sick, or dead. (By and large, for statistical purposes, 'sick' means 'in hospital or losing time from work'; we have no good statistics on sickness other than those.) Of course, the data have changed since 1974, but not enough to lessen the interest of the Green Paper.

The Analysis of Health Policy in *A New Perspective*

The Green Paper proposed the broad lines of health policies that might promise a better return on investment, focusing on the health impact of various economic and political choices made institutionally or collectively ('Environment'), on the effects of more or less free and conscious personal choices governing individual behaviour ('Lifestyle'), and on strategy for fundamental and applied scientific research ('Human Biology'). These alternative policy orientations overtly aim at the same targets as do advocates for a variety of currently popular special causes: industrial and other pollution of earth, water, and atmosphere, the nutritional value of marketed food, the safety of work and other environments, the safety of cars and roads and the practices of car drivers, smoking, alcohol abuse, etc. These policy orientations have in common two significant characteristics: the most relevant skills are not medical or surgical, and the individual is not cast in the role of patient.

Policy for Wellness vs. Policy for Illness

Another way of putting it is that the alternative policies proposed in the Green Paper and elsewhere are not essentially policies for the care of the sick; they are, rather, policies for the more or less well. They are concerned with the enhancement of health and the prevention of illness and accident.

This insistence on health rather than illness is basic to much criticism of current health policy. To be sure, in principle no one is *against* improving health and preventing illness. Still, the political choice to spend less money on hospitals and doctors, and more on other health programs, is a difficult one to make. It is idle to discuss the subject without acknowledging the massive social support enjoyed by the hospital and the medical profession. To appreciate this support, one has only to observe the outcry that greets any governmental decision to close a hospital. Candour compels the observation that few of the radical critics of the over-allocation of health funds to hospital and medical care are prepared openly to support government cutbacks in hospital and medical services budgets.

Health Policy in the Context of Social Criticism

So strong and pervasive have the institutions of health care become, says Ivan Illich in his *Limits to Medicine: Medical Nemesis*, that individuals in 'advanced' societies have virtually lost all sense of responsibility for the well-being of their own bodies. Whether that responsibility has been willingly surrendered by individuals or has been seductively appropriated by other interests, as Illich contends, may matter less than the simple fact that, to some degree, it has been lost. One may appreciate the pertinence of Illich's central thesis without agreeing with his further conclusion that medical domination has been actually harmful to health.

A different political-economic twist is given to the basic thesis of the alienation of Western societies from control of their own health by such critics as Donald Swartz and Vincent Navarro (most notably in *Medicine under Capitalism*). In this view, health care has become an instrument of class domination and the individual merely an exploited consumer, as any consumer in capitalist society is exploited, albeit benignly at times. Needless to say, Illich, Navarro, Swartz, and others adhere to ideological lines that are not pursued in the Green Paper, an official document of the Government of Canada. Their analyses tend to confirm, however, that there is nothing superficial about the reordering of health policy proposed in the Green Paper.

The Role and Responsibility of the Individual

In any discussion of health policy, the place given to the individual is problematic. Alternative proposals, equally critical of the status quo, may be diametrically opposed to each other in terms of the role they provide for the individual.

On the one hand, it may be argued that the individual is relatively help-less against factors beyond his/her control, managed by powerful organiza-tions that have an interest in keeping people ill informed about their plight—factors such as economic deprivation, adulteration of food, pollu-tion of the environment, unsafe products produced and marketed on a global scale. So besieged, the individual is not to blame for his/her plight. Pursued with force, such an approach might lead to the conclusion that we can only be rescued by the mobilization of equally powerful contrary forces, typically governmental, equally far beyond the individual's power to control. (Governments themselves do not have perfectly clean hands in such matters as environmental pollution.)

On the other hand, it may be argued that, in large part, the individual *is* responsible for crucial decisions that would affect health and *can* act effec-tively—he or she can eat well, avoid noxious substances and their fumes, temper ardour with prudence in sexual behaviour, drive only when sober, exercise appropriately, demand environmental protection both through the market (avoiding the products of anti-social producers, investing in the enterprises of responsible ones) and through political means (supporting politicians and interest groups with positive programs). Pushed to an extreme, this argument may lead to the conclusion that we are to blame for much of our own ill health, and may be seen as a variant of 'blaming the victim'. An appropriate balance must be struck between individual and collective responsibility.

The Mediating Social Environment of the Individual
In between the society and the individual comes his/her immediate social environment: family, private social circle, workplace, community. In 1995 an advisory committee to the Ontario Premier's Council submitted a report entitled *Optimizing Resources for Health*, in which there is barely a mention of illness, medication, treatment, or medical or hospital service. Apart from what a supportive social network can do to enhance access to and delivery of professional health care, there is abundant evidence that the emotional security that comes with social support is a positive health fac-tor in itself. Programs may accordingly be built around the health-main-taining potential of an adequate social life.

Before presenting material concerning two kinds of health programs that move away from the hospital and the doctor's office, a word of cau-tion. The remarkable rise to social eminence of medical and hospital prac-tice in recent history was not an accident. Vested interests promote that eminence, no doubt, but its basis has been the growth in technical ability to locate and neutralize the effects and often the causes of specific illnesses. Once a disease has been defined and its causes isolated, treatments address themselves to it; results are subject to test; success can be seen and felt by everyone. Diseases can not only be cured, they can be all but eliminated. Many of the dreaded killers of a couple of generations ago have all but disappeared from our mortality statistics, for example, diphtheria and scarlet fever. To be sure, grim new reapers—e.g., HIV/AIDS—appear on the scene from time to time. And bacteria and viruses are maliciously resistant; some infectious diseases once thought to have been conquered, such as tuberculosis, have been reappearing in new forms. On the whole, however, progress against disease has been remarkable, and medicine has contributed indisputably to this progress, even if conceivably the public mind gives credit to hospitals and doctors for improvements in health that have really originated elsewhere, as in effective sewer systems, better nutrition, and personal hygiene.

Once one addresses the broad causes of generalized ill-health in populations, however, one enters a complex realm of multiple interrelated causes, difficult to trace, and of diffuse effects, difficult to measure, sometimes difficult even to perceive. It is no longer a matter of *this* specific intervention leading in linear fashion to *this* desired outcome. (According to observers like Rachlis and Kushner [1989], that popular perception is not borne out in the world of medical practice either.) This is even more true when one moves from the causes of ill-health and addresses the factors conducive to good health, which cannot be traced, as illness usually can, to a specific condition. A health policy that purports to take account of environmental and social factors moves into areas where certainty is elusive, the relevant time span is long, and results often cannot be demonstrated conclusively. Society's reluctance to reduce its reliance on the familiar health care services is therefore not surprising, even if possibly wrong-headed.

ALTERNATIVE POLICY #1: COMMUNITY HEALTH

What Is It?

In practice, 'community health' is an umbrella term of vaguely defined circumference. Meaning no denigration, the safest working definition of 'community health' is 'Whatever is done by people who say they are doing community health'. Even authoritative attempts at definition tend less to define community health than to list the characteristics that its advocates want to see realized—non-bureaucratic, de-professionalized, decentralized, controlled by the consumer, preventive, 'holistic' (which defines one vague expression by another equally vague) (CCSD, 1985, p. 23; also Banta, 1979).

At the minimal limit, some self-styled community health programs have consisted merely of systematic efforts to make the professional services of doctors and nurses more readily accessible to the population: convenient locations, working evenings and weekends, not insisting on appointments, and organized, supportive home care. While highly valued, such programs embody little of a distinctive *community* dimension, even when matched to the social milieu and the lifestyles of their clienteles.

Taking into account the actual content as well as desirable characteristics of programs, the following are more clearly identifiable as community health functions:

- the detection, tracking, and control of contagious diseases;
- fitness programs for the general population or for special groups such as expectant mothers or the elderly;
- self-help groups for sufferers of specific diseases (arthritis, multiple sclerosis) or injuries, not aimed at *cure*, but at accommodation to normal life;
- local health-related educational programs, whether aimed at improving health in general by the dissemination of information (e.g., on child nutrition) or at the prevention of health problems (e.g., anti-smoking programs, programs concerned with sexually transmitted diseases);
- promotion of social and recreational activities, especially for such vulnerable groups as the elderly, on the principle that mental, emotional, and physical health are interrelated;
- advocacy and organizational activities conducted in the spirit of encouraging people who are well to 'take charge of their own health';
- education and promotion around locally relevant health issues, e.g., waste disposal and storage, control of noxious substances, accident prevention.

In all of the above, close association among people will enhance the success of the activities, as will an element of self-determination by the clientele; they are essentially *community* programs, whether the community is based on a neighbourhood, an occupational group, or some other natural affinity. All call for the active participation of people; to elicit this participation demands a special skill. The skills involved range from conventional medical and nursing skills through other health specialties like nutrition, epidemiology, and physical education, to skills not essentially related to health at all, such as communications, counselling, group work, and community organization.

Who Does It?

Community health services are provided by a myriad of private, voluntary, religious, public, union-based, and community-based agencies. In Quebec, within the *public* network of health and social services agencies authorized by the Health and Social Services Act (initially passed in 1972, substantially amended in 1991), two classes of agencies have explicit mandates in community health: the regional Directorates of Public Health (known as the DSPs—Direction de la santé publique) and the Local Community Service Centres (known as the CLSCs—Centres locaux des services communautaires). Though their services overlap to some extent, the Directorates of Public Health tend to work at the level of applied research into environmental conditions and the dissemination of information, and the CLSCs tend to work directly with individual and group clienteles.

Other provinces have their own institutional arrangements, variously known as Community Health Centres, Area Health Centres, etc. The Hastings Report, a Canada-wide survey of community health centres published in 1972, reported much activity along this line across Canada through the 1950s and 1960s. It seems, however, that no other province has gone as far as has Quebec in the systematic support of public community health institutions.

Community Health as Research and Information

The standard Canadian approach is to entrust to municipalities such public health and community health functions as fall outside the ambit of provincial departments. In Quebec, the Public Health Directorate is a department of a regional authority known as the Regional Health and Social Services Board ('Régie régional de la santé et des services sociaux'). The DSP works through public health units in designated general hospitals serving the region (some rural regions may have only one general hospital) (Marchand et al., 1996). Quebec's Regional Boards are totally outside the municipal government system. The Regional Boards, of which there are seventeen in the province, are responsible for program development and resource allocation within their territories, including budgetary authority over hospitals and CLSCs. The governing bodies of these regional authorities are made up of representatives of service-providing agencies, professional groups, community organizations, and the public at large, some elected by their respective constituencies, some appointed.

The evolution of the DSP may be seen as the continuation of an interesting administrative experiment. 'Community Health Departments' were originally located in general hospitals in all regions, in such a way that the whole province was covered. Hospitals are not really 'community' organizations, but they are well known and highly respected. But a community health department remained something of a foreign body in a hospital—it had no patients, was not preoccupied with healing, did almost of its work outside the hospital building, and collaborated more with agencies not linked administratively with the hospital than with other hospital departments. Thus the links between Community Health Departments and their host hospitals remained tenuous. On the other side, for much of their work, they did not relate particularly closely with organizations in the community. Moreover, to the extent that it implies a territorial dimension, the idea of 'community' was quite irrelevant to many of the mandates exercised by Community Health Departments, as the 1985 program of one Community Health Department of the time shows:

DENTAL HEALTH: Instruction and information on dental hygiene; identification and referral of urgent cases.

HOME CARE: Nursing and physiotherapy services to post-hospital and short-term patients, and to chronic and handicapped patients not served by a Local Community Services Centre (CLSC).

MOTHER AND CHILD HEALTH: Prenatal and postnatal visits; prenatal classes; Well Child clinics.

NUTRITION: Support service for all programs of community organizations.

OCCUPATIONAL HEALTH: Identification of local industries whose workers are at risk; surveillance of health of workers; development of health programs in workplaces.

SCHOOL HEALTH: Screening of general health, especially vision, hearing, tuberculosis, scoliosis; immunizations; health education.

SEXUALLY TRANSMITTED DISEASES: Support services for physicians; follow-up of contacts of patients coming to school; STD clinics, educational programs for schools, community clinics and groups, industry, etc.

GERONTOLOGY: Co-ordination of programs and services to meet the needs of the elderly; organization of in-service training programs for professionals; participation in public education.

INFECTIOUS DISEASES: Immunization programs; investigation of reported cases of infectious disease; control of epidemics.

SEXUAL ASSAULT: Co-ordination of resources, program development and evaluation; public education.

RESEARCH: Identification of the health needs of the population; determining priorities; evaluation of preventive health programs. (Community Health Department, Montreal General Hospital, 1985)

Some of these functions fit with the limited territorial and/or population boundaries associated with the idea of 'community'—e.g., the participative and the educational components of 'Gerontology', 'Sexually transmitted diseases', 'Nutrition', and 'Mother and Child Health'. Others, however, clearly call for action on the very large scale associated with public health (e.g. 'immunization programs' and 'control of epidemics'), transcending community boundaries, though no doubt community mobilization would be helpful.

The Health and Social Services Act of 1991 transferred the administrative responsibility for all such functions from the hospitals to the regional boards. 'Public health units' remained physically in the hospitals. The list of the public health units' responsibilities has not changed appreciably from those listed above (cf. Régie régionale de la santé et des services sociaux de Montréal, 1995, p. 18).

The Quebec experience may be seen as an attempt simultaneously to decentralize responsibility for public health functions to the regional level, and to leave resources useful to community health programs in hospitals with community contacts. In the nature of things, considerable information about the health of the people in any area is concentrated in the hospital. And the continued association of the hospital with the work of community health may improve the hospital's ability to carry out its role as (still) the major health institution in society.

All provinces have public health and community health authorities at the provincial and municipal levels, carrying out many of the functions of Quebec's regional public health units. All rely to a greater or lesser extent on collaboration with voluntary community agencies, particularly with respect to disease-specific programs of prevention and education: heart and lung disease, arthritis, AIDS, etc. All struggle with the prevalent bias in the public mind in favour of hospitalization and curative medicine, reflected in their budgets. The Quebec experience has been described here at length as one example of an effort on the part of government to broaden health services beyond their conventional conceptual boundaries. The experience of Quebec's Local Community Services Centres also merits attention as an attempt to further decentralize implementation.

The CLSC: Community Health as Direct Service

The second agency created by Quebec's 1972 Health and Social Services Act to work in community health, and confirmed in its functions by the new Act of 1991, is the Local Community Services Centre, which has substantial mandates in both health and social services.

The CLSC (the French-language acronym) has been regarded as the most striking innovation of the Health and Social Services Act: a publicly funded agency created to carry out health and social programs that would meet the expressed needs of the local community, operating with substantial client and community input, designed to be highly accessible (in principle, there is supposed to be a CLSC within thirty minutes' travel time of every resident, allowing for exceptions in sparsely settled regions), and to be unencumbered with the bureaucratic/professional apparatus that is known to be a barrier to real access to service for some people. Before the passage of the Act in 1972, there had been a few community clinics in the province, operating under various auspices, but except for these the CLSCs had little organizational precedent in Quebec. (The other types of agencies set up by the Act—social service centres, hospital centres, reception centres—all took over the functions and the establishments of pre-existing agencies.) The network of CLSCs covers the entire population and territory of Quebec.

The CLSC is designed to be the first entry gate into the health and social service system for anyone feeling the need for either kind of service. As noted earlier, most CLSCs, particularly those in rural areas where hospital clinics and emergency departments are not as readily available as in the cities, provide front-line health services of a non-specialized nature. They all have a mandate to provide home care services. In addition, they are authorized to carry out health programs of an informational, preventive, or otherwise positive character, in keeping with the expressed needs and wishes of the local population (Act respecting Health Services and Social Services, *Revised Statutes of Quebec,* Ch. S-4.2, Articles 80, 131).

One inevitable consequence of the *community* orientation of CLSCs is that they do not provide uniform services. Communities do not all have the same needs; community centres do not all provide the same programs. The autonomy of CLSCs to customize their programs has been considerably qualified by the reform embodied in the 1991 Act, however; the CLSCs have all been mandated to carry out certain programs that implement provincial government policies, particularly in the field of front-line health services, including medical, home care, and school social services, and health and social services for youth at risk, but there is still variation from one CLSC to another.

Of particular interest are the clauses of the Health and Social Services Act outlining the provision for the specific representation, on the Board of Directors of CLSCs, of users of the services and representatives of the local community. In fact, the Act requires that users of services and the local community be explicitly represented on the boards of directors of *all* health and social service establishments in the public network (including Hospital Centres and Youth Centres and also, as noted above, the Regional Boards), but the proportion of such representation is highest in the CLSCs. Regular elections take place to fill the positions of users' representatives and public representatives; the elections are sometimes hotly contested, eliciting a fair degree of electoral participation, sometimes not.

The health authorities of other provinces have been sympathetic to community health in principle. It is seen as a way of enlisting populations into useful preventive programs, and of making good use of health care personnel other than (relatively expensive) physicians. But the barriers and the pitfalls are considerable. The principal barrier is the combination of resistance on the part of the medical profession and continued public preference for conventional medical treatment. In Quebec, successive governments have committed themselves to the CLSCs as a vehicle for health policy; doctors are obliged in the early years of their practice to devote a stated proportion of their time to service in public agencies, including CLSCs. Even so, doctors have easily succeeded in retaining their position as the most frequented providers of unspecialized, front-line primary health care. A common phenomenon in Quebec, as everywhere else in Canada, is the grouping of doctors' offices in accessible 'polyclinics' where a range of services is available relatively free of delay—and, of course, free of charge under medicare (above, p. 210; CCSD, 1985, pp. 32-42; Lomas, 1985, ch. 10; York, 1987, p. 187).

ALTERNATIVE POLICY #2: OCCUPATIONAL HEALTH

Work accidents and unhealthy working conditions inflict casualties in Canada on the scale of a small guerrilla war: over 800 deaths a year in recent years, and over 1,000 recorded injuries and illnesses a day. Just as at the dawn of the twentieth century, when much legislation, litigation, and professional and public attention was focused on the carnage rampant in mines, mills, and factories, occupational health has once again become a major social issue, attracting considerable professional and public attention and giving rise to important new legislation and regulations in most provinces. A disaster like the 1992 explosion in the Westray coal mine in Stellarton, Nova Scotia reminds us of the lethal dangers associated with certain occupations, but even aside from such large-scale, highly publicized, and fortunately infrequent tragedies as Westray, workplace conditions pose a threat to the lives and the health of many Canadians.

Human suffering is the first cost of dangerous conditions of work. To this must be added the cost of care and compensation for injured and ill workers, borne in the first instance by employers but largely passed on to consumers in the price of the product, plus the heavy economic burden borne by society in terms of reduced production. The economic, as distinct from the financial, cost is shared, in the end, *by society at large*, even though it shows up in the first instance in the accounts of employing organizations. Interest in and responsibility for safety in the workplace are shared by employers, workers, and the public; in our era, 'the public' is typically represented by the state. Policy must take account of the interests of these three parties, and of their interactions.

The passage of the Occupational Safety and Health Act (OSHA) by the Congress of the United States in 1970 both testified to growing concern over and stimulated further interest in the issue throughout North America (Ashford, 1975). It is more than a coincidence that many provinces of Canada subsequently enacted major legislation of their own. Here we shall pay particular attention to programs in Ontario and Quebec.

We have paid some attention to work accidents and work-related illnesses in Chapter Four on Workers' Compensation. In addition to providing health care to injured or sick workers after the fact, Workers' Compensation Boards/Commissions in all jurisdictions have long exercised some powers to regulate safety in workplaces. In so doing they pursue the compatible goals of reducing *accidents* and reducing *claims* for compensation and for health services.

In a few provinces, including Quebec, responsibility for work safety rests principally with the Workers' Compensation authority; in most, the responsibility lies with the Departments of Labour or their equivalents. As we have seen, the Quebec legislature in 1979 changed the name of the responsible administrative body in Quebec from Workmens' Compensation Commission to the broader Commission on Health and Safety at Work (Commission de la santé et de la sécurité du travail, abbreviated to CSST), and at the same time enlarged its mandate in the fields of regulation, inspection, prevention, and research. In the section immediately preceding this one, we saw that regional public health units also have certain responsibilities with respect to work safety; and while it was not mentioned specifically in our description of CLSCs, they too may take on the provision of direct health-related services to working people in their territories.

In Ontario, the Occupational Health and Safety Act (now Chapter O-1 of the *Revised Statutes of Ontario, 1990*) gives broad responsibilities for the maintenance and improvement of safe and healthy conditions in workplaces to the provincial 'Workplace Health and Safety Agency'. This agency is financed by an allocation from the revenues of the Workers' Compensation Board, but is more closely linked with the Ministry of Labour, which carries out safety-related worksite inspections. Similar work is done in Manitoba, as we have seen, by the 'Workplace Safety and Health Division' of the Department of Labour, also paid for out of the revenues of the Workers' Compensation Board.

In both Quebec and Ontario, a crucial role in the maintenance and surveillance of safety conditions is assigned to joint health and safety committees. These committees must be organized in all workplaces employing more than a certain number of workers, with equal representation of the management and the employees; in an industry with several workplaces, none very large, in a given geographic location, 'sectoral' health and safety committees are authorized. The laws of the two provinces give these committees extensive powers to investigate conditions, to inspect worksites, and to make recommendations to which the employer(s) must respond. In Quebec, the public body, the CSST, provides technical advice to these workplace safety committees, as do also the public health units of the Regional Boards, but the committees are essentially relied on to prepare the safety programs for their own workplaces.

The underlying philosophy of this structure for the administration of the work safety legislation is twofold: first, it devolves substantial responsibility upon the people directly involved in the workplace; second, it assumes a large measure of recognized common interest in safety: the committees can only succeed if the atmosphere is more collaborative than confrontational.

Both of these aspects of the policy give rise to controversy. Some degree of devolution is inevitable whether it is part of the law or not, for no regulatory body can possibly be attuned to all the variations that exist from one workplace to the next. The devolution of responsibility to people on the site is in keeping with ideals of decentralization of authority in society, and of personal and group responsibility with respect to health. Some argue, however, that where things do not work out well, the devolution of responsibility provides a convenient route by which the public authority can escape blame.

As to the assumed common interest of all parties in improved work safety, no doubt some employers are convinced that it is in their long-term material interest to invest in improved safety conditions for workers; as a matter of fact, there are employers who insist that the care they take to put in place a superior health and safety organization gives them a cost edge over their less careful competitors. There are, however, notorious instances of employers who are not in the least enthusiastic about the work of the safety committees. Even in Quebec, where the law dictates that, in the absence of a health and safety committee, the employer must organize one at the demand of one employee, it cannot be said that there are active committees in all workplaces. The presence of an active union with informed and interested leadership is a favourable factor. And, with or without a safety committee, it is observed that workers themselves may aggravate certain problems by neglecting to take safety measures available to them; a commonly cited example is the neglect of devices that protect the ears in workplaces with dangerous noise levels. There is not much information in circulation concerning the actual functioning of health and safety committees.

Each observer of such situations will come to his/her own assessment of how the burden of responsibility should be apportioned: on the employer, to reduce the level of risk, though it may cost money to do so; on the government, to enforce standards rigorously, so that all employers will have to meet the same safety costs and the less scrupulous will have no advantage over competitors on that particular score; on the worker, to exercise vigilance where safety is concerned.

One somewhat paradoxical phenomenon has been an *increase* in the number of reported industrial accidents since 1980 (Ontario Task Force, 1987, p. 4). This may reflect a number of things: a real increase in the rate of accidents; the increase in number of persons employed as the economy has recovered from the recession after 1983; or better reporting of accidents, following the improvement of safety administration and better dissemination of information concerning the rights of injured workers.

Clearly, it is important to pay attention to the protection of health and the maintenance of safety in the locations where people spend as much as one third of their waking hours.

OTHER ALTERNATIVES FOR HEALTH POLICY

In addition to community health and occupational health, there are other kinds of programs that aim at improving health in ways that no amount of hospital care and medical care could reach: controls over industrial uses of technology; regulation of pollution and the disposal of waste; nutritional education programs; driver safety education and enforcement of safe driving laws; enforcement of automobile safety standards; attention to safety features of road design; fitness programs; preventive programs aimed at health-destroying social ills such as abuse of alcohol, tobacco, and drugs; prevention of sexually transmitted diseases; and many others. Some of these overlap with community health and occupational health, some call for interventions at other levels.

Threats to health sometimes arrive from unexpected directions, for example, the transmission of HIV/AIDS through transfusions of inadequately tested blood in the 1990s, a tragedy that gave rise to the Krever Inquiry into Canada's system of blood donation and distribution. A current health policy debate rages, not over whether tobacco is potentially harmful or not, but over the limits to be set to the advertising of tobacco products through the widely publicized sponsorship of sporting and cultural events by tobacco companies—an intervention that requires legislation. And bodies like the National Council of Welfare remind us, if a reminder is necessary, that poverty is bad for one's health, that mothers who are poor materially are likely to have babies who are poor physically, that poor children have a high probability of growing up to live shorter-than-average lives as unhealthy adults, that the poor people congregated in poor neighbourhoods live shorter, less healthy lives than their fellow-citizens congregated in higher-income neighbourhoods. A program aimed at providing adequate incomes is a health program—a point made in the federal government's 1974 Green Paper, *A New Perspective on the Health of Canadians* (National Council of Welfare, 1990; Health and Welfare Canada, 1974).

It is difficult, however, to locate responsibility in the most effective hands, to design the agencies and instruments most likely to do the job well, and to identify and mobilize the kinds of expertise needed. Progress along any of these lines requires much more than well-founded convictions and good intentions. Absolutely needed are large investments by private and/or public organizations, intelligent planning, and changes in behaviour on the part of masses of people.

Motivating people to behave in such ways as to prevent ill health, whether by diet, exercise, avoidance of behaviours known to be risky, and so forth, is naturally more difficult than motivating the same people to go to the doctor or to the hospital when illness arrives. With respect to prevention, people are asked to make an effort that will *probably* (but not certainly) prevent the occurrence of some harm that otherwise *probably* (but not certainly) will befall them. This hardly compares with the degree of certainty we have, rightly or wrongly, that medical treatment will deal effectively with something that certainly has happened to us. A 1996 Statistics Canada inquiry found that, in spite of the abundant promotion of better health habits over the last three or four decades, the physical health of the now middle-aged first wave of 'baby boomers' shows no striking improvement over that of their parents. Nobody ever said it would be easy.

LEGISLATION

Ontario. Occupational Health and Safety Act. *Revised Statutes of Ontario, 1990*, Chapter O-1.

Quebec. An Act respecting Occupational Health and Safety. *Revised Statutes of Quebec*, Chapter S-2.1 (was Chapter 63, *Statutes of Quebec, 1979*). Most of the Act concerns rights and obligations of employees and employers, health and safety committees, safety representatives, sector-based health and safety associations, and the regulatory mandate of the Commission.

————————. An Act respecting Health Services and Social Services. *Revised Statutes of Quebec*, Chapter S-4.2. Local Community Services Centres (CLSCs) are referred to in Articles 80, 131.

BIBLIOGRAPHY

General

Burgess, Michael (1996). 'Health Care Reform: Whitewashing a Conflict between Health Promotion and Treating Illness?', in Michael Stingl and Donna Wilson, eds, *Efficiency vs. Equality: Health Reform in Canada*. Halifax: Fernwood.

Canada. Health and Welfare Canada (1974). *A New Perspective on the Health of Canadians.*

Canada. Health Services Review 1979 (1980). Report of Hon. Emmett Hall, Special Commissioner. *Canada's National-Provincial Health Program for the 1980s: A Commitment for Renewal* (1980). (English and French, back-to-back.) This Report reviewed health care programs in Canada since the advent of hospital insurance and medicare. According to Malcolm Taylor, who was research consultant to Justice Hall, 'Funding [for this Review] was inadequate, lack of staffing absurd, and the targeted date for reporting wholly unrealistic' (Taylor, 1988, p. 429). See Chapter 6, which gives some attention to lifestyles and health care, preventive health care, mutual aid groups, and community health.

Davis, D.L., and H.P. Freeman (1994). 'An Ounce of Prevention', in *Scientific American*, Vol. 271, No. 3.

Demby, A. (1980). 'Preventive Intervention in an HMO', *Social Work in Health Care*, Vol. 5, No. 4 (Summer). Some mention has been made in the text of the so-called 'Health Maintenance Organization' (HMO). An HMO is a consortium of health practitioners, not necessarily limited to physicians, that contracts to look after a range of health services for its subscribers over a period of time. The subscribers do not pay fees for individual services rendered; they pay a fee for services over the period of the contract. In the United States, HMOs are the vehicle for many employment-based health insurance plans, partly because of their perceived cost-controlling capability.

Dreitzel, H.P., ed. (1971). *The Social Organization of Health: Recent Sociology* No. 3. New York: Macmillan. See especially 'The Social Causes of Disease', by Alfred Katz, pp. 5–14 and 'The Health Crisis of the Poor', by Rodger Hurley, pp. 83–122. (Hurley's data relate to the United States, where as we all know there is no universal public health care insurance; but that does not affect the relevance of his work, for the 'health crisis' he analyses is so deep and wide that hospitals and doctors could not resolve it.)

Dubos, René (1971). *Mirage of Health*. New York: Harper and Row. See especially Chapters V and VIII. Philosophical speculations on the history, mythology, science, and technology of the pursuit of health and the treatment of illness.

Evans, Robert G., M.L. Barer, and T.R. Marmor, eds (1994). *Why Are Some People Healthy and Others Not? The Determinants of the Health of Populations*. New York: Aldine de Gruyter.

Fuchs, Victor R. (1983). *How We Live: An Economic Perspective on Americans from Birth to Death*. Cambridge, Mass.: Harvard University Press. A highly regarded American health economist traces the life cycle of the general North American population from a number of linked points of view, notably including work, income, and health.

Gorin, Stephen, and Cynthia Moniz (1996). 'From Health Care to Health: A Look Ahead to 2010', in Paul Raffoul and C. Aaron McNeese, eds, *Future Issues for Social Work Practice*. Toronto: Allyn and Bacon.

Hayes, M.V., L.T. Foster, and H.D. Foster, eds (1994). *The Determinants of Population Health: A Critical Assessment*. Victoria, B.C.: Western Geographical.

Health and Social Work (1985). Tenth Anniversary Issue, Vol. 10, No. 4.

Health Values. A journal that gives attention to non-medical approaches to 'wellness'. See especially Vol. 8, No. 5 (Sept.–Oct. 1984).

Higgins, Agnes C. (1972). *A Preliminary Report of a Nutrition Study on Public Maternity Patients, 1963–1972*. A path-breaking longitudinal study of the effects of pre-natal nutrition on the long-term physical *and mental* health of children. The study was carried out at the Montreal Diet Dispensary, of which Mrs Higgins was the director for many years. An excellent example of a non-medical contribution to health, and of a non-medical health service need. Regrettably, not easily accessible, but of historic significance.

Illich, Ivan (1975). *Medical Nemesis: The Expropriation of Health*. Toronto: McClelland and Stewart. The consequences for society in general of the professional and technological domination of health care. (No references to Canada.)

Langlois, Kathy (1997). 'A Saskatchewan Vision for Health: Who Really Makes the Decisions?', in Robin Ford and David Zussman, eds, *Alternative Service Delivery: Transcending Boundaries*. Toronto: KPMG Centre for Government Foundation and Institute of Public Administration of Canada. Saskatchewan decentralized to better pursue 'Wellness'—or did it?

Leeson, Howard (1996). 'Health Care Reform in Saskatchewan', *Management* (Institute of Public Administration of Canada) 6/4. The reform Leeson describes was promoted by the Saskatchewan government under the banner of 'Wellness'.

Lesemann, Frédéric (1984). *Services and Circuses*. Translated from *Du Pain et des services*. Montreal: Black Rose. An account of the social and health services reforms of the

1970s in Quebec, strongly suggesting that bureaucratic influence was reinforced by the process.

Levin, L.S. (1987). 'Every Lining Has a Cloud: The Limits of Health Promotion', *Social Policy* 18/1. A useful corrective to over-enthusiasm.

Mustard, J.F., and J. Frank (1991). *The Determinants of Health*. Toronto: Canadian Institute for Advanced Research.

National Council of Welfare (1990). *Health, Health Care and Medicare*. Ottawa: The Council.

Navarro, Vincent (1976). *Medicine under Capitalism*. New York: Prodist. Navarro sees medicine, as practised in capitalist societies, as an instrument, witting or unwitting, of class domination.

Offord, D., M. Boyle, and Y. Racine (1990). *Children at Risk: The Ontario Child Health Study*. Toronto: Queen's Printer for Ontario.

Rachlis, Michael, and Carol Kushner (1994). *Strong Medicine*. Toronto: HarperCollins.

Rachlis, Michael, and Carol Kushner (1989). *Second Opinion: What's Wrong with Canada's Health Care System and How to Fix It*. Toronto: Collins.

Swartz, Donald (1977). 'The politics of reform: conflict and accommodation in Canadian health policy', in Leo Panitch, ed., *The Canadian State: Political Economy and Political Power*. Toronto: University of Toronto Press.

Taylor, Malcolm (1988). *Health Insurance and Canadian Public Policy*. 2nd edn. Toronto: University of Toronto Press.

Vakil, Thea (1997). 'Bringing Health Closer to Home: Reform in British Columbia', in Robin Ford and David Zussman, eds, *Alternative Service Delivery: Transcending Boundaries*. Toronto: KPMG Centre for Government Foundation and Institute of Public Administration of Canada.

Community Health

Anctil, Hervé (1982). 'Les Dix Ans des CLSC', *Carrefour*, Vol. 4, No. 4 (September). A thumbnail evaluation of the first ten years' experience of Quebec's CLSCs. (See also Duplessis, cited below.)

Banta, James E. (1979). 'Definition of Community Health', in National League for Nursing (U.S.A.), *Community Health—Today and Tomorrow*.

British Columbia (1974). *Consumer Participation, Regulation of the Professions, and Decentralization*. A Special Report prepared by J.T. McLeod, appended to *Health Security for British Columbians* (the Foulkes Report). Victoria, B.C.: Queen's Printer.

Canada. Department of National Health and Welfare (1970). *Task Force Reports on the Cost of Health Services in Canada, Volume 3: Health Services*. Appendix 3, Part C: 'The Nurse and the Social Worker as Multipurpose Workers in Community-Oriented Practices', pp. 138–56.

Canada. Health and Welfare Canada (1974). *A New Perspective on the Health of Canadians*.

Canada. Report of the Community Health Centre Project (1972). *The Community Health Centre in Canada* (the Hastings Report), Vol. 1, pp. 1–59.

Canada. Senate of Canada (1970). *Proceedings of the Special Senate Committee on Poverty* (1969–70 Session), No. 45, Statement submitted by and testimony of Pointe St Charles Community Clinic, pp. 9–21, 39–40. This pioneer community clinic was a direct predecessor of the CLSCs in Quebec. Part of the initiative for its foundation came from the Faculty of Medicine of McGill University. The testimony given by its representatives to the Senate Committee on Poverty laid heavy emphasis on the specific health deficits attributable to an environment of deprivation.

Canadian Council on Social Development (1985). *Community-based Health and Social Services: Conference Report.* Ottawa: The Council. Note Chapter 2, an attempt to define community health. More of a wish-list than a concrete definition.

Demers, Andrée, and Deena White (1997). 'The Community Approach to Prevention: Colonization of the Community?', *Canadian Review of Social Policy*, No. 39 (Spring). When health professionals sit down with community folks, who takes over whom?

Duplessis, P. (1982). 'Opinions on Community Health Departments Ten Years after the Reform', *Administration Hospitalière et Sociale,* Vol. XXVIII, No. I. (See also Anctil, cited above.)

Lomas, Jonathan (1985). *First and Foremost in Community Health Centres: The Centre in Sault Ste. Marie and the CHC Alternative.* Toronto: University of Toronto Press. See especially Chapter 10, 'The Role of the Community'.

Mable, Ann L., and John F. Marriot (1994). *Comprehensive Health Organizations: A New Paradigm for Health Care.* Kingston: Queen's University School of Policy Studies.

Marchand, Robert, Jacques Durocher, and Pierre Tousignant (1996). 'Assessing the Effects of Health Service Reform on Population Health: A New Role for the Public Health Directorate in Quebec', in Michael Stingl and Donna Wilson, eds, *Efficiency vs. Equality: Health Reform in Canada.* Halifax: Fernwood.

Montreal General Hospital, Department of Community Health (1985). *Health Priorities 1985–1987.* This document gives an idea of one hospital's community health program; as noted in the text, the work of the 'public health units' that replaced the 'Community Health Departments' is quite similar. (See 'Régie régionale. . .', cited below.)

Ontario. The Premier's Council: Final Report of the Resource Management Committee (1995). *Optimizing Resources for Health: Are we tackling the real issues?*

Quebec. Report of the Commission of Inquiry into Health and Social Services (1968–70), Volume IV, *Health,* Tome II, Chapter 5, 'The Health Plan', especially pp. 37–57 on health centres and regionalization, and Chapter 6, pp. 93–109. This Report presents the idea of the health centre as it was conceptualized before it took form in the Health and Social Services Act. (Paging is different in the original French version.)

Régie régionale de la santé et des services sociaux de Montréal-Centre (1995). *Rapport d'activités 1994–1995.* Pp. 19–20 describe the work of the public health directorate of the regional board, including the work delegated to the two public health units in hospitals in Montreal. (See 'Montreal General Hospital', cited above.)

Villedieu, V. (1976). *Demain la Santé.* Québec: Le Magazine Québec-Science. Chapters 7 and 10 include passages on the DSCs and CLSCs; see pp. 171–9, 234–9.

Young, T. Kue (1975). 'Lay-Professional Conflict in a Canadian Community Health Centre', *Medical Care* 13 (November). An examination of the failure of a community clinic in Regina. The blame is shared.

Occupational Health

Ashford, N. (1975). *Crisis in the Workplace: Occupational Diseases and Injury.* Cambridge, Mass.: MIT Press. The final report of a Ford Foundation study. This is an American book about American industry, and should be read with due reservations by Canadians. The Summary, pp. 1–37, however, outlines many of the organizational, economic, political, and social issues surrounding the regulation of health and safety conditions in industry in a social system much like Canada's.

Canada. Human Resources Development Canada (1994). *An OSH Program in Your Work Place.* (OSH Stands for Occupational Health and Safety.)

Canada. Labour Canada (1991). *Occupational Safety and Health Concerns of Canadian Women.*

Canada. Labour Canada (1989). *The Selective Protection of Canadian Working Women.*

Doern, G. Bruce (1977). 'The Political Economy of Regulating Occupational Health: The Ham and Beaudry Reports', *Canadian Public Administration* 20/1 (Spring). The two reports referred to in the title studied conditions in the mining industry in Ontario and in the asbestos industry in Quebec. The article analyses the political-economic environment of occupational health regulation in general.

Leyton, Elliott (1987). 'Dying Hard: Industrial Carnage and Social Responsibility', in D. Coburn et al., *Health and Canadian Society: Sociological Perspectives.* 2nd edn. Toronto: Fitzhenry and Whiteside.

Ontario Task Force on Vocational Rehabilitation (1987). *An Injury to One Is an Injury to All.* A summary of a Report submitted to the Minister of Labour.

Perceptions (1982). Canadian Council on Social Development. One issue of this magazine largely devoted to health and safety at work.

Québec. Commission de la santé et de la sécurité du travail. *Annual Report* (current). Note especially passages concerning prevention activities, organization of Safety Committees in workplaces, etc.

Québec. Commission de la santé et de la sécurité du travail (1984). *The Québec Occupational Health and Safety Plan: Elements of comparison with other Canadian plans.* 2nd edn.

Québec. Commission d'enquête sur les services de santé et les services sociaux (1988). *Rapport.* Especially pp. 445–72, on health goals. English *Summary* (1988).

Quebec. Minister responsible for Social Development. *Health and Safety at Work* (1978). (French title: *Santé et sécurité au travail.*) This document was a so-called White Paper, meaning a fairly firm statement of government policy preceding the presentation of related legislation. It reviewed the historical and legal background of occupational health and safety in Quebec (painting a rather dark picture), and presented the main lines of the policy underlying the Act respecting Health and Safety at Work of 1979.

Reasons, C., L. Ross, and C. Paterson (1981). *Assault on the Worker.* Toronto: Butterworths. Chapters 2 and 6, and Appendix D. Canadian content, with a strong Western Canadian emphasis.

PART THREE

Retrospect and Prospect

The New Millennium:
Sunset or New Dawn of the Welfare State?

OVERVIEW

Over the past three-quarters of a century, the history of the welfare state in Western democratic countries presents a fairly widely accepted generalized picture. To thus limit the time frame omits important earlier developments, such as those undertaken in Germany under Bismarck more than a century ago and in Britain during the Liberal Party ascendancy before World War I; the limitation to democracies passes over in silence the misadventures of twentieth-century fascist states, some of which did, with all their follies, have elaborate welfare-like programs. This limited history runs roughly as follows.

First was the response to the crisis posed by the Great Depression of the 1930s. Second, there was a steady growth in the size and importance of welfare programs throughout the decade or so after World War II. Third, there followed a considerable expansion of programs in the 1960s and early 1970s, the era of 'New Frontiers', 'Wars on Poverty', 'Great Societies', and 'Just Societies' (Kennedy and Johnson in the United States, Pearson and Trudeau in Canada). Fourth, this flood was followed by an ebb tide of restraint and reappraisal through the later 1970s and the 1980s, an ebb that has, if anything, accelerated into the 1990s. This latter period has been characterized by widespread economic difficulties, of which Canada's share was the serious recession from about 1982 to 1986 (Guest, 1987; Moscovitch and Drover, 1987), another recession in the early 1990s, and the uneven recovery thereafter. The changes that welfare programs have undergone in the last two decades have in some cases stepped up from changes in degree into changes in kind.

In Canada's case, the student will not go far wrong in using the dates of the major federal interventions as markers on Canada's path towards a welfare state. As markers only: by no means has progress in social policy in Canada come about always through the leadership of the federal government. Provincial welfare and health departments would rightly resent any such inference. The original programs for the financial support and the health care of the indigent were necessarily the work of the provinces, within whose jurisdiction these matters lay, and still lie. And in the later history, as we have seen, other major initiatives have come from the provinces, as in hospitalization insurance and medical care insurance; federal interventions have followed commitments by provinces as often as they have stimulated them.

With that important proviso, the stages along the road may conveniently be marked by the following legislation: the Old Age Pensions Act of 1927; the Unemployment Insurance Act of 1940; the Family Allowances Act of 1944; the Old Age Security Act of 1951; the Hospital Insurance and Diagnostic Services Act of 1957; the Medical Care Act of 1966; the Canada Assistance Plan Act, the Canada Pension Plan Act, and the Guaranteed Income Supplement amendment to the Old Age Security Act, all of 1967; and the Unemployment Insurance Act amendments of 1971 and 1976.

Steps that many regard as backward turns in the road, discussed in previous chapters, are (at the federal level) the far-reaching Unemployment Insurance amendments of 1981 and, still more so, the further-reaching change to employment insurance of 1995–6; the income limits on Child Benefit and (projected) Seniors Benefit recipients and the only partial indexing of the benefits paid through those programs; and the overall reductions in transfers to the provinces, which cut into welfare and health care programs (though, hopefully, only temporarily). But the federal level has no monopoly on backward movement, any more than on forward. Since the late 1970s, the provinces, too, have contributed to the revision of the welfare state: provinces have altered welfare eligibility rules so as to reduce payments to 'employable' recipients (Quebec); have redefined 'other income' to include income from universal programs, thereby reducing welfare payments (Saskatchewan); have broadened their resort to user charges for hospital services, and so on. An independent-minded student will of course take the trouble to examine such policy shifts, and the reasons by which they have been justified, before concluding that they are all necessarily retrograde steps. At the very least, however, these shifts suggest that the Canadian welfare state, as it has been understood, is highly vulnerable.

Throughout almost the entire democratic world, the economic difficulties of the 1980s and 1990s have prompted a wave of reaction against direct government intervention in general, not only with respect to social welfare. In Canada in the 1980s, self-styled Conservative governments were elected resoundingly at the federal level, followed in 1993 by a Liberal government fairly characterized as socially conservative. Re-elected in 1997, it faced an official Opposition party, the Reform Party, which espoused a clear and explicit platform of reducing all kinds of government intervention. Small-c conservative governments, whatever their party labels, have won and held power in almost all of the provinces. Somewhat against the tide, mildly left-wing New Democratic Party governments have come to power in British Columbia and Saskatchewan. The Saskatchewan government has proudly combined successful deficit-cutting with some fidelity to its traditional welfare philosophy.

This observed swing to the right might have been otherwise: in the past, economic distress has usually led to more rather than less social intervention by the state. Some observers argue that the 'wave of neo-conservatism' is the result of the tightening grasp of the corporate élite on the political process. But few if any of the same observers had been heard to argue that the corporate grasp had been loosening in the preceding thirty years of expanding government. At least as plausible as the corporate élite theory is the thought that the conservative wave on the part of non-élite electorates expresses a certain scepticism toward the interventionist policies of the preceding three decades, not least in the field of welfare: a certain disappointment over the limited fulfilment of bright promises to the unemployed, the young, the old, the poor, in return for the costs; a growing fatalism about what can reasonably be expected in the way of effective social change; around the world, a spreading—and alarming—cynicism about democratic politics; and, as a result of all this, a falling back on familiar private values. As always, facile exaggerations must be avoided: Canadians may be sceptical about government's capacity to deliver, but they have not withdrawn their support in principle for the social safety net (*Time* Canada, 1995; *Maclean's*, 1995; Crane, 1994, ch. 8).

Both governments and electorates have also been persuaded, rightly or wrongly, that the deficit-and-debt situation described in this book's introductory chapter requires a general tightening of the public belt, at least temporarily. More mordant critics see no mere attrition, but rather a systematic attack on the very basis of the welfare state, a deliberate turning away from the priorities assigned to the mitigation of poverty and to universal access to health care; and they are inclined to see the heavy emphasis on deficit- and debt-reduction as 'The Great Deficit Hoax'—a handy, all-purpose excuse for a radical reduction in the role of government (Canadian Centre for Policy Alternatives, 1996; Cameron and Finn, 1996).

Whatever the mix of causes, there is ample evidence in Canada both of limited attainment of some of the goals of the welfare state and of deliberate retrenchment in the social policies of most of our governments. Whether or not one concludes that the welfare state has 'failed', one must at least acknowledge that many things are not working out as hoped.

As to the attainment of goals, incomes remain highly unequal; the poor remain numerous; the number of Canadians dependent on public assistance, after a brief period of stabilization, has risen again (but may fall back, not because there are fewer in need, but because rules governing eligibility have tightened); unemployment, especially among the young, remains at a high level (even as the number employed full-time continues to increase); and there is a persistent increase in such manifestations of poverty as homelessness and resort to food banks.

As to retrenchment in social policy, Unemployment (now Employment) Insurance has pulled back sharply from the levels of provision offered even a few years ago. The number of recipients of benefits has dropped noticeably, despite continuing high unemployment. No universal income support programs remain—the last such program, Quebec's Family Allowances, expired in 1997 (Québec: Ministère de la securité sociale, 1997). The financially squeezed public pension programs are under critical scrutiny on the part of the public, to the point where large numbers of young people apparently do not expect ever to receive pensions from them. Federal financial support of provincial public assistance programs is becoming less generous, and the provinces are responding by simply reducing the benefits and/or by stepping up their surveillance of the able-to-work. It must be noted that restrictive measures have been taken by provincial governments of all partisan stripes, from the harsh cuts of the proclaimed right-wing conservatives of Alberta and Ontario to the milder ones of the professed socialists and social democrats of British Columbia, Saskatchewan, and Quebec (Mishra, 1987). A severe brake has been applied to the further growth of every province's medical care and hospital services programs; and, as we have remarked, government health departments have been driven to impose widespread hospital closures, as much to contain costs as to rationalize services (Rosenberg and James, 1994).

Every reduction in any social program cannot be considered necessarily a change for the worse, but such coast-to-coast, across-the-board restraint in social programs constitutes a critical juncture for social policy in Canada. It is highly unlikely that an economic recovery, even a strong one, will see Canada retrace the steps of the last few years, taking us back to where we were.

THE LIMITED ATTAINMENT OF GOALS

Inequality

Even though equalization has never been an explicit goal of Canadian social policy or a widely proclaimed ideal, measures of income inequality may be adduced as indicators of the progress, or lack of it, of the welfare state.

Using annual incomes as the key measure, Canada remains a highly unequal society. The proportion of the annual national income going to the top stratum of the population, after taxes and after transfer payments to individuals, remains almost as high as it ever has been—by some measures, it is increasing, as a result of a subtle upward and downward polarization of incomes from employment (Morissette, Myles, and Picot, 1995, Table 3). The proportion going to the lowest stratum is only slightly larger than several decades ago (Ross, Shillington, and Lochhead, 1994, p. 43; National Council of Welfare, 1985). The effect of the recession of the early 1980s is instructive. Real incomes declined right across the spectrum, as befits a recession, but the highest income earners, as a class, cushioned the shock better than the middle and lower income groups (National Council of Welfare, 1985, p. 57).

One may put the best face on it one can: there has been some increase in the real national income; the lowest quintile's share today is a slightly larger slice out of a much larger pie—but only after taxes and transfers are taken into account. The real purchasing power of average family incomes increased by nearly thirty per cent between 1969 and 1984, but has drifted downward since then; it is noteworthy, however, that the incomes of elderly-headed families in that period increased by over one-half in real terms. Moreover, any given distribution of income has a more favourable meaning in terms of overall well-being in a society with free health care and an assured income floor in old age than in a society lacking those features. The 'annual value' of health care insurance is not counted as part of anyone's income after taxes and transfers, but it is surely worth a great deal financially to those who may need health services at any time, as is more likely for low-income people.

The recent modifications to family benefits and to benefits for older people will make some difference to the after-tax, after-transfer distribution of income after 1997. Lower-income people with dependent children and lower-income older people will have a somewhat reduced tax burden and much-enhanced transfer incomes; middle and higher-income people in those categories will receive no transfer incomes at all from these programs, and will henceforth gain little or nothing from related tax credits. How greatly this affects final income distribution, however, remains to be seen; it will depend on whether upper-level incomes from earnings and investment continue to increase in relative terms, as apparently they have in recent years. Incomes from paid work are highly unequal, and incomes from investment—that is, from 'ownership of the means of production'— are even more unequal. (And, as noted earlier in the text, the distribution of *wealth* is still more steeply unequal than that of income.)

How much priority do Canadians really give to material equality as a value? On the one hand, the fairness and indeed the economic rationality of the highest incomes are widely questioned. On the other hand, our values probably allow a generous reward to the successful achiever. (There are, however, questions about those who grow rich on the achievements of others.) And even many egalitarians profess exasperation when accused of promoting literal equality of income (Tawney, 1964; Ryerson, 1981). What cannot be specified with confidence is the degree of inequality that a national consensus would endorse nor the degree of equality that would be regarded as essential to the character of a welfare state. Nor, most emphatically, can anyone specify the sacrifice that Canadians with above-average incomes would willingly make in order to achieve some greater measure of equality.

It is necessary to distinguish the goal of equalization of incomes from that of assurance of *adequacy*, or the reduction of poverty. The maintenance of an adequate minimum living standard for all is a much higher priority for Canadians than any particular degree of material equality. This may be inferred from many indications: the general tenor of income support programs that have proven politically viable; the platforms of the major political parties; the thrust of major public inquiries; and the critiques of observers ranging as far to the left as the authors of *Inequality* (Moscovitch and Drover, eds, 1981) and *The 'Benevolent' State* (Moscovitch and Albert, eds, 1987). Continued poverty is considered much more relevant than continued inequality in evaluating the performance of the welfare state in Canada. To poverty we now turn.

The Persistence of Poverty

In assessing Canada's success or failure in providing minimally adequate material standards for all, four kinds of evidence are commonly put forward:

- the number and proportion of Canadians living under some variant of what we commonly if not quite accurately call the 'poverty line';
- the extent of poverty among children, a phenomenon fraught with long-term menace;
- the increasing numbers of Canadians seeking assistance at voluntarily operated food banks; and
- the increasing numbers of homeless people in Canada in recent years.

Canadians of Low Income

Even though all programs of the 'welfare state' are not for the poor alone, the number and the plight of the poor are primary data for a judgement of Canada's welfare state performance. If poverty is to be expressed in terms of a meaningful dollar–income 'poverty line', then the ultimate objective of a Canadian welfare state must be to have zero Canadians living in poverty. We are far from attaining any such objective, whatever current 'poverty line' is used as the measuring instrument. In the discussion of 'poverty lines' in Appendix A, a distinction is drawn between 'market-basket' and 'relative' poverty lines. In Canada there is no 'market-basket' poverty line in general use as a reference point. The provinces' public assistance benefit scales are generally based on the market-basket cost of meeting basic needs, but they cannot be used as 'poverty lines'. For one thing they vary considerably; and for another, in order to arrive at the 'minimally adequate' dollar–income standard of living implied by these scales, one would have to factor in the dollar values of other items that also vary, like housing allowances, permitted earnings, and tax credits.

Public assistance recipients in this country are called upon to live at a level at or not far below a 'market basket' definition of long-term minimally adequate subsistence. A discussion of poverty, in any case, must not be confused with a discussion of public assistance: welfare recipients account for a substantial percentage of Canadians counted as 'poor', but not a majority, as we have known at least since the 1970 Senate report on poverty.

There are many other Canadians whose annual incomes are in much the same range as the incomes of welfare recipients: older people, mostly older women, with little income other than their Old Age Security and Guaranteed Income Supplement (or Seniors Benefit), low-paid workers, irregular workers, workers depending on part-time jobs, people marginal to the labour market, like subsistence farmers and some classes of immigrants. And there is, in addition, a statistically elusive, heterogeneous category whose incomes, at any given time, are even further below the poverty level because they have temporarily or permanently fallen through what Sherri Torjman has dubbed 'the tangled safety net' of the public welfare system: transients, dropouts, dischargees from institutions of one kind or another, Native people living at odds with the dominant culture, lone parents whose wages or salaries may appear adequate but whose expenses gouge holes in their disposable incomes, victims of accident or disabling illness, survivors of broken families.

The poverty line most often referred to in Canadian discourse is that developed by Statistics Canada. The allowances that any welfare program in Canada provides are well below the StatsCan low-income cut-offs (LICOs)—so much so that in 1988 the Ontario Social Assistance Review Commission proposed as a medium-term goal the raising of the total incomes of families on welfare in that province to 85 per cent of the LICOs. As explained in Appendix A, the StatsCan standard of 'low income' depends on the standard of living of the overall population. Average income has risen over time, though with occasional reversals, as in 1981–4 and again in 1989–92, and LICOs have risen accordingly—much as though the standard height that defined 'short' were to rise as the average height of a population increased. In 1984, depending on the size of community of residence, the LICOs for a family of four ranged from $14,720 (rural) to $20,010 (metropolitan) per year; in 1996, the same LICOs ranged from $21,690 to $31,862, in 1996 dollars (National Council of Welfare, 1997). The nature of the LICOs must be borne in mind in interpreting most of the reported trends in the numbers of the poor in Canada.

Under the StatsCan definition, Canada's success in abolishing poverty has been barely perceptible. The percentage of the population under the StatsCan LICOs has stayed consistently between 14 per cent and 18 per cent. During the recession of the early 1980s it appeared to rise somewhat, as was to be expected, from a low of 14.7 per cent in 1981 to a high of 18.2 per cent in 1983; that meant nearly four and a half million people. In

1989, the rate dropped to 13.6 per cent. In 1995, it rose to 17.4 per cent, as the lagging effects of the recession of the early 1990s continued to weigh upon the poor (National Council of Welfare, 1997). Improvement on this score is slow and irregular. (With the population now approaching 30 million, each one-tenth of one per cent represents roughly 30,000 people.)

As we have seen, transfer incomes do not all go to the poor, but to the poor, transfer incomes are vitally important. For older Canadians with incomes below the LICOs, over 90 per cent of income comes through transfers, principally OAS/GIS and including as 'transfers' Canada/Quebec Pension Plan pensions. Even those older couples whose incomes are above the LICOs realize, as a group, about 40 per cent of their incomes from transfer programs. For the approximately three million who are dependent on welfare, transfer incomes (public assistance, child benefits) will amount to very nearly 100 per cent of their total incomes.

Critics of low levels of public assistance sometimes express their outrage by alleging that welfare is designed to keep people poor. Welfare will certainly not raise people out of poverty, but government transfer programs make up nearly half of the incomes of *all* poor families with children, and nearly 100 per cent of the incomes of families on welfare. It is a truism, of course, but it is difficult to imagine the plight of the lowest fifth of the Canadian population without the transfer programs that are in place (Moscovitch, 1985, pp. 20–1).

The simple number or percentage of people with inferior incomes is a rather crude measure of national poverty. Some people will have incomes just a little below the line, others far below. Both types are counted equally as 'poor', but their respective poverties are far from equal. The National Council of Welfare's series of reports entitled *Poverty Profile* presents, among other analyses of poverty in Canada, some statistical measures to show just how poor the poor are. For the year 1995, for example, the National Council segregated the population with incomes below the StatsCan LICOs; among the poor so defined, standard families (couples under sixty-five with children under eighteen) had incomes, on average, $8,564 less than the low-income defining lines applicable to them (National Council of Welfare, 1997, pp. 51–2). As always, one must bear in mind what is implied by an 'average': if the average poor family of this description had an income $8,564 below the LICO, then many poor families must have had incomes even lower than that.

Poverty Among Children

Old age was long observed to be a principal correlate of poverty. It is less so now than ever before; today the very young are more at risk. To emphasize poverty among the very young is not to minimize its ravages among the old, but in Canada, as elsewhere, public policy addresses the income problems of the elderly fairly substantially. The National Council of Welfare has tracked a steady decrease in the proportion of older Canadians whose incomes are below the StatsCan LICOs—from 34 per cent in 1980 to 16.9 per cent in 1995 (National Council of Welfare, 1997, p. 13). In 1994–5, about 3.5 million older Canadians received well over $20 billion in the Old Age Security package of transfer programs (OAS, GIS, and Spouse's Allowance), excluding C/QPP pensions and all provincial top-ups. In 1995, a couple, both eligible for OAS and GIS, with no other resources at all, were assured of a combined income of $16,642 from that source alone; and since the vast majority do have other income, almost all older couples did have incomes higher than that (National Council of Welfare, 1997, p. 64). That OAS/GIS income of $16,642 actually exceeded the StatsCan LICO for rural areas and almost reached the LICO for small towns, falling more seriously short for medium to large communities. At worst, the distance an older couple or an older person can fall below the poverty line is limited. In comparing the material support of the old and the young, it must also be taken into account that older people receive more 'in-kind' benefits.

With children, the universally available safety net is far lower and looser. (We distinguish here, as one must, between what is 'universally *available*' and what is 'universally *provided*'; as we have seen, there are no more universally provided income benefits, for children or anyone else, but the Child Tax Benefit is available to everybody, should their incomes fall below the statutorily determined level.) Child Tax Credits and supplements in 1993–4 afforded maximum annual benefits of a little over $1,200 per child nationwide. Total federal benefits were a little over $5 billion. Provincial child supplements added a little over $500 million, mostly via Quebec's family allowances and allowances for young and for new-born children. Thus, families with low incomes through earnings, poor by the StatsCan standard, got much less relief for the support of their children than for the support of their elders.

Poor families mean poor children. Families on welfare are poor, and their children are poor. All commentators have observed, with some anxiety, the increases in the numbers of younger people on welfare, including young married people and, most notably, young lone parents. The additional welfare allowances given by provinces for additional children, devised as they are with one eye on the federal Child Tax Benefit, are such that, with each child, the family will fall a little further below the LICOs.

Other family units with children, that fall into certain vulnerable categories, swell the number of children living in conditions of poverty:

- single-parent families (now known to Statistics Canada as 'lone-parent families'), most often the product of divorce or separation; where female-headed, as is nearly always the case, lone-parent families have a 60-40 (or 3 to 2) chance of being poor; where the mother is young, as a growing number are, and/or where there are two or more pre-school children, the chances of being low-income are 9 to 1;
- families with heads vulnerable to unemployment;

- families with heads in low-income jobs, especially if only one parent is earning an income;
- families with heads in the early years of their work careers, when their incomes are at their lowest.

The upshot is that, according to a National Council of Welfare estimate, over one-fifth of Canadian children—just under 1,500,000—were poor in 1995, the worst record since Statistics Canada began publishing analyses of low income (National Council of Welfare, 1997, pp. 75-6; Caledon Institute, 1995, p. 1). Sad to say, the percentage is even higher among children under seven. Grimly but inevitably, poor children have a higher mortality rate, a greater prevalence of developmental difficulties, and generally worse health, on all indicators, than other children (Canadian Council on Social Development, 1989). These deficits translate into lifelong disadvantages, inflicting a heavy price upon the individuals concerned, and upon society.

Poverty among children has become the focus of attention of many observers of social policy in Canada in recent years (e.g., Caledon Institute, 1995; Battle and Muszynski, 1995). As stated in the chapter on public assistance, concern over the welfare of children in the families of the working poor as well as in families on welfare inspired the recommendation of Ontario's Social Assistance Review Commission (1988) that the province create a new program of enhanced income-tested benefits available to *all* children.

The conception behind the recommendation is intrinsically interesting. It seeks to deal with two problems at once: first, the well-documented extent of low incomes among working family heads, one result of which is deprivation for their children; and second, the equally well-documented 'welfare trap': the perverse effect of the scaling of welfare benefits to family size, as a result of which many family heads receiving welfare will be better off on welfare than if they were working for insecure earnings that take no account of the number of dependent children (Naylor, Abbott, and Hewner, 1994). The proposal was to make public assistance benefits entirely independent of the number of children in the family—'to remove children from the welfare system', by introducing a new income-tested benefit for *all* low-income families. The proposal attracted favourable attention from one (NDP) government of Ontario (Ontario Ministry of Community and Social Services, 1993, p. 17), but the succeeding (Conservative) government has had a very different approach to welfare. Broadly speaking, this idea lies behind the new federal Child Tax Benefit and Quebec's current revision of its package of benefits for families with children.

The Ambiguous Experience of the Food Banks

Feeding the hungry has always been an honourable pursuit. A number of years ago, a new movement developed in several North American cities to rationalize the somewhat hit-or-miss voluntary giving of food to hungry people. The method is to mobilize the food production and distribution industries so as to collect systematically and to make available to poor people food that might otherwise be wasted by retail food outlets and restaurants, producers, and wholesalers. Graham Riches, the closest student of this phenomenon in Canada, arrived at a careful estimate that 'between 1981, when the first food bank was started in Edmonton, and the end of 1984 seventy-five organizations in Canada calling themselves food banks were set up' (Riches in Ismael, 1987, p. 126). Riches certainly did not overstate the numbers. Centraide Montreal reported in 1988 that 'the number of soup kitchens in the Montreal area had increased from six (6) to about forty (40) in the past four years, and the number of food depots has risen from fifteen (15) to more than sixty-five (65) over the same period' (Centraide Montreal, 1988). The increase in the numbers of outlets for free food was outstripped by the increase in the numbers of their clients, as Riches's evidence shows. The phenomenon persisted through the recession of the early 1990s and the subsequent recovery, which left many unemployed, even though according to some indicators the economy improved.

Through the 1980s in particular the operation of food banks grew to so large a scale as to provoke uneasy second thoughts on the part of those involved in the experience. Riches observed that in Saskatchewan the surge in the use of food banks coincided with an increase in unemployment, nation-wide but very marked in Western Canada, that was accompanied by somewhat paradoxical reductions (rather than increases) in public assistance caseloads (Riches and Manning, 1989, p. 19). Patterns of usage in most places showed an increase in resort to the food banks toward the end of each month, when welfare incomes would likely have been exhausted—especially in months with a fifth weekend, meaning late arrival of an end-of-month cheque. A notorious incident in 1996 gave yet another indication of the near-institutional importance that food banks have taken on: one overburdened community food bank in British Columbia felt obliged to limit its provision of food packages to local residents—in other words, to impose a residence requirement, something that public assistance authorities are prohibited from doing by the terms of federal cost-sharing!

One tenable explanation for the rise of food banks is the face-value one: they have met, efficiently and economically, a high-priority need that is ever present, more than usually so during a period of recession. According to this view, if the food banks have added something from private good will to the lives of needy people, both receiver and giver have been blessed and there is little more to be said.

Riches and others propose a more disturbing alternative explanation. The number of clients is an inverse measure of the adequacy of the public safety net for people in hardship: if so many thousands line up for *food*, surely that indicates that the programs in place to help the needy are inadequate to the purpose. And as a second-order effect, it is at least conceivable that public welfare authorities would see in the food banks a rationale for not expanding public programs to meet the expanded needs. This view of matters is in keeping with the normal propensities of governments in a period when the political-economic climate is restricting government revenues and putting demands on government in many directions. It is consistent also with the widely observed trend among governments to 'privatize' functions. In this case, the privatization changes the claim the needy person has for support from one of defined public obligation to one of undefined and unenforceable private charity, a change of no small practical and philosophical significance.

Homelessness

Among the most distressing phenomena in Canadian society in the last decade or so has been the rise in the numbers of homeless people.

In the nature of things, the homeless are a difficult population about which to gather precise, reliable information. They have become highly visible on the streets of all of Canada's cities, but they cannot be counted easily. The increasingly hard-pressed voluntary organizations and municipalities that operate emergency shelters are usually able to say how many people they have assisted; but since there is a great deal of movement from place to place, in and out of such facilities, as well as movement in and out of homelessness, and since an undetermined proportion of the homeless seldom or never make use of shelters, an accurate census of the homeless is next to impossible. On the basis of the best information available to them, municipal officials in greater Montreal have estimated that there were as many as 20,000 homeless people in that city in 1995–6—and that the number had been rising by 3,000 a year.

Some homeless people, especially among the young, are no doubt only temporarily rootless while in transition from one place or life stage to another. Many others are the casualties of a well-intentioned but inadequately planned movement for the de-institutionalization of people with mental illness and other behavioural problems. But what is strikingly different about today's homeless is the number of family units among them, with a high proportion of lone-mother families. The increase in the number of homeless in the 1980s and 1990s has coincided broadly with the two recessions during that time, with continuing high levels of unemployment through the post-recession recoveries, with the contraction of unemployment insurance benefits, and, lately, with cuts in the incomes of welfare recipients. A frequent complaint of those homeless who are reached by voluntary social service organizations is that they have lost their homes because they have been unable to pay their rent with their reduced welfare cheques. The current problem of homelessness is in large part the consequence of current social policy decisions.

Deterioration of Hospital Services

Finally, in this review of evidence of the declining performance of the welfare state in Canada, hospital services have been under great pressure. 'The crisis in health care' has been something of a cliché in the Canadian media for several years. There is nothing mysterious about the problem: the demand for hospital services is growing, even if less explosively than some media presentations would have it, with the aging of the population, with a still-continuing increase in total population, and with continuous advances in techniques that make more readily available many interventions that require some hospitalization—by-pass surgery, organ transplants, etc.—all of this in the context of a system whereby the cost of services is met by the public, for the most part, without charge to the patient. The demand is growing, unrestrained by price, but the supply—the public resources to build the facilities, provide the equipment, and pay the people who do the work—is limited by the constraints on the public purse.

Technology mitigates the problem at the same time that it contributes to it. Hospital stays are becoming shorter for many interventions, e.g., the birth of a child, and for post-operative follow-ups in general. Shorter stays mean that the same number of patients can be served with fewer beds. Whatever the benefits may be of a shorter stay in hospital following surgery, they will be negated unless appropriate home care is provided, an aspect of health care that remains problematic. And there is a strong suspicion that length of stay in hospital is conditioned by hospital budget realities as well as by treatment indications.

In such a context, there is little wonder that some form of rationing will take effect, whether it is called that or not. The implicit rationing device in Canadian hospitals is the waiting period, or 'rationing by queue' (Naylor, 1986, p. 13). Thus we hear and read well-documented stories of people waiting months for operations to deal with cardiac problems, and so on.

Not to diminish the resulting hardship, an evaluation of the system must give due weight to the services that *are* provided to general satisfaction (Taylor, 1988, p. 499). Such 'crisis' as there is arises from the pressures generated by universal access to services which, fifty years ago, were far from universally accessible. What is in question is what mix of different kinds of service is needed. Governments defend the imposition of some limits by arguing that the demand for services has grown out of proportion to our capacity to pay for them; critics of the health system charge that the problem lies with skewed priorities of governments (and perhaps of the population), and with intra-system conflicts in which the interests of patients are squeezed out.

CHANGES IN POLICY AND ADMINISTRATION

The *performance* of the Canadian welfare state, therefore, is seen as somewhat spotty. The other source of concern—perhaps deeper, in the eyes of defenders of the welfare state, than actual performance—has been the response of public *policy* and *administration* to the challenges of economic and social change.

Even so watchful an observer as Allan Moscovitch cautions against associating the Canadian governmental response to the challenges of recent decades with that of certain other governments. That we are experiencing an unmistakable wave of conservatism is true, but in social policy it is not of the same order as the changes pursued in the 1980s by Thatcher in Britain or by Reagan in the United States. By certain measures, our social expenditures have continued to increase, if more slowly than in the 1970s and 1980s (Moscovitch, 1985; Moscovitch and Drover, 1987; Human Resources Development Canada, 1994). As the always-critical Caledon Institute acknowledges, 'the trend is upward, no matter what measure used' (Battle and Torjman, 1993). Whether the upward trend will continue when the budget-driven restrictions of the last few years take hold remains to be seen, as does the public response to be expected when governments feel that the fiscal situation is under satisfactory control.

There have been many indications, however, of a general move toward containment and possible reduction of public commitments in the field of social welfare, of which three shall be mentioned here:

- in public assistance, the erosion of welfare benefit levels, the tightening of administrative controls, and the greatly increased attention to incentives to work;
- the shrinkage of unemployment insurance;
- the virtual abandonment of universality, in favour of income-tested selectivity.

Public Assistance

The Erosion of Welfare Benefit Levels
On the face of it, one sees many indications that public assistance is far from contracting as a budget item. Total expenditures, federal and provincial, continued to grow, even as the recession of 1981–3 receded (Health and Welfare Canada *Inventory*, 1987, pp. 40–1), and spending increased rapidly again through the recession of 1991–3 (Human Resources Development Canada, 1994, pp. 241–52). By dint of benefit reductions or severe eligibility restrictions, or both, Alberta and New Brunswick began to record reductions in their public assistance outlays in the early 1990s, as did Ontario more recently; more such reductions may be expected in other provinces as well.

All provinces make some provision for adjustment of welfare rates in terms of the cost of meeting needs; the long-term trend of upward increases in rates has been reversed in many provinces as of 1995–6. As always, the details are significant. Adjustments of rates may be made in two ways: automatically, or at the discretion of the responsible authorities. Only two provinces, Quebec and Nova Scotia (for its provincial program), have automatically adjusted benefits to the cost of living at regular intervals; other provinces have adjusted rates on an *ad hoc* basis, usually annually (Human Resources Development Canada, 1993). Even such automatic cost-of-living adjustments might not mean a proportional improvement in the living standards of people living on welfare incomes, because the composition of the cost-of-living indices might not conform to the mix of items that determines the cost of living of the poor.

A number of bodies have analysed changes in the real purchasing power of welfare benefits over time, reaching conclusions that are far from reassuring. The most frequently cited study showed that the buying power of benefits in Ontario was reduced in the years between 1975 and 1982 by between 25 per cent and one-third (Social Planning Council of Metropolitan Toronto, 1982; Levens and Melliship, 1987; Riches and Manning, 1989, pp. 11-17; National Council of Welfare, 1987, pp. 75-6; see Riches in Ismael, ed., 1987, p. 147 for a thorough list of studies in this vein). Recent cuts in the dollar amounts of welfare benefits in the two largest provinces, Quebec and Ontario, have not been in effect long enough to be analysed, but they cannot fail to reverse a favourable trend in the buying power of welfare benefits up to 1993. In fact, only in British Columbia and, very marginally, in New Brunswick did benefits improve after 1993; notably, New Brunswick slightly improved benefits while, as stated above, reducing total outlays (National Council of Welfare, 1995a, pp. 34–5).

Tightening of Administrative Controls

As discussed in the chapter on public assistance, policing of beneficiaries has always been a component of public welfare administration. Care is necessary at all times in the disbursement of public funds, for welfare as for any other purpose, but a consistent pattern of heightened surveillance may be taken to represent a shift in policy. And there has been much of this heightened surveillance in Canada lately.

In Quebec, the Liberal government undertook a special program of inspection of recipients and their benefits in 1986-7, requiring over 100,000 home visits and the services of many specially hired employees. The governmental initiative aroused the ire of social workers and community groups, elicited admonitions from the Catholic hierarchy and from the province's own Human Rights Commission, provoked numerous adverse editorials in the press, and gave rise to legal proceedings on behalf of some persons whose benefits were withdrawn (Borovoy, 1988, p. 170). The Quebec government somewhat smugly publicized a number of instances where benefits were increased as a by-product of the process. What must be added is that, despite the hue and cry from the circles mentioned, the government's program was not badly received by the general public, according to opinion polls: welfare recipients are a politically vulnerable minority. Riches and Manning (1989, pp. 19, 24-32) provide exuberant detail on similarly intensified mechanisms in the welfare system in Saskatchewan.

Lately, in the context of deficit elimination and a perceived need for tax restraint, nearly all provinces have made widely publicized overtures in the direction of helping, or pushing, apparently employable people, especially young employable people, out of public assistance.

Incentives to Work; 'Workfare'

Statistically speaking, the recent changes in the characteristics of the population of welfare recipients in Canada that catch the attention are the increase in the numbers of younger people and, linked to that, the increase in the numbers who are judged to be employable. It is a commonplace that, over a period of twenty or thirty years, the relative proportions of unemployable and employable recipients in the welfare caseload have reversed themselves: employable recipients now dominate, by far.

It is certainly appropriate that policy should respond to such a significant shift in clientele, and, by implication, in the nature of need. The response could be one of positive assistance and encouragement to people to equip themselves to make the transition from the privations of welfare (accompanied by some measure of security) to the material possibilities of employment (accompanied by some risk). Such a positive policy response would provide training, counselling, supportive services (e.g. child care), and income back-up, all within realistic consideration of the capacity of the labour market to make available the necessary jobs. But the response could just as well be simply a tightening of the screws on the employable recipient: limits on cash benefits, time limits on benefits, ultimatums for entry into work, demands for more frequent self-reports, heavy pushes into low-end jobs (any job will do, especially if you're young), etc., all of which neutralize any bargaining power the individual might have *vis-à-vis* a potential employer. The contrast between the two concepts is drawn, with some nuance, in Andrew Johnson's comparison of Ontario's Work Incentive program and Quebec's Work Income Supplement (now Parental Wage Assistance) (Johnson in Ismael, 1987).

Suffice it to say that there is cause for concern that governments will be so focused on expenditure reduction that employable people will become hostages to provincial budgetary politics. The terms of the Canada Assistance Plan denied federal cost-sharing to any public assistance program that imposed a work requirement as a condition of receiving assistance (so-called 'workfare'—show up for work or training or you're refused benefits). Even within that prohibition, several provinces have shaped their welfare programs in ways that reinforce both the push of implicit penalties for unreadiness to work and the pull of incentives to engage in work or work-related activity. As with many current developments in social policy, governments of various professed political orientations, left, right, and centre, are pursuing very similar paths. It remains to be seen whether the new block funding arrangements for federal support of welfare under the Canada Health and Social Transfer will dilute somewhat the federal government's influence over provincial public assistance practices (Armitage, 1996, pp. 30–42; Séguin, 1987; Institute for Research on Public Policy, 1995; National Council of Welfare, 1995b).

The Shrinkage of (Un)Employment Insurance

(Un)Employment Insurance is not just another social program. As Canada's oldest strictly federal social program, as an exceedingly large program (a little smaller than OAS, GIS, and Spouse's Allowance put together, but larger than Social Assistance), (Un)Employment Insurance is a Canadian institution. Revisions of UI/EI may therefore be taken seriously as indicators of trends in the social temper.

In the early 1970s, UI shared in the expansive glow of the times; in the later 1970s, as the economic shadows began to lengthen, UI was somewhat contracted but still represented a very large commitment of publicly managed funds to broadly social purposes. At the end of the 1980s, the program was contracted in ways consistent with the then current *Zeitgeist*. The amendments in 1995, described in Chapter Three—including the symbolic change of name to 'Employment Insurance'—went even further in the same direction.

To repeat some of what was said in Chapter Three: the government as such now pays no part of EI benefits, all of which are paid out of the premiums collected from employers and workers; in that sense, the scheme is totally privately financed. Where it once paid for 'extended benefits' out of general taxation, the government now pays for only work retraining and job creation efforts for unemployed people. Less of the apparatus for carrying out the program is public. The work requirement to qualify for benefits has been increased; the duration of benefit periods has been shortened; weekly benefits have been reduced as a percentage of normal income. The possibility that an unemployed person might receive repeated long-term benefits for short-term work—a controversial feature of UI since 1971—has been virtually eliminated, with serious consequences for seasonal workers; it will be much more difficult henceforth to have repeated recourse to benefits as an income source over an extended period.

The spirit of these changes is entirely consistent with the tendencies discernible in public assistance policy. To restrict unemployment insurance benefits at a time when the unemployment rate remains relatively high has had the inevitable consequence of increasing the number of people who need public assistance; and since this, too, comes at a time when the federal government is reducing its support of provincial welfare programs, a new and difficult situation is created for the provinces, mitigated for some of them, it is true, by the promised increase in equalization payments to the 'have-not' provinces.

The Abandonment of Universality

In the past, the existence of universal programs—Old Age Security, Family Allowances—was regarded as a hallmark of Canadian social policy. Now, the tendency, visible for many years, to nibble away at universal programs has led to their complete abolition.

It might be said that the principle of universality has been outflanked rather than overrun, by giving something else in place of a universal benefit. In their turn, the Guaranteed Income Supplement (1967) and the Child Tax Credit (1979), both income-tested and therefore selective, were evidently seen by government as preferable to increases in Family Allowances and OAS. The more recent Child Tax Benefit and anticipated Seniors Benefit have proceeded logically along the same path.

Health care insurance, for both hospital and medical care, remains universally accessible. Hitherto, policy has firmly resisted all forms of co-payment by users, which in the judgement of most analysts would create a barrier to access. 'Extra billing' for medical services and user charges for hospital services faded as a live issue after the amendments to the Canada Health Act in 1984 legislated dollar-for-dollar reductions in federal fiscal transfers to provinces that allowed them. But there is simply no question that hospitals are financially hard-pressed, that medical services cost the provinces a good deal, and that alternative sources of income are being looked at. One alternative source is the patient. The attenuation of central government interest in the operation of provincial programs resulting from the switch from 50-50 cost-sharing to EPF and then to the Canada Health and Social Transfer has removed one line of defence against the ingenious introduction of financial participation by patients.

The issue of universality is not one of angels vs. devils. Many sincerely feel that the taxpayer's money should go more to those in need and not at all to the relatively affluent. Richard Titmuss, himself an influential advocate of the principle of universality, wrote perceptively of developments in the welfare state in Britain (1963, p. 229):

> Many of us must also now admit that we put too much faith in the 1940s in the concept of universality as applied to social security. Mistakenly, it was linked with economic egalitarianism. Those who have benefitted most are those who have needed it least.

If the existence of tax avoidance opportunities for the wealthy is offensive, so too might be tax-supported income transfers to the wealthy, though admittedly the tax avoidances cost the state a great deal more than the transfers to the wealthy ever did.

The issue of universality vs. selectivity in income maintenance programs has already been discussed in Chapter Two. To summarize briefly, the arguments for universality are moral (solidarity, avoiding a givers–receivers split in society); political (the underlying policy, having many beneficiaries, will have many defenders); economic (need is sufficiently widely felt that little of the transfer actually goes to those who are wealthy to superfluity, especially net of taxes); and administrative (with no eligibility testing, universal programs are cheap to manage).

As Canada shifts away from universal income benefits of any kind, the idea has begun to enjoy a modest resurgence of interest and support in Europe. The moral case for universal social programs is restated vigorously by A.B. Atkinson (1995) and by Philippe Van Parijs (1995). The moral and political arguments retain considerable force, founded as they are on the collective mutual responsibility of members of a modern society, where interdependencies are complex and dense.

The economic argument remains viable, too. Old Age Security is still 'universal' for the vast majority of older people, in that the income test comes into play at the level of about $55,000, above which it is gradually reduced until it disappears for those of upper-level incomes. (The Seniors Benefit, as we know, will operate similarly, but at lower income levels, benefitting fewer people.) Without OAS, Canada's elderly as a class would require vast financial support in some other way; without the certainty of OAS, more Canadians not yet old would face the risk of a future of dependency. A similar case for benefits both universal and selective for families raising children has been discussed under 'Poverty Among Children', above.

SOCIAL PROGRAMS: THE WELFARE STATE

What has been

At the beginning of this book, it was affirmed that a grasp of the ways in which social programs operate is necessary to a critical understanding of social policy.

For the most part, what we know about social issues are the facts that emerge from the operation of social programs. As we have seen, public assistance policy is currently being revised because of concern with the 'employable unemployed'; in fact, most of what is actually known about the employable unemployed is known through the operations of public assistance programs, and comes to us classified in ways that are relevant to public assistance programs. Much of what we know about employment and unemployment we know through data on unemployment insurance. The Forget Commission on Unemployment Insurance obtained information from a wide range of sources, but the bulk of the data in the Commission's report came from the records of the UI program, gathered for the purposes of the program. Much of what we know about the older population we owe to the records of Old Age Security, the Guaranteed Income Supplement, the Spouse's Allowance, and the Canada and Quebec Pension Plans. And so on. We see the issues of social policy mostly through the prism of social programs. Whoever fails to understand the programs will fail to understand the facts—including the shortcomings of the facts. The student of policy will be grateful for the information, but at the same time will be critically aware of its source. This important aspect of our knowledge of the social world is too easily overlooked.

For good reason, most program changes are incremental, and incremental changes will invariably be based on the data that emerge from the experience of the programs. On occasion, to be sure, a more radical change will take place, based on external input. An example is the current fiscally driven re-examination of all programs. Even in such a case, the information accumulated in the records of the programs affected exerts major influence over what is done. By contrast, in the absence of concrete programs addressed to particular social problems, reformers and legislators may find themselves looking at issues without much information at all. The members of the Boucher Commission on public assistance in Quebec in 1963 complained that the absence of any articulated program of assistance to employable people meant that they were being asked for policy advice when they knew very little about the problem.

In new problem areas, there may well be no relevant experience upon which to draw, and a lot of informed guessing must be done; the issue of HIV/AIDS is an example. And there can be a darker side to the illumination that comes from already existing programs: society can effectively blind itself to facts that do not come to light because the prevailing program does not focus on them. Up to the late 1950s, welfare programs had little room for employable people and none at all for the employed; the working poor were simply not in the poverty picture. It was the independent work of the Economic Council of Canada and of the Croll Report, *Poverty in Canada*, in the 1960s, that brought the working poor to the attention of the public and to that of policy-makers, as no analysis of then-existing welfare programs could have done.

What might be

The point that the means are as important in their way as the ends is often lost on reformers, conservative as well as radical. Reformers lament that the brave new legislation they enacted has failed to usher in the brave new world they wanted, even within the narrow compass of a single program (in support of children, the elderly, addicts, etc.). 'Grand designs' may make for gripping slogans and rousing speeches, but it is necessary to understand the working parts in order to execute the grand design.

Whatever the fundamental moral, political, and economic choices underlying a social program, it must be legislated in a certain form. Rule-making and administrative authority must be allocated. The population covered must be defined. The terms upon which a person becomes eligible to enjoy its benefits must be stated clearly, because the criteria are sure to be tested repeatedly. The money that the program spends must come from an indicated source, taxation or other; there will probably be some logical link between the source of funds and the administration. The choices that are made at each of these stages tell us more of the true character of the program than all the noble promises of its advocates, or the furious denunciations of its opponents.

Whether the ideology or political perspective underlying a particular program be liberal, conservative, socialist, or other, the choices of operating mechanisms are limited, as comparative studies of social policy tend to show. Even when basic policy orientations are different, there are bound to be similarities in programs, as well as differences. An approach that gives attention to the components of programs attunes the student to discerning where the real differences lie. A focus on program structures and program data has nothing to do, therefore, with a bias towards either 'conservative' or 'progressive' orientations. A study of policy that takes seriously the working machinery of social programs will not conceal from students the need for change where the need exists, and it *is* likely to alert them to real-life necessities, ignoring which dooms change to failure.

Sometimes, certainly, incremental changes will be seen as insufficient. New objectives will have to be formulated. But objectives do not achieve themselves. Without giving in to the cautious instincts of the middle manager, feasible ways to achieve the alternative objectives must be sought. The feasible *modus operandi* even of a radical change must rely heavily on the experience of actual programs.

From the point of view of the users or beneficiaries, the policy *is* the program. If, for instance, unemployed persons respond to a program positively, as enhancing solidarity, or negatively, as stigmatizing, they do not do so on the basis of the intentions of the legislator; what makes it stigmatizing or not for them is the terms upon which they are made to interact with the program. In this respect, too, program operations take on great importance as *parts of the policy*, not as mere mechanical details.

No one questions that the gritty details are relevant. The pragmatic questions in social policy and social program debates are important: Would it cost more to do it this way or that? Which kind of administration would demand the greatest effort? How will the program affect the behaviour of clients and of others (families, employers) who can affect outcomes? But a meaningful understanding goes beyond the pragmatic. Public discussion of issues is typically framed in terms of grand social choices, such as the common good vs. the pursuit of private interest. The nuts-and-bolts approach provides an essential component of understanding even at that more abstract, ideal level.

The higher-level value issues—solidarity and non-stigmatization, the individual interest and the social interest—can in fact be meaningfully discussed in terms of the structure of programs; perhaps they can be most meaningfully discussed in those terms. For example, people get family benefits when they are raising children, and they get old age benefits when they are older. Some participants in debates over universal family allowances and old age pensions seemed to assume that an attachment to social solidarity necessarily inclined one towards universal payments and away from income-related tax credits or supplements. That is a defensible position, but it is not the only position. It is relevant to look at how the options compare through the life cycles of people. Over the long run, how many people would be served according to the different proposals? Putting the tax and transfer systems together, does a non-universal approach actually split the population into two groups, givers and getters, as universalists fear? Or do most of us take turns as net givers and net receivers?

Having looked at the case in such terms, one might still come down on either side, for ultimately choices are made on the grounds of values. But judicious attention to the facts of the case will contribute to the implementation of values that are shared and reduce the depth of disagreement on values that are difficult to reconcile.

There is no doubt a further level of resolution of social issues where power is the arbiter. At that point the issue enters a realm where welfare, like any other issue, relies on the play of social values and social forces, working through the institutions of government and the other social institutions identified in the opening chapter as decisive factors in the game.

AT THE GATE OF THE TWENTY-FIRST CENTURY

As we approach the year 2000, the challenges enumerated in the introductory chapter promise to continue to dominate the social policy agenda in Canada.

• The financial difficulties of Canada and the provinces have the proponents of expansive social programs very much on the defensive. Most of the provinces (Quebec the notable exception) have balanced their annual budgets, but in doing so they have reduced public commitments in health, education, and welfare, their largest expenditure fields; of the three, welfare is politically the most vulnerable by far. Meanwhile, the provinces are obliged to go on spending large shares of their annual revenues to pay the interest on their debts. The federal government, for its part, has for many years realized a surplus on its annual operations—it has spent less on its various programs than it has collected in revenues—but the burden of interest on its debt has kept it at an overall deficit, and will continue to do so for a number of years, despite the contractions in transfers to the provinces and in transfers to individuals through some of its own programs.

In such a context, one may well fear that health and welfare cuts, once made, will not be restored, even when fiscal difficulties have been overcome. A pessimistic view is that hostile interests will succeed in making welfare-state policies and programs the whipping-boy for the fiscal issue; the rhetoric of those who promote efforts to 'downsize government' has that very flavour. It is neither pessimistic nor optimistic to be wary lest decisions made in the fervour of deficit- and debt-reduction create huge costs for Canadians in the near and especially the distant future. Here as elsewhere, the tax side of the equation is just as important as the spending side, and serious students of social policy must give attention to revenues as well as to program benefits. Proposals for simple solutions of matters as complex as taxation must be looked at as sceptically as simple solutions for anything else.

• There is little need to repeat warnings about the problems Canada will face in the twenty-first century arising from the age composition of the population. Those in the unusually large 'boomer' generation, born between about 1950 and 1965, will begin reaching the conventional milestone age of sixty-five around the year 2015, while those in the unusually small cohorts that followed will soon dominate the so-called active adult population. Working-age adults will comprise a smaller part of the total population than ever before.

If one accepts the picture of a retired population drawing off a share of the economic wealth produced by a working-age population, one sees a social problem that cannot be shirked. In most of the second half of the twentieth century, there have been five Canadians of working adult age for every Canadian over sixty-five; on a national basis, those five have been supporting themselves, the children and the youth of the country, and the non-working elderly. By about 2030, there are expected to be only three working-age adults for every person over sixty-five. Those three will need some help from social policy to carry the same load that their five predecessors carried in the past. Prophets of doom see the interests of the two age groups grinding against each other like the tectonic plates whose frictions cause earthquakes. (It should be recalled from Chapter Five that working Canadians provide the output that supports non-working Canadians no matter what forms the incomes of the latter take; a retired older person relies on the output of the Canadian economy if he or she lives off a private pension just as much as if he or she relies wholly on public income transfers.)

To claim with righteous indignation, as many do, that the elderly have earned the support allotted them does not refute the analysis; it confirms it, for it implies that someone is going to produce enough to be able to allot to the elderly in the present what they have earned in the past. Enough wealth must be generated to make a share available to them; and if the active population does not produce sufficiently, it will not matter how thoroughly the elderly have deserved their share. The 'inactive' most obviously depend upon the 'active', if incomes are simply transferred to them each year on a pay-as-you-go basis; but the 'inactive' depend on the 'active' even if the 'inactive' are paid from the fruits of investments they made from their own savings in their income-earning years. In either case, the current productive population must deliver the product out of which the incomes of all are generated, including the incomes of the elderly.

Obviously it would help if each 'working' generation, in its turn, were to save enough consistently to make investments that would enhance national productivity sufficiently to assure continued adequate incomes for the elderly. That is the crux of current discussions about the future of the Canada and Quebec Pension Plans, about tax support of retirement savings plans, and about the appropriate roles of public and private pension-providing vehicles. At the heart of the issue is the encouragement of sufficient savings to pay for the investments needed to enhance productivity so as to assure sufficient returns to provide satisfactory financial support for the 'inactive' population—the old and the young.

As has been noted, the generational divide affects policy in health care provision as well as in income provision.

• Along with a potential age fissure, Canada faces a potentially growing income fissure in the twenty-first century. As observed in the introductory chapter, a tendency toward a polarization of earned incomes has been noted: the hitherto large middle strata have been shrinking, as a proportionately large number of earners have increased their incomes, and a yet larger number and proportion of earners find themselves in the lower strata. Because annual incomes and lifetime incomes play such a large part in our income security system, such a change would have far-reaching consequences (Banting and Beach, 1995).

• Social policy in the next century must confront the consequences of the powerful world trend toward the globalization of economic relations. The irrepressible emergence of Third World countries into world markets will raise incomes and living standards in those parts of the world, and will make many goods cheaper for consumers in the 'advanced' countries, like Canada, but it will also force the developed countries to adapt their patterns of investment and production to the vastly enhanced competitive capacities of the emerging nations. Canadians will produce a different array of goods and services; they will do different jobs; quite possibly the nature of employment relationships in Canada will change. Even moderate commentators see a sharp decline in the kind of long-term career with a few employers that has characterized the labour market experience of many typical Canadians. Already the percentage of Canadians who describe themselves as 'self-employed' has risen sharply. 'Self-employed' can mean many different things.

Whether, on balance, more market opportunities, jobs, and incomes will be lost or gained in Canada through globalization is a hotly debated issue. Canada has had a few years of experience with the liberalization of trade in North America through the North American Free Trade Agreement (NAFTA). The net effects of NAFTA are also hotly debated. Canada is a Pacific country as well as a North American one, and the scale of the possibilities of trade with the heavily populated countries of the Asian Pacific is quite staggering. The task of economic and social policy will be to see that the benefits do not go mostly to one part of the population and the costs to another.

• Finally, as has been remarked, the waning decades of the twentieth century saw, in virtually every country in the industrial world, a growing scepticism about the capacity of government to achieve all the tasks that it had been asked to undertake. This weakening of confidence in the state could be observed even in countries with the historically longest, most profoundly rooted welfare state establishments, like Sweden and New Zealand. Left and Right share in 'widespread cynicism, or at least scepticism, about the confident large-scale designs which previously characterized social welfare programs, and state intervention generally' (Leonard, 1994, p. 57). Canada is no exception: downsizing of government, divestment of government functions to private bodies, deregulation of previously regulated fields of enterprise—all have been popular political programs in Canada. Certainly, those interests that expect to do well for themselves if government is weak provide much of the enthusiasm for the contraction of the state—but not all of it.

The challenge for social policy will be to see that we maintain the social machinery necessary to continue to enlist the resources, material and moral, of society in the support of the disadvantaged, and that we do not abandon the weak and the victims of the risks of modern life. And who can tell, from moment to moment, whether they may find themselves among the victims? Throughout most of the twentieth century, we learned to rely on the state to lead society in that effort. It would be a serious mistake to weaken the state's capacity to manage welfare without giving that capacity to other institutions, and there are at present no other plausible contenders for that role.

BIBLIOGRAPHY

Atkinson, A.B. (1995). *Public Economics in Action: The Basic Income/Flat Tax Proposal.* London: Oxford University Press.

Banting, Keith, and Charles H. Beach, eds (1995). *Labour Market Polarization and Social Policy Reform.* Kingston: Queen's University School of Policy Studies.

Battle, Ken, and Leon Muszynski (1995). *One Way to Fight Child Poverty.* Ottawa: Caledon Institute of Social Policy.

Battle, Ken, and Sherri Torjman (1993). *Federal Social Programs: Setting the Record Straight.* Ottawa: Caledon Institute of Social Policy.

Borovoy, Alan (1988). *When Freedoms Collide: The Case for Our Civil Liberties.* Toronto: Lester & Orpen Dennys.

Caledon Institute of Social Policy (1995). *Colloquium on Child Poverty, January 26, 1995.* Ottawa: The Institute.

Cameron, Duncan, and Ed Finn (1996). *10 Deficit Myths: The truth about government debts and why they don't justify cutbacks*. Ottawa: Canadian Centre for Policy Alternatives.

Canada. Health and Welfare Canada (1987). *Inventory of Income Security Programs in Canada*.

Canada: Human Resources Development Canada (1994). *Basic Facts on Social Security Programs*. This invaluable compendium, updated at intervals, gives essential data on all *federal* income programs—numbers of beneficiaries, dollar outlays, recent year-by-year comparisons.

Canada. Human Resources Development Canada (1993). *Inventory of Income Security Programs in Canada*. A highly informative survey of *all* income security programs in effect in Canada, federal and provincial. Like *Basic Facts. . .*, *Inventory* is published at irregular intervals. An essential source for students.

Canadian Centre for Policy Alternatives (1996). 'Large public debt no excuse to cut social programs', CCPA *Monitor* 2/8 (February).

Canadian Council on Social Development (1989). *A Choice of Futures: Canada's Commitment to Its Children*. Fact Sheets #1 and #3.

Centraide Montreal (1988). *Brief to the Minister of Manpower and Income Security on Social Aid Reform*.

Courchene, Thomas J. (1995). *Redistributing Money and Power: A Guide to the Canada Health and Social Transfer*. Toronto: C.D. Howe Institute. The financial, administrative, and political implications of the latest variation in federal transfers to the provinces in support of social programs.

———— (1993). 'Path Dependency, Positive Feedback and Paradigm Warp', in Elisabeth Reynolds, ed., *Income Security in Canada: Changing Needs, Changing Means*. Montreal: Institute for Research on Public Policy. Courchene argues powerfully that Canada's social policy was constructed to satisfy the conditions of the 1960s, and that Canada is now paying the price for failing to keep up with changes in the world economic environment.

Crane, John A. (1994). *Directions for Social Welfare in Canada: The Public's View*. Vancouver: UBC Press. A deeply concerned empirical exploration of the character of public debate over social programs in Canada, led by a professor in the School of Social Work of the University of British Columbia. Chapter 8, 'Public Support for Comprehensive Social Programs in Canada', presents findings of opinion research (including opinions from organizations known to be hostile to social programs) conducted in British Columbia. Opinion was found to be more than moderately favourable to all current social programs *except* tax shelters for the encouragement of investment.

Dilnot, Andrew (1995). 'The Assessment: The Future of the Welfare State', *Oxford Review of Economic Policy* 11/3 (Autumn).

Drache, D., and A. Ramachan, eds (1995). *Warm Heart, Cold Country: Fiscal and Social Policy Reform in Canada*. Ottawa: Caledon Institute and Robarts Centre for Canadian Studies. See especially articles by Kitchen on the passing of universality, Torjman and Battle on the growing influence of the federal Department of Finance on social policy, and Part IV on the deficit-debt issue.

Freeman, Richard B. (1995). 'W[h]ither the Welfare State in an Epoch of Rising Inequality?', in Banting and Beach, eds, *Labour Market Polarization* (cited above).

Guest, Dennis (1987). 'World War II and the Welfare State in Canada', in Allan Moscovitch and Jim Albert, eds, *The 'Benevolent' State: The Growth of Welfare in Canada*. Toronto: Garamond.

Hills, John (1995). 'Funding the Welfare State', *Oxford Review of Economic Policy* 11/3 (Autumn).

Institute for Research on Public Policy (1993). *Federal Transfers to the Provinces*. Montreal: The Institute. Expert discussions of many aspects—administrative, economic, political—of intergovernmental transfers. Dated 1993, this volume makes no references to the CHST.

Ismael, Jacqueline, ed. (1987). *The Canadian Welfare State: Evolution and Transition*. Edmonton: University of Alberta Press.

Johnson, Andrew F. (1987). 'Ideology and Income Supplementation: A Comparison of Quebec's Supplément au revenu de travail and Ontario's Work Incentive Program', in Jacqueline Ismael, ed., *The Canadian Welfare State* (cited above).

Leonard, Peter (1994). 'The Potential of Post-modernism for Rethinking Social Welfare', in Leslie Bella, ed., *Rethinking Social Welfare: People, Policy and Practice: Proceedings of Sixth Biennial Social Welfare Policy Conference, 1993*. St John's, Nfld.: Memorial University.

Levens, Bruce, and Kaye Melliship (1987). *Update: Regaining Dignity*. Vancouver: Social Planning and Research Council of British Columbia.

Maclean's (1995). 'Can Canada Survive? The 12th Year-end Poll'. 25 December, 1995/1 January, 1996, pp. 20–2.

Mishra, Ramesh (1987). 'Public Policy and Social Welfare: The Ideology and Practice of Restraint in Ontario', in Jacqueline Ismael, ed., *The Canadian Welfare State* (cited above).

Morissette, René, John Myles, and Garnett Picot (1995). 'Earnings Polarization in Canada, 1969–1991', in Keith Banting and Charles Beach, eds, *Labour Market Polarization and Social Policy Reform*. Kingston: Queen's University School of Policy Studies.

Moscovitch, Allan (1985). *The Welfare State since 1975*. Occasional Paper Series, No. 3, Social Administration Research Unit, University of Regina.

Moscovitch, Allan, and Glenn Drover (1987). 'Social Expenditures and the Welfare State: The Canadian Experience in Historical Perspective', in Allan Moscovitch and Jim Albert, eds, *The 'Benevolent' State: The Growth of Welfare in Canada*. Toronto: Garamond. This article, and that of Guest (cited above) sum up the development of the welfare state in Canada, broadly confirming the schematic representation of that development given in the text. With ample statistical and financial data, Moscovitch and Drover show that welfare has unquestionably grown, especially since the 1940s, though neither they nor Guest, who confines himself to post-World War II history, have much praise for Canada's performance.

National Council of Welfare (1997). *Poverty Profile 1995*. Ottawa: The Council.

——— (1995a). *Welfare Incomes 1994*. Ottawa: The Council.

——— (1995b). *The 1995 Budget and Block Funding*. Ottawa: The Council. The Council points out that public assistance burdens rise when GNP growth either slows or stops, whereas at such times, in principle, CHST block funding will slow down; this contradiction will put provincial programs in greater difficulty. The Council also argues that the only national standard that prevailed under the Canada Assistance Plan that survives under CHST is the no-residence-requirement rule; the 'needs test' rule, once regarded as essential, has been given up, in the name of allowing provinces greater flexibility in their welfare programs. The Council sees this as a grave threat.

——— (1985). *1985 Poverty Lines*. Ottawa: The Council.

——— (1975). *Poor Kids*. Ottawa: The Council.

Naylor, C. David (1986). *Private Practice, Public Payment: Canadian Medicine and the Politics of Health Insurance, 1911–1966*. Kingston and Montreal: McGill-Queen's University Press.

Naylor, Nancy, Ruth Abbott, and Elizabeth Hewner (1994). *The Design of the Ontario Child Income Program*. Ottawa: Caledon Institute of Social Policy. The reader must understand that this paper describes the *proposed* Ontario Child Income Program. Due to the vicissitudes of politics and public finances, the program never got off the ground. It had some influence, however, on later developments in child benefits in Canada.

Ontario. Ministry of Community and Social Services (1993). *Turning Point: New Support Programs for People with Low Incomes*. A policy statement issued by the NDP government of Ontario. Its spirit was not much respected by the succeeding Progressive Conservative government.

Oxford Review of Economic Policy (1995). Special issue, Reform of the Welfare State, 11/3 (Autumn).

Pulkingham, Jane, and Gordon Ternowetsky (1996). 'The Changing Landscape of Social Policy and the Canadian Welfare State', in Pulkingham and Ternowetsky, eds, *Remaking Canadian Social Policy*. Halifax: Fernwood.

Québec. Ministère de la sécurité sociale (1996). *La réforme de la sécurité du revenu, un parcours vers l'insertion, la formation et l'emploi*.

Riches, Graham (1987). 'Feeding Canada's Poor: The Rise of the Food Banks and the Collapse of the Public Safety Net', in Jacqueline Ismael, ed., *The Canadian Welfare State* (cited above).

——— (1986). *Food Banks and the Welfare Crisis*. Ottawa: Canadian Council on Social Development.

Riches, Graham, and Lorelee Manning (1989). *Welfare Reform and the Canada Assistance Plan: The Breakdown of Public Welfare in Saskatchewan, 1981–1989*. Working Paper Series, No. 4, Social Administration Research Unit, University of Regina.

Rosenberg, Mark W., and Amanda James (1994). 'The End of the Second Most Expensive Health Care System in the World: Some Geographical Implications', *Social Science and Medicine* 39/7 (October). The 'System' in the title is Canada's. The authors show that hospital closings in Ontario have damaged equality of access to hospitalization, and may not have reduced total costs appreciably.

Ross, David, E. Richard Shillington, and Clarence Lochhead (1994). *The Canadian Fact Book on Poverty, 1994*. Ottawa: Canadian Council on Social Development. Chapter 2, 'Working Definitions of Poverty'. Offers a thorough explanation of various 'poverty lines' and 'low income lines' in use in Canada.

Ryerson, Stanley (1981). 'A future for equality in Canada?', in A. Moscovitch and G. Drover, eds, *Inequality*. Toronto: University of Toronto Press.

Séguin, Gilles (1987). 'Descriptive Overview of Selected Provincial Income Supplementation and Work Incentive Initiatives', in Jacqueline Ismael, ed., *The Canadian Welfare State* (cited above). Now dated in the details, this article casts light on the ways in which income programs attempt to cope with the issue of incentive to work.

Social Planning Council of Metropolitan Toronto. *Social Infopac*. A periodical report of the Council. A number of issues, 1982 through 1984, dealt specifically with the dwindling adequacy of welfare benefits.

Tawney, R.H. (1964). *Equality*. 4th edn. London: Allen and Unwin. A reasoned, humane statement of an egalitarian position.

Taylor, Malcolm (1988). *Health Insurance and Canadian Public Policy*. 2nd edn. Toronto: University of Toronto Press.

Tester, F. (1991). 'The Globalized Economy: What Does It Mean for Canadian Social and Environmental Policy?', *Canadian Review of Social Policy*, No. 27 (May). To the author, the answer is 'Nothing good'.

Time (Canada) (1995). 'A Nation Blessed, a Nation Stressed', 146/21, 20 November, pp. 30–40.

Titmuss, Richard (1963). *Essays on the Welfare State*. 2nd edn. Boston: Beacon Press. The chapter entitled 'Social Welfare and the Family' is an exploration of social needs created by industrialism, calling for a collective response.

Van Parijs, Philippe (1995). *Sauver la Solidarité*. Paris: Cerf.

Webb, Steven (1995). 'Social Security Policy in a Changing Labour Market', *Oxford Review of Economic Policy* 11/3 (Autumn).

Appendix A: Poverty Lines

WHY A POVERTY LINE?

A 'poverty line' is simply a dollar income, above which a person or family is considered by someone to be not poor, below which to be poor. 'Simply' is perhaps the wrong word.

Any poverty line, no matter how ingeniously devised, has limitations. For one, it is limited to dollar income, though many would insist that 'poverty is more than a matter of money.' The poor themselves, however, give high priority to the money part. For another, no way has been developed to arrive at a poverty line that is completely free of some arbitrary input; in the calculation, there is unavoidably a crucial point at which one chooses, arbitrarily, a factor that yields a higher poverty line (more people 'poor') or a lower one (fewer 'poor'). Some dislike the expression 'poverty line', perhaps because it suggests that poverty is just another interesting social phenomenon to be measured. Statistics Canada prefers the expression 'low-income cut-offs' (LICOs).

Subject to those limitations, a poverty line may be used for statistical purposes, as a guide to policy, as a measuring stick for the evaluation of the outcomes of specific programs, or as a yardstick for overall social progress (the number under the poverty line today versus ten years ago).

It may be objected that one could do all of these useful things by defining poverty with any reasonable income figure, even one picked out of a hat. But the way in which a poverty line is calculated determines the way it registers changes in the distribution of income, that is, it will affect the measurements made when one uses it. Moreover, the soundness of the method by which a poverty line is worked out affects the credibility of the result and of any conclusions drawn from applying the poverty line to real-life conditions. Finally, the choice of method may reflect basic value positions about the nature of society (Johnston, 1993; Spector, 1992).

When we talk about poverty lines or low-income lines, we are not talking about the benefits that will be given through any program or combination of programs; there is no relationship between the two. As far as governments are concerned, poverty lines are, at best, guidelines. No government promises that all its citizens (and certainly not all its welfare recipients) will have incomes equal to or higher than any specific poverty line. Despite occasional misinformed statements to the contrary, *there is no 'official' poverty line in Canada.* The fact that the defining measure most in use is one developed by Statistics Canada may mislead some into thinking of it as in some way official, but Statistics Canada is not a mouthpiece for the Canadian government.

There are two ways to approach the definition of a poverty line income: the 'market basket' or 'absolute' approach, and the 'relative' approach, of which there are a number of variants.

THE MARKET BASKET APPROACH

This approach sees poverty as the lack of a sufficient income to meet certain specified needs. Somebody lists the goods and services that a person or family needs and adds up their prices. The total becomes the poverty line. Whether or not they actually spend their incomes on the prescribed 'market basket', persons or families with incomes below the line are counted as 'poor'.

Needless to say, people differ on what items should be included in the list of needs. Past a certain core—food, clothing, shelter—differences in values and tastes will influence the choice of items. Even within the core needs of food, clothing, and shelter (abbreviated as 'FCS'), people disagree as to how much money it takes to satisfy real 'need'. When an agency makes up a list of needs, it is not telling low-income people how they ought to spend their money (though some commentators will predictably accuse it of doing so), but the list, and therefore the poverty line, may well reflect the ideas of 'need' of the people making up the list, not necessarily those of the poor themselves. Absolute poverty lines therefore vary in the 'minimum' standard at which they implicitly aim.

To mitigate this arbitrary element in the identification of needs, the actual consumption practices of low-income people may be taken into account. The Montreal Diet Dispensary, for instance, bases its low-income family budgets on a combination of its own expertise in nutrition and its observations of the actual spending patterns of low-income people. Its budgets accordingly include such items as newspapers, church attendance, and entertainment. The Metropolitan Toronto Social Development Council adopts a similar approach.

Another market-basket poverty line was developed in 1992 by economist Dr C. Sarlo. Having been published by the Fraser Institute of Vancouver, widely and proudly recognized as a proponent of right-wing thought in economic and social matters, Sarlo's poverty line has attracted much attention (Sarlo, 1996). It is indeed a *poverty* line, allowing for a good deal less consumption than the Montreal Diet Dispensary's budgets. While dollar figures should not distract from the underlying conception, examples do prove illuminating. The Diet Dispensary sets the annual income required to meet its 'minimum adequate standard' for a family of four at $19,960. For a family of four in Quebec, Professor Sarlo set the poverty level at $13,848. Not to belabour the obvious, if our understanding of poverty was represented by an income as low as this, the percentages of poor people, poor elderly, poor single mothers, and children living in poor families would fall to extremely low levels. Sarlo's estimate is that about 2 per cent of Canadians are poor.

The key characteristic of absolute poverty lines is that they focus on what people need in order to live at some standard—'subsistence', 'minimal adequacy', or whatever. They are not essentially concerned with what other people earn or consume. Later on, this Appendix suggests that absolute poverty lines cannot completely ignore the living standards of others, and conversely that purveyors of relative poverty lines do not ignore the market basket; theoretically, however, the absolute approach pays no attention to where a poverty-line income would stand in relation to other people's incomes.

With the above-noted exception of recent public-assistance practice in Quebec, public welfare authorities have all followed 'market-basket' approaches when determining the levels of income to be allotted under public-assistance programs. (But again, welfare levels must not be confused with poverty lines.) Professionals and academics have shown more interest in so-called 'relative' poverty lines.

THE RELATIVE APPROACH

The relative approach to the fixing of a poverty line begins with a concept of poverty as the lack of an income sufficient to provide a standard of living reasonably close to what is common in a society. This is conceptually different from defining sufficiency in terms of capacity to purchase a given list of goods and services. If the prevailing standard of living in the society is low, a relative poverty line income will yield a standard of living that is still lower, low enough perhaps to fail to meet subsistence requirements. If the standard of living is high, the relative poverty line will be high, more than meeting basic needs. Poverty conceived in this way is not, then, a matter of being unable to buy certain specified things. Rather, it is a matter of being unable to buy what most other people are able to buy. There are two main variations of the relative approach:

(1) *In direct relation to the prevailing standard.* First, the prevailing standard of living is expressed in terms of some annual dollar income. The obvious expressions of this are either average annual income or median annual income (the average will always be higher). The poverty line is then fixed in some direct arithmetical relationship to this income, with appropriate adjustments for dependants. One says, 'In order to be considered poor, a person's income must be less than some stated percentage of the average (or median) annual income.' There is a tendency to settle upon *half* the average or median income, but the 'half' is chosen largely on the basis of its arithmetical convenience, i.e., it is easy to divide by two. The percent-

age chosen is up to the chooser. It is a good exercise to pause and think what percentage you would choose, and why. The poverty line propounded by the Canadian Council on Social Development is of this nature—one-half of median income.

(2) *In relation to income distribution strata.* A second way is to rank all the incomes in society, from the highest to the lowest, and then to say that all incomes below a certain *rank* are to be considered 'poverty' incomes. In practice, the ranked incomes of all persons or families are divided into deciles (tenths), quintiles (fifths), or quartiles (fourths). The judgement is made that all those whose incomes are in, let us say, the two lowest quintiles (40 per cent of the population), or the three lowest deciles (30 per cent), are poor. The poverty line itself is the income at the top of the stratum identified as poor. There will always be a bottom 30 per cent, or 40 per cent, or whatever percentage is chosen, so the proportion considered poor will remain *exactly the same* at all times and under all conditions except total equality of incomes, in which case there would obviously be no top or bottom strata. Again, it is instructive to reflect on just what factors one would take into consideration when determining the poverty line in this way.

Clearly, a relative poverty line, however determined, may be low or high, in dollar terms, according to the decision of the person who determines it. In theory, it might be so low as to fail to provide even a subsistence income, but in a so-called affluent or developed society it would be absurd to designate as 'poor' only those unable to survive. On the other hand, there has to be some upper limit; it would be a contradiction in terms to say that people were 'relatively poor' whose living standard was as high as that of the ordinary members of society. They might be poor in some sense, but they are not 'relatively poor'.

There is one important difference between the two types of relative poverty line. The first type, relative to the common standard of living, allows for the possibility that poverty could be eliminated; it is theoretically possible that income could be distributed so that even the poorest enjoyed at least 50 per cent of the national average or median income (or 40 per cent, or whatever percentage is chosen). The second type, based on income strata, does not allow for that possibility: as has been pointed out, even if everybody had an income that was more or less adequate, there would always be a bottom 30 per cent or 40 per cent, *unless incomes were virtually equal across the board.* Following this second approach, therefore, it would be difficult to measure any progress in the alleviation of poverty.

Published relative poverty lines will be observed to be invariably higher than published absolute poverty lines. It would be hasty to conclude that the proponents of relative poverty lines are necessarily more generous of spirit than those who propound absolute poverty lines. As we have noted, in principle a relative poverty line can be set at any level whatsoever; there is nothing in the relative approach itself that determines the level of poverty.

THE STATISTICS CANADA LOW-INCOME CUT-OFFS

In Canada, the poverty line most commonly referred to in public discussion is one produced by Statistics Canada. To repeat a word of caution, Statistics Canada notes that, although its low-income cut-offs 'are commonly referred to as official poverty lines, they have no officially recognized status nor does Statistics Canada promote their use as poverty lines.' (Statistics Canada, 1994, p. 178). And according to a 1997 press report, a Statistics Canada document went so far as to say, 'One could argue that had Statistics Canada known that those original LICO measures would be used so widely as poverty lines, it is doubtful that the agency would have published them as extensively as it has' (Beauchesne, 1997). StatsCan's computation ensures that the standard of living it represents will indeed be low relative to the standards of Canadians in general, but would necessarily exceed subsistence, an interesting combination of features.

Statistics Canada prefers the expression 'low-income cut-offs' (LICOs). The plural is necessary because StatsCan works out differentiated low-income lines for metropolitan, urban, and rural communities. Their derivation is logically consistent with the basic definition of poverty proposed by StatsCan: 'Inadequate access to the goods and services that are accessible to the general population'. To that extent, this is clearly a relative definition of poverty, but StatsCan's calculation incorporates an important qualifying constraint: subsistence must be covered—in fact, more than subsistence. Without that qualification, as we have seen, a purely relative poverty line might imply a standard of living below subsistence. The Statistics Canada poverty line was first developed in a study by Jenny Podoluk (1958). It has been updated, revised, and refined a number of times since then, but the concept remains the same.

The low-income cut-off is worked out in a series of steps:

1. An answer is sought to the question, 'What goods and services are accessible to the general population?' The best way to answer it is empirically: to observe actual experience. A survey is therefore made of the goods and services that Canadians of all income classes actually consume. Since a population survey would be much too expensive, a Canada-wide sample survey is done, at intervals of a few years.

2. Certain goods and services are identified as indisputably 'needs': food, clothing, and shelter (FCS). The percentage of gross (before-tax) income that all Canadians devote to these needs is calculated. No judgement is made as to the percentage that people *should* spend on FCS, nor of what items of FCS they ought to buy (fruit juice vs. soft drinks, generic products vs. name brands). In keeping with the basic definition of poverty, the question is, simply, what percentage of income *do* Canadians, on average, spend on these things?

This becomes the pivotal quantity in the computation: *the percentage of income of all Canadians that is spent on FCS*. This is an empirically observed measure, not an arbitrary one. Some judgements may have to be made as to whether a particular solid or fluid that is swallowed counts as 'food' (a hot dog at a hockey game? liqueurs after a meal?) and whether a particular piece of raiment counts as 'clothing'.

3. Significantly, it is assumed that people have needs other than FCS, varying from person to person. And since the general population consumes many items that are not needs in the sense that FCS are needs, a person who is not to be considered poor must be able to afford some of those non-need items: the low-income cut-off must allow for consumption beyond the level of subsistence.

A family has 'inadequate access' to the normal Canadian bundle of goods and services if the percentage of income it must spend on FCS is larger than the average. How much larger? *There is no definite answer.* Those not far below the average income, not easily considered relatively poor, may spend on FCS 5 per cent more than the average percentage; those far below the average income, decidedly poor, may spend on FCS a percentage that is 40 per cent more. As Podoluk explained, the determination of that additional percentage spent on FCS that will define 'low income' is a matter of judgement, *with an inescapable arbitrary element.* If too large, it would push the poverty line close to the level of sheer survival; if too small, it would identify as poor many who do not think of themselves, nor are thought of by others, as poor. This arbitrary factor in the computation of the LICOs has always been openly acknowledged.

StatsCan, in effect, has said: 'Whatever the average percentage spent on FCS needs may be, if you have to spend that percentage of your income *plus 20 per cent or more* on FCS, you are of relatively low income, because after providing for FCS, what is left over is not enough to give you adequate access to the goods and services that are accessible to the general population.' We shall refer to this percentage—the average percentage spent on FCS plus 20 per cent—as the 'LICO percentage'. If Canadians, taken all together, spent 55 per cent of their incomes on FCS, the LICO percentage would be 75 per cent, leaving at most 25 per cent for everything else; if Canadians all together spent 40 per cent of their incomes on FCS, the LICO percentage would be 60 per cent, leaving at most 40 per cent for everything else.

4. Finally, StatsCan figures out, on the basis of its survey of people's actual expenditures, what FCS costs in dollars. Whatever the LICO percentage is (average percentage spent on FCS plus 20 per cent), it is equated to the dollars required for FCS. The low-income line is then that income of which the LICO percentage equals the dollar cost of FCS.

To illustrate: suppose the expenditures survey showed that, on the whole, Canadians spent 55 per cent of their incomes on FCS (a fictitious figure, much higher than the actual). Adding 20 per cent, StatsCan would say that if you have to spend 75 per cent or more of your income on FCS, you are low-income. Now suppose the expenditures survey showed that

FCS for a family of five in a metropolitan area—Montreal, Toronto, or Vancouver—cost $12,000 a year (also fictitious). That $12,000 would then be 75 per cent of the poverty line income. By simple arithmetic, the poverty line income for such a family would be $16,000 a year (if 75 per cent = $12,000, then 100 per cent = $16,000).

The first StatsCan low-income line, calculated in 1958, was based on a survey that was too limited in extent to include a significant number of large families and was therefore, on StatsCan's own admission, insensitive to the impact of large family size on poverty line incomes. Somewhat larger, more reliable surveys have been carried out since then (they cost too much to be done frequently). The later surveys have differentiated between metropolitan, urban, small-town, and rural consumption patterns, and have taken better account of family size. Each time, a change has been observed in the critical percentage, the average percentage of income spent on FCS by all Canadians, and therefore the low-income cut-offs have changed.

In between these *revisions*, the LICOs have been *updated* in accord with the Consumer Price Index. This is what is meant by references to the 'revised' and 'updated' LICOs. Changes in the proportion of the population living under this poverty line must be interpreted in light of the fact that the line itself is shifting. In terms of real purchasing power, the StatsCan low-income cut-off is not fixed, like the freezing point on a thermometer; it shifts. The more prosperous Canada becomes, the higher the LICOs rise. The buying power of the LICOs in 1997 is greater than it was in 1960; a standard of living that would have put a family a little above the LICO in 1960 (not poor) would put them a little below it in 1997 (poor).

As noted, the StatsCan low-income cut-offs are usually what is meant when a Canadian writer refers to 'the' poverty line, unless otherwise specified. The distinct conceptual feature of the StatsCan technique is its combination of market basket and relative approaches.

StatsCan also calculates, but does not widely circulate, low income cut-offs applying the same approach to after-tax income instead of gross income, on the grounds that it is after-tax income that people actually have at their disposal to spend (or save) (Geddes, 1997). The percentage of after-tax income spent on FCS will, of course, be higher than the percentage of gross income, and so the LICO percentage (the FCS percentage plus 20 per cent) will also be higher. Since the same dollar amount (FCS) will now represent a higher percentage of the LICO, the LICO itself will be lower, and fewer people will fall below it. In 1996, with the LICO based on pre-tax income, about 17 per cent of Canadians were judged to be of low income; on the after-tax basis, 12 per cent. A preference between the two approaches ought to be based on something more profound than which one shows more or fewer Canadians to be 'poor'.

In attempts like this to quantify the extent of poverty, it is useful to speculate on the differences that would ensue from different approaches to the quantification. While the StatsCan criterion of low income is the one most commonly referred to in Canada, that proposed by the Canadian Council on Social Development (CCSD) is also prominent in discussion. The CCSD line, as described above, is simply one-half the median income of Canadians, classed according to number of dependants. This calculation pays no attention to what goods and services that particular annual income would buy. This way of defining poverty would therefore be of no interest whatever in a poor society, where the average income, let alone half the average, might be too low to provide what Canadians regard as the minimal standard of living. The CCSD approach is more sensitive to changes in the *equality* of incomes than to changes in their *adequacy*. It could happen that, over a given time span, *all* incomes increased, but the lower incomes increased proportionately less than the higher. Since all incomes have risen, the median income has risen. Then, by the CCSD definition, more Canadians would be counted as poor, even though their incomes had risen, because 'half the median' would now be at a higher level, with more people below it. The outcome does not pay any attention to the buying power of the income defined as the poverty line.

MARKET BASKET OR RELATIVE: HOW REAL IS THE DIFFERENCE?

No matter who makes it up, an absolute or market-basket poverty line income in an affluent society is sure to be high enough to pay for some items that would not be included in the context of a more deprived society. Writing in the eighteenth century, Adam Smith included the 'customary decencies' among the necessities of life, along with food and shelter. A market-basket poverty line in Canada today will certainly buy things that a similar poverty line devised fifty or twenty-five years ago would not have bought. It is inconceivable that the contents of the market basket will not be influenced by the prevalent standard of living. That is another way of saying that a market-basket poverty line will be in some degree implicitly relative.

Similarly, anyone who propounds a relative approach to determining a poverty line must at some point fix some quantity as the defining point: some percentage of the average income, some point in the ranked distribution of incomes, or, as in the case of StatsCan, some margin for expenditure above bare subsistence. This choice is inescapably arbitrary. Advocates of the strictly relative approach are not likely to make themselves look ridiculous by choosing a measure that will yield an income that is *below* subsistence level—so they do, in practice, take into account the market basket that a particular income will buy. In this way, each approach implies a little of the other.

BIBLIOGRAPHY

Beauchesne, Eric (1997). 'Low-Income Lines Poorly Understood: StatsCan', Montreal *Gazette*/Southam Newspapers, 27 August.

Canada. Economic Council of Canada (1968). Fifth Annual Review, *The Challenge of Growth and Change*. Chapter 5, 'The Problem of Poverty'.

Canada. Senate of Canada (1971). Report of the Special Senate Committee on Poverty, *Poverty in Canada/La Pauvreté au Canada*. Section I, Chapter 2. An alternative 'relative poverty line', never widely used, perhaps because it is quite complicated.

Canada. Statistics Canada (1994). *Income Distributions by Size in Canada*. Catalogue No. 13-206. An annual series.

Geddes, John (1997). 'Counting the Poor', *Financial Post*, 15 April. This article deals with the StatsCan after-tax LICO computation. Since StatsCan does not publish this computation for general circulation, references to it are exceedingly rare.

Johnston, Patrick (1993). 'The Threat to Canada's Poverty Lines: Implications and Strategy', *Canadian Review of Social Policy*, No. 31 (Spring).

National Council of Welfare (1985). *1985 Poverty Lines*. Good explanation of the StatsCan 'low-income cut-offs'.

——— (1994). *Poverty Profile 1992*. Both of these National Council of Welfare publications are re-issued from time to time.

Phipps, Shelley (1993). 'Measuring Poverty among Canadian Households: Sensitivity to Choice of Measure and Scale', *Journal of Human Resources* 28 (Winter). What difference does the choice of measuring instrument make to estimates of the prevalence of poverty?

Podoluk, Jenny (1958). *Incomes of Canadians*. Chapter 8, pp. 179-94. Ottawa: Statistics Canada. The original presentation of the Statistics Canada approach. Historically an important document.

Ross, David (1994). *The Changing Face of Poverty*. Ottawa: Canadian Council on Social Development.

Ross, David, E. Richard Shillington, and Clarence Lochhead (1994). *The Canadian Fact Book on Poverty, 1994*. Ottawa: Canadian Council on Social Development. Chapter 2, 'Working Definitions of Poverty', offers a thorough explanation of various 'poverty lines' and 'low-income lines' in use in Canada.

Sarlo, Christopher (1996). *Poverty in Canada*. 2nd edn. Vancouver: Fraser Institute.

Spector, Aron (1992). 'Measuring Low Income in Canada', *Canadian Social Trends* (Summer).

Appendix B: The Redistribution of Income

It is taken for granted that in a free enterprise economic system, with a substantially free labour market and with only modest controls on the private accumulation of capital, incomes will be unequal. Most defenders of capitalism and free enterprise, while firmly denying that there is anything fundamentally wrong with such inequality, concede some place for transfers of income to people who are unable to take part in the labour market, and for measures to protect people in the labour market from some of its chronic misadventures. Opponents of capitalism do regard its systematic inequalities as evidence that the system is unsound at its foundations. To such critics, attempts at redistribution of income within the system are at best palliatives, likely to have little effect. (We shall leave aside discussion of inequalities in systems other than our own.)

Chapters Three to Six have dealt with an array of public programs in effect in Canada whose end product is the distribution of incomes to various people under various conditions. The avowed purpose of some of these programs is to improve the distribution of incomes, by transferring incomes from some part of the general public—basically the taxpaying part—to certain people.

Redistribution is a two-way process. First, people *pay* money to the government in taxes; that alone affects income distribution, in that the after-tax distribution of income will differ from the before-tax distribution. Second, people *receive* money from the government, further changing the distribution of income. Many people both pay and receive. Such final redistribution as takes place is the net effect of the paying and the receiving. The purpose of this appendix is to heighten awareness of the redistributive factors on both the paying and receiving sides.

Programs of an 'insurance' character—EI, WC, and the Canada/Quebec Pension Plans—are not essentially aimed at redistribution at all. As we have seen, Unemployment Insurance was often accused in the past of having been converted largely into an income supplementation program for the frequently unemployed; on the other hand, it was also accused of giving too much to some highly paid workers in thriving industries, whose stretches of unemployment were predictable, indeed controllable. The Canada and Quebec Pension Plans, while dressed up at least in large part as savings-and-investment plans, have been so managed, at least until the late 1990s, as to have become virtual pay-as-you-go programs, in effect redistributing money from the working-age to the elderly. The scheduled sharp increases in contribution rates are intended to put the plans on more of a 'saved-and-funded' and less of a 'pay-as-you-go' basis.

Whatever they do in practice, in principle these 'insurance' programs mobilize parts of the incomes of certain classes of people so that they continue to receive incomes if and when they fall victim to some of the risks of life in industrial society. The programs may, however, have a redistributive effect if the risks themselves systematically fall more heavily upon some kinds of people than others. Such redistribution is incidental to their main purposes, but it is substantial enough to merit discussion (see below).

Here, our attention will be given to factors that affect the power of income transfer and income security programs, in combination with the tax system, to *redistribute* the incomes realized through the labour market. As we know, there are other forms of income besides incomes earned through work, but by far most personal income in Canada is earned through the labour market. (The labour market includes, of course, public sector workers; though they are paid out of taxes, their wages and salaries are 'market' incomes.)

The most important factors to keep in mind are: (1) the real rates at which people contribute to government in the form of taxes and to government agencies in the form of special contributions; (2) the real effect of special concessions by which taxes payable are reduced on certain conditions; and (3) the net amounts of income people receive from income programs, that is, the amounts received less any tax they pay on that income.

REDISTRIBUTION THROUGH TAXATION: WHO PAYS HOW MUCH?

Progressive, Proportional, Regressive: Income and Other Taxes

The distribution of the burden of many taxes is obscure. Almost the only tax whose 'incidence' is fairly transparent is the personal income tax, and even there parts of the picture are cloudy. With other taxes, like the federal sales tax, the various provincial consumption taxes, and corporation income tax, to mention only the simplest, there can be at best only estimates, albeit well thought out, of how much different classes of people finally pay.

The key concepts in the discussion of the burden of taxes are *progressivity, proportionality,* and *regressivity.* These concepts relate tax paid to the ability to pay. 'Ability to pay' is almost always taken to mean annual income, which is probably a satisfactory though not a perfect index of ability to pay. Wealth—the value of one's possessions—is also a measure of ability to pay. Indeed, property taxes, still important, though less so than seventy-five or a hundred years ago, are wholly based on one form of wealth. But wealth is much more difficult to keep track of than income, and recorded wealth does not translate into ability to pay as directly as recorded income. For example, the owner of farm land of high market value may have a very modest income over a period of years. The progressivity or regressivity of a tax is therefore usually measured in terms of the taxpayers' incomes.

A *progressive* tax is one that falls more heavily on those better able to pay. That is, people with higher incomes not only pay larger *amounts* of tax, they pay at higher *rates*. The personal income tax is about the only tax that we can confidently say is progressive overall, and there are some qualifications to that. The higher the tax rates imposed on upper-level incomes, for instance, the stronger the temptation to try to avoid taxes and thus diminish the effective progressivity.

A *proportional* tax is one whose burden is evenly spread, without regard to relative ability to pay. That is usually taken to mean that all classes pay at about the same rate in proportion to income. There is much discussion currently of the virtues of a flat-rate income tax. Such a proposal obviously appeals to high-income earners, by contrast with a tax structure that taxes higher incomes at higher rates, but the defenders of a flat-rate income tax argue that it would have advantages for the whole fiscal system, not just for upper-income recipients. Notably, a flat rate would make tax avoidance less rewarding for the rich.

A *regressive* tax is one that falls more heavily on those less able to pay. There is general agreement that sales taxes hit low-income people proportionately harder than higher-income people. Wealthy people pay sales tax of the same percentage of the price of a product as do poor people, and because they buy more things, and more expensive things, they pay larger *amounts* of sales tax; but the wealthy devote lesser percentages of their total incomes to buying goods and services than do the poor (they save and invest more), so the amounts that they pay in sales tax work out to lesser percentages of their incomes. Some sales taxes are made less regressive by being limited to luxury articles, and in practice most sales taxes exempt at least some articles that loom large in the spending of low-income people; and, not to get ahead of our story, the regressivity of sales taxes is mitigated by such measures as Canada's Sales Tax Credit, which refunds a certain amount to low-income people, to pay them back for the sales taxes they presumably paid over the counter during the year.

There is a natural inclination to look favourably on progressive taxes. Strong feelings are aroused when any decrease in progressivity is proposed. Even if there were no programs that transferred incomes to poorer people, progressive taxes would contribute positively to the redistribution of income from rich to poor: in return for much the same ostensible level of government services, those with lower incomes will have given up a smaller proportion of their incomes than the better-off. (There may be some argument as to whether all classes of citizens *do* benefit equally from the general services of governments.) Through progressive taxation, redistribution via transfers from rich to poor is made more feasible. It is worth considering, however, that there might be policy reasons, e.g., disincentives to consume certain substances, for which regressive taxes would be more effective than progressive taxes.

The picture with respect even to the generally progressive personal income tax, however, is not simple. Until recently in Canada, a basic *exemption* from taxation of a certain amount of income was allowed to all income taxpayers. The overt purpose of this exemption was to protect low-income people from having to pay any, or much, tax. Older people and people with dependent children used to be granted additional exemptions. These special exemptions seemed humane. Yet, as we have seen, when such exemptions were combined with a progressive income tax, the tax saving was greater for a wealthy person than for a poor person, and the more progressive the tax structure, the greater the relative tax saving to the wealthy. To avoid this perverse effect, straightforward *exemptions* have therefore disappeared from the Canadian income tax structure, having been replaced by *tax credits* of equal dollar value to all, and, as a proportion of income, of greater value to lower-income people (see below).

Canada's personal income tax system allows *deductions* whereby certain expenditures that one has incurred may be deducted from income before calculating one's income tax. Examples are private pension plan contributions and contributions to Registered Retirement Savings Plans (discussed in Chapter Five), union and professional dues, expenses for child care and attendant care, and certain health care and educational expenses. Again, each of these, by itself, seems humane. Many of the deductions offered, however, are more likely to be used, and will be used to greater effect, by higher-income people (e.g., RRSPs); and, as with exemptions, an equal tax deduction results in a greater tax saving for a higher-income than for a lower-income person. The outcome is an instance of two rights making, if not a wrong, at least an anomaly. Finally, deductions for certain expenses yield no benefit to a person whose income is low enough to incur no income tax at all, a further reason for greater reliance on tax credits (see below).

A somewhat similar analysis might be made of various other kinds of exemptions, deductions, and tax shelters used by wealthy people and by corporations to shelter their incomes against taxation. The more steeply progressive an income tax is, the more powerfully will high-income earners and corporations be motivated to avoid taxation of parts of their incomes. But the defenders of tax shelters are not exclusively rich capitalists, corporation executives, and right-wing ideologues. They include spokespersons for workers in depressed regions, in vulnerable industries, and/or in industries deemed to be 'in the national interest', such as shipping or energy exploration, whose jobs may depend on tax-privileged investments; representatives of organizations supported by foundations, by charities, and by private donations, including social agencies, universities, medical and social research institutes, art galleries, community theatres, etc., whose good works would be threatened by the removal of tax privileges enjoyed by foundations and donors; and so on. These and other features of the tax system make the personal income tax less progressive in its final impact than might appear from the schedule of income tax rates.

Canada has recently moved strongly in the direction of replacing the *exemptions* and *deductions* by a variety of personal tax credits. We have described the federal Child Tax Benefit and the proposed Seniors Benefit, both of which operate as income-tested 'refundable' tax credits. The tax credit device can be ingeniously designed to combine the reduction of taxes for people in certain defined categories (e.g., parents) with the distribution of cash benefits for people in those categories whose incomes are low. 'Non-refundable' and 'refundable' tax credits are discussed below.

Special Contributions: EI, C/QPP

(a) The Individual's Share

The analysis of linkages with the personal income tax system is equally relevant with respect to contributions to Employment Insurance, as it is now known, and the Canada/Quebec Pension Plans. Some commentators, such as the National Council of Welfare, insist on using the word 'taxes' for EI, pension plan, and, where applicable, health insurance contributions, on the grounds that they are compulsory levies to support legislated programs. The federal government's accounting practice lends credence to this usage by counting EI surpluses as contributing to its current success in deficit reduction. In this text, we have consistently distinguished *contributions* from *taxes*, on the grounds that contributions are not normally available to government for its general purposes. (Governments may borrow temporary surpluses in such funds as those of EI and the pension plans, but the money belongs to the funds, and must be made available to them if needed for their purposes.) In the context of a discussion of redistribution, the words make no difference.

It is often pointed out that both EI and C/QPP are financed by flat-rate contributions on income up to a certain ceiling; above those ceilings, the contributions stay the same, so that as income increases, the percentage of income represented by the contributions gradually decreases. It may be said, therefore, that EI and C/QPP are financed by special levies that are *proportional* up to certain income levels and *regressive* beyond them. In saying this, one must also point out that the pensions and other cash benefits payable also decrease as a percentage of income for those whose incomes are above those levels.

The tax treatment of EI and C/QPP contributions has been changed. To understand the tax treatment, remember that in both cases contributions are certain percentages of income up to ceilings; above the ceilings, everybody contributes the same dollar amount. Until the mid–1990s, such contributions were simply deducted from income for tax purposes, yielding the above-noted perverse effect of a higher tax saving to a high-income, high-tax-rate person than to a low-income, low-tax-rate person. Now, these contributions are lumped in with a number of other tax concessions as part of a (non-refundable) tax *credit*. The credit is such that a taxpayer in the lowest tax bracket (17 per cent) will be credited with (will get back) *all* the tax he or she would have paid on the amount of income that paid the EI and C/QPP contributions. People in higher tax brackets must be content with the credit based on the lowest tax rate; that is, the maximum tax saving in dollars (17 per cent of the maximum contributions) is the same no matter how high one's income. In the cases of both EI and C/QPP, the net contributions (contributions minus tax savings) of higher-income persons are now exactly the same as those of lower-income contributors; they enjoy no advantage. And to round out the picture, it must be noted that the wealthy person, once receiving the pension, will probably be paying more income tax on it than the less wealthy pensioner, which again affects their relative positions.

The tax credit linked to EI and C/QPP contributions has been characterized as *non-refundable*. From the earlier discussions of tax credits, the reader may recall that this means that the credit may be claimed only up to the taxpayer's total tax liability; if the taxpayer's income tax is so low that it is less than his/her non-refundable tax credits, the individual will have no tax to pay but will not get a refund for the difference. (With a *refundable* tax credit, the individual receives the difference in cash.) The non-refundable tax credit for C/QPP contributions will grow in importance to the individual taxpayer as these contributions gradually increase, as they are scheduled to do over the forthcoming years.

(b) The Employer's Share

The 'employer's share' of the financing of EI is far larger than the 'employee's share'. Who finally pays the employer's share? There are four possible answers:

- the *owners* of the enterprise (in the case of a corporation, the shareholders), whose surpluses, and therefore profits, are diminished by their contributions to EI;
- the *workers* themselves, perhaps, because the employers' EI contributions are a direct addition to expenditures on payroll, and they take them into account when they bargain with their employees over salaries and wages;
- the *consumers*, because of course the employer tries to get back in the price of his/her product all of his/her expenditures and then some;
- and—easily overlooked—the *taxpayers* in general, because it must be assumed that the taxpayers in general must make up for the tax reductions granted to employers for their contributions to EI. At this point we enter into what economists call the thickets of tax incidence; the employers and the workers *are* taxpayers in gen-

eral—between them they account for most of the taxes collected by government—so if they get away with some share of the country's total tax load because of the deductions allowed for EI contributions, they are the ones who have to make up most of it. Let us not try to unravel this any further.

It is difficult to say with finality who pays what part of the employer's share of EI, and it is therefore difficult to say how much of the money that is distributed as EI benefits is supplied by the rich, the average, or the lower-income. It is safe to say that little is actually contributed by the poor, nor do they receive much in the way of EI benefits, since those who earn very low incomes are not included in EI at all. The burden of contributions to the other 'insurances', WC and the two pension plans, might be analysed in much the same way.

To sum up, the personal income tax looms large enough, by comparison with those other major tax sources whose incidences are, on the whole, regressive, to make the whole tax system mildly progressive. With a truly progressive tax system, the wealthy would be left with less of their income to dispose of, while the non-wealthy would be left with more of theirs. Contributions to the major social insurances, EI and C/QPP, have become much less regressive than they were, thanks to the way they are now treated by the tax system. But whether the overall revenue system is progressive or regressive, a *de facto* redistribution of resources is achieved, on the collecting side, even before the benefits of income-distributing programs are taken into account.

Before proceeding to redistribution through the payment of benefits, bear in mind that the main function of a tax system is not to redistribute income but to pay the expenses of government. For better or worse, whether taxes improve the distribution of incomes is only one of many factors to which governments give weight in developing their tax systems.

REDISTRIBUTION THROUGH BENEFITS: WHO GETS HOW MUCH?

Tax-Supported Benefits Programs

The picture is almost as mixed on the *receiving* side of income redistribution as on the paying side. Public assistance is the program that most clearly redistributes income in the direction of the poor. The net benefit may be reduced a little by the taxes that welfare recipients do pay, notably sales taxes and their share of property taxes (as tenants), but their tax burden is lightened by the federal Sales Tax Credit. Depending on the province in which they live, the net taxes that they pay may be lessened by other tax credits. In Quebec, for example, tenants with low incomes, welfare recipients, and others receive a rebate on their imputed shares of their landlords' property taxes. Public assistance is undoubtedly highly redistributive.

A tax credit, unlike an exemption, has the same dollar value for any recipient. Moreover, a tax credit can be rigged so as to cut out at a given income ceiling, so that high-income people will receive no benefit from it at all. The Child Tax Benefit and the Seniors Benefit are similar in this respect.

All told, programs are now more and more slanted toward low-income people, while still giving diminished benefits to people with not-so-low incomes. (Whether one considers program benefits adequate or not is another matter.) The Child Tax Benefit yields something to families with incomes up to a level that is close to the average. Since it is a tax credit, none of it is taxed back. The current Guaranteed Income Supplement gives income-tested benefits to older people of low income, but it does not reach people high enough in the income spectrum to be taxed. The proposed Seniors Benefit would give maximum benefits to older people with incomes up to a modest level ($20,921 in 1997) and gradually scaled-down benefits to people with personal incomes higher than that, but will give nothing to people whose incomes are much higher than the average for older people. This benefit will not be taxed at all.

Among the income-providing programs, in theory, the rich-to-poor redistributive potential was weakest where benefits were provided universally. For that reason, universal programs were easy targets in the federal government's recent drive to reduce its deficits. But a logical deduction of that nature must always be tested against the facts of experience. Old Age Security used to be given to a class of people, the elderly, not defined in terms of income, but whose incomes have been relatively low. In 1994, 40 per cent of the recipients of OAS had total incomes under $24,000, not much above the StatsCan low-income cut-off, qualifying them to receive full or partial GIS in that year; and many of the other 60 per cent were not far above the LICOs (Human Resources Development Canada, 1994, pp. 47–8). The conclusion is that, even before the clawback cut higher-income Canadians out of OAS entirely, no great proportion of OAS benefits was going 'upward' in the income spectrum. The Seniors Benefit would be— and its predecessor programs, OAS (clawed back), GIS, and Spouse's Allowances are—net distributors of income to the poor, without question.

There are now, fortunately, many older Canadians who will be quite comfortable without the new Seniors Benefit: fortunately, because moderate-to-upper-income older Canadians will soon (as of the year 2001) receive it no longer. It may still be true, however, that a few working people with *low* incomes are paying federal taxes that will go in part to provide the Seniors Benefit for a small number of old people with *higher* incomes than their own.

A similar empirical analysis might be sketched to evaluate the actual redistributive effects of various family-oriented income provision programs, like Canada's and Quebec's child tax benefits. How are the benefits to be distributed across income classes? Granted, all but the upper-income classes are eligible to receive them, but what are the characteristics of those who *do* receive them? During their child-rearing years, families are more likely to be somewhat low in the household income scale—overall, family incomes increase with the age of the 'head of family'; poor families, like others, therefore have to meet the costs of bringing up children over a period while their incomes are probably at their lowest. Do lower-income families actually have more children—something once accepted as fact without question? If so, downward redistribution will be enhanced; if not, not. (The Quebec program has had an anti-poverty objective, to be sure, but it openly aims benefits at the non-poor for other objectives.)

Contributions-Supported Benefits Programs

Finally, we have said that the 'contributory' or 'insurance' programs—EI, WC, and C/QPP—are not essentially redistributive programs. Do they redistribute income, all the same? If so, how? Here we risk plunging into so many complications that the reader may feel he or she will never emerge.

An ordinary insurance program—fire, theft, life insurance, etc.—does not redistribute income explicitly at all, and certainly not in inverse relation to the recipient's current income. People pay a premium for protection against the financial consequences of the occurrence of some risk. Some people suffer the risk and draw cash benefits; some do not, but do not feel unfairly treated because they paid premiums for years without ever getting any benefits. This paying of insurance premiums by all and collecting of insurance benefits by some cannot seriously be called 'income redistribution'.

A *social* insurance is naturally a little different. The risk covered affects primarily the financial position of the individual, to be sure, but it also has consequences affecting the realization of some other social values probably compatible with some redistribution. EI is explicitly intended to replace income when employment is interrupted, so its benefits go to people whose incomes are somewhat depressed, at least temporarily; if their spells of unemployment are prolonged, their annual incomes will certainly be low.

Moreover, certain characteristics of EI have been designed to work in favour of workers of low income. One of the most significant (and permanent) changes made in the 1971 revision of UI was, in fact, to remove the income ceiling on eligibility, thereby including millions of better-paid, more secure workers, who had previously been excluded. The revenues of the Unemployment Insurance program were swollen by the relatively large contributions of the better-paid workers. Unemployment is more widespread, and total EI benefits probably larger, in industries that characteristically pay low wages and salaries than in those with higher pay (with some exceptions, notably in seasonal industries, where *weekly wages* are sometimes fairly high but *annual incomes* much less so). Higher-paid people regularly pay their (relatively high) EI premiums, but they are less likely ever to get EI benefits, being on the whole more secure in their jobs (with certain notorious exceptions). Finally, unemployment is most serious, seemingly permanently, in precisely those regions of the country where incomes are low, making EI a conduit for substantial interregional redistribution of income. In those regions, as we have seen, benefit periods are longer, and the periods of work required to qualify for benefits are shorter. The 'clawing back' of a percentage of benefits paid to people whose annual incomes in the year of the claim exceed a stated amount (quite high in the personal income scale) limits the net flow of benefits to upper-income people. Finally, as we have seen, the new EI provides for increased benefits to low-income EI recipients with dependent children. In all of these ways, EI is 'redistributive downwards'.

At the same time, there are aspects of Employment Insurance that do *not* work so as to redistribute income from richer to poorer. Below the level of annual income, well above average, at which EI benefits are subject to the clawback, benefits are not 'income-tested'. Nor are benefits ever reduced in accordance with the incomes of others in one's household. Thus an eligible unemployed person receives undiminished benefits even if living in the bosom of a high-income family. Below the level at which the clawback begins to take effect ($48,750 in 1996), benefits are not affected by one's total income in the year in question. EI covers virtually all wage and salary earners, including those with the highest wages, some of whom are in seasonal jobs where unemployment recurs regularly; the benefits go to them when they are unemployed just as to poorer workers. And since benefits are proportioned to 'normal' weekly earnings, the higher the normal earnings, up to the maximum insurable, the higher the benefits.

There is a minimum weekly income below which workers are not eligible to be part of the EI plan. Workers who are frequently unemployed (hence probably poor) are unlikely ever to receive benefits at all because of the rules requiring a certain number of hours of work since the last claim; increasing the work attachment needed to qualify for benefits, as was done in 1995, excludes more such workers. Workers with the lowest total incomes therefore receive *no* income through EI.

We have, therefore, one set of factors enhancing the 'downward' redistributive capacity of EI benefits, and one set limiting it. To arrive at one's own judgement, it would be necessary to look at the statistics of EI benefits to determine what proportion of benefits went to lower-income workers. Reasonable people may judge differently, but past experience would lead one to expect that among participants in EI, the higher earners as a class contribute more than they receive, while lower earners receive more than they contribute. (The same applies as between wealthier and poorer regions of Canada, but one must be careful not to mix up interregional and interpersonal redistributions.)

In 1986 the Commission of Inquiry into Unemployment Insurance, chaired by Claude Forget, issued a Report that made many controversial recommendations. Among the most controversial were some aimed at returning UI more closely to its original intended function of insuring relatively permanent workers against the consequences of temporary, irregular interruptions of income through unemployment. The Report emphasized that we need fair and effective income redistribution for the benefit of low-income Canadians, but argued that UI, or EI, as a contributory scheme, could not redistribute income fairly or effectively. Its proposals for unemployment insurance were explicitly designed to limit its redistributive functions. The Report got a stormy reception, but it gives an invaluable analysis of the operation of UI. The 1995 changes to EI evidently took seriously many of the criticisms expressed in the Forget Report (Caledon, 1995).

Less controversy has emerged over the redistribution, or lack of it, achieved through Workers' Compensation and the Canada and Quebec Pension Plans. Workers' Compensation adheres closely to the 'insurance' model. For there to be any consistent redistribution via WC, the firms in certain industries would have to be consistent net payers of premiums and the workers in certain other industries consistent net receivers of benefits; and since the contribution rates of industries vary according to their claims experience, it is most unlikely that any such systematic redistribution takes place.

Similarly, the basic structure of the pension plans is not designed to achieve redistribution. The more one earns, the more one pays in; and the more one pays in, the more one receives as a pension, always up to a point. For lower-income workers, the principal advantage of the pension plans, in theory, is not any redistribution to them of the resources of the better paid, but the fact that a retirement pension, disability benefit, and survivors' benefits are made securely available to them.

As has been noted, the effect of the actual financial performance of the plans is that income has been transferred from today's working people to today's retired people; and among the latter, those who had the highest incomes during their earning careers receive the highest pensions. The current increases in contribution rates, while primarily intended to assure the solvency of the plans, will ultimately lessen somewhat this intergenerational redistribution (see Chapter Five).

In assessing the pension plans from the point of view of redistribution, having taken into account the above-mentioned tax treatment of employee and employer contributions, one must bear in mind as well the relation between C/QPP benefits and the proposed Seniors Benefit. Recall that for each $100 of personal income above a fixed amount ($20,951 in 1997), one will give up $20 of benefit, so that the benefit is gradually reduced to zero (at about $52,000 for an individual, $77,000 for a couple). Therefore, if one's income after sixty-five is going to be low enough to qualify for the partial Seniors Benefit, then for every five dollars one has built into one's C/QPP pension, one's Seniors Benefit is reduced by one dollar, somewhat blunting the redistribution to moderately lower-income seniors. (Of course, one may not know in advance how low one's after-sixty-five income is going to be. The Seniors Benefit provides a secure guaranteed annual income for all. That a great many Canadians will exceed the guarantee through their own devices is all to the good.)

At present, the maximum C/QPP pension, approaching $9,000 per year for one who takes the pension at sixty-five, is much less than double the maximum GIS for an individual, which is about $5,600 per year, and will be less than double long into the foreseeable future (both figures are subject to change through indexation). That is, even a *maximum* C/QPP pension will not by itself wipe out one's entitlement to GIS/Seniors Benefit. A pensioner with no personal income other than C/QPP will receive a substantial Seniors Benefit. Those with no pensions or other income, not a desirable situation, get the full benefit of Seniors Benefit redistribution; the redistributive effect is somewhat reduced for those with middle to maximum C/QPP pensions but little else.

Finally, C/QPP and EI benefits are subject to taxation, if the recipient's annual income is high enough to be taxed; and EI is subject to a further clawback for recipients with fairly high total annual individual incomes. This lessens any redistribution achieved through these plans for all but the poorest of older people.

REDISTRIBUTION: TAXATION AND BENEFITS COMBINED

The foregoing may persuade the reader of two propositions:

(1) It is difficult to say with assurance how much of the money for income programs comes from the rich, how much from middle-income people, how much from the poor, though there is some doubt that the overall burden is apportioned in close accordance with ability to pay.

(2) It is naturally easier to say how the benefits are distributed among income classes. The replacement of the once-universal OAS by the new income-tested Seniors Benefit, and the replacement of federal Family Allowances by the income-tested Child Tax Benefit, mean that the rich get little or nothing. The benefits of programs that are income-tested, as most now are, or needs-tested, as public assistance is, are substantially limited to the poor and to those with low incomes. Virtually no part of the net benefits of C/QPP goes to the genuinely poor, some part goes to low-income people, and most goes to those of middle and higher incomes. A goodly proportion of net EI benefits goes to low-income working people; more goes to working people with modest to mid-level incomes; and those few rich people whose employment experience qualifies them to draw EI benefits, will get at most a residue from EI, because of the clawback. Together, these effects incline the benefits paid through the whole transfer and insurance system sharply downward.

Combining propositions (1) and (2), one concludes that the net effect, after taxes, of the whole package of income maintenance and income security programs is redistributive downwards, more so than before the 1990s, but still less than one might imagine.

That the system had (and probably still has) internal contradictions has never been a dark secret. The 'Introductory Paper' to the Quebec Government's 1984 *White Paper on the Personal Tax and Transfer Systems* said flatly (p. 19):

> ... Under the existing system, the multiplicity of transfer programs (some of which are universal) has created a situation in which persons who have difficulty meeting their essential needs are taxed. ... The tax and transfer systems are complex and illogical; they force the taxpayer who is less well off to finance a portion of the transfers he requires to cover his essential needs. ...
>
> At the high end of the income scale, households which are more than able to meet their essential needs receive transfers which are superfluous to their requirements and which, to a certain extent, they are paying to themselves by paying taxes.
>
> The integration of the tax and transfer systems therefore brings the issue of universality of certain transfer programs such as family allowances into particularly sharp focus.

Clearly, many of the changes in taxes, contributions, and benefits over the past few years have been designed to answer this criticism—to 'integrate the tax and transfer systems' to a degree. The 'assault on universality' has ended in total victory. When the tax system, including the tax credits we have described, and the benefits of all the income programs together are analysed, as in the Caledon Institute's 1995 publication, *Government Fights Growing Gap between Rich and Poor*, the wealthy emerge with incomes somewhat reduced after taxes and transfers, and the poor with incomes somewhat higher. But the gap is still wide and, in the judgement of most current students of income distribution, it is growing. The changes in both sides of the system—the collecting of revenues and the payment of benefits—have been so recent and so far-reaching, however, that we must wait for a few years' experience before reaching a conclusion about their redistributive effect.

Two concluding 'door-knobbers': (1) Inequality in the distribution of *wealth* (what is owned) is far greater than inequality in *income*; we have looked only at the latter; and (2) we have not looked at many other 'transfer payments', in the form of public subsidies to various kinds of organizations, business corporations, and others, that are not transfers to *individuals*, but that undoubtedly influence the distribution of income (National Council of Welfare, 1977; Canadian Centre for Policy Alternatives, *seriatim*).

BIBLIOGRAPHY

Caledon Institute of Social Policy (1995). *Government Fights Growing Gap between Rich and Poor*. Ottawa: The Institute.

───── (1995). *Critical Commentaries on the Social Security Review*. Ottawa: The Institute.

Canada. Human Resources Development Canada (1994). *Basic Facts on Social Security Programs, November 1994*.

Canada. Statistics Canada. *Income Distributions by Size in Canada*. Catalogue No. 13-206. Annual.

─────. *Income after Tax, Distributions by Size in Canada*. Catalogue No. 13-210. Annual.

Canadian Centre for Policy Alternatives. *Monitor*. Almost any issue of this lively publication will point to the inequalities in income and wealth in Canada, and to the avoidance of taxes by Canadian corporations, both of which it considers scandalous.

Djao, A.W. (1983). *Inequality and Social Policy: The Sociology of Welfare*. Toronto: John Wiley and Sons. Chapter 7, 'Social Inequality II: Income Distribution', and Chapter 9, 'Welfare Institution I: Income Security'.

Falconer, K. (1990). 'Corporate Taxation in Canada: A Background Paper', *Canadian Review of Social Policy*, No. 26 (November).

Gillespie, W. Irwin (1980). *The Redistribution of Income in Canada*. Ottawa: Carleton University Press. A heroic attempt to estimate the redistribution achieved not only through transfer payments and taxation, but also through all operations of government. Dated, of course, because taxes and transfers have both changed since 1980.

National Council of Welfare. (1994). *Poverty Profile 1992*. Ottawa: The Council. Not just an update of the 1985 version. Pp. 49–56 show the impact of public income programs on the incomes of the poor.

───── (1985). *Poverty Profile 1985*. Ottawa: The Council. The Distribution of Income, pp. 63–9. Note especially Table 52, 'The Impact of Taxes and Transfers', 1981 to 1983.

───── (1978). Ottawa: The Council. *Bearing the Burden—Sharing the Benefits*. A detailed analysis of the net redistribution effects of a number of programs. The pervasive 'perverse effects' of the combination of income tax exemptions and the progressive income tax are highlighted. The authors are particularly critical of UI and CPP on this ground.

───── (1977). *The Hidden Welfare System*. Ottawa: The Council. A critique of federal subsidies and tax concessions.

Quebec. Ministère des Finances (1984). *White Paper on the Personal Tax and Transfer Systems: Introductory Paper*. Criticism by the provincial government's advisers of the joint effect of existing taxes and transfers.

Ross, David (1995). *Canadian Fact Book on Poverty in Canada 1994*. Ottawa: Canadian Council on Social Development. Highly recommended. This book continues the series *Fact Books on Poverty*, published by the Canadian Council on Social Development.

Appendix C: Federal Sharing in the Costs of Health Care

Hospital insurance and medicare are examples of provincial programs, to help meet the costs of which the federal government makes contributions, subject to conditions that are agreed on by both federal and provincial governments. In describing the evolution of the federal cost-sharing of these programs, a sketchy outline was given of the original 50-50 formulas, in force until 1977–8, and of the subsequent financial arrangement known as Established Programs Financing (EPF), which was in turn superseded in 1997 by the Canada Health and Social Transfer (CHST). In 1997, as well, the Canada Assistance Plan, whereby the federal government paid one-half of covered costs for the provinces' needs-tested public assistance programs, came to an end; henceforth the federal contribution to public assistance programs will be wrapped in with its grants for health and post-secondary education as part of the CHST, calculated in the same way.

This appendix will seek to explain EPF and CHST more fully, in consideration of the importance of federal transfers to provinces, especially with regard to social policy. The names are self-explanatory: Established Programs Financing was a new vehicle for the federal share of the financing of established provincial programs in post-secondary education and health. The Canada Health and Social Transfer is the one omnibus transfer which from 1997 on embodies federal transfers to the provinces for health, social welfare, and post-secondary education.

THE HEALTH INSURANCE 50-50 COST-SHARING FORMULAS

Nation-wide hospital and medical care insurance both began with a 50-50 division between the two levels of government of the costs of services. One form of 50-50 division is to grant to each province an amount equal to 50 per cent of its actual costs. This simple pattern prevailed in the former categorical welfare programs (Old Age Pensions, Blind and Disabled Persons Assistance, etc.), and through most of the life of the Canada Assistance Plan (until the federal government imposed a limit on the annual rate of increase to the wealthiest provinces). But the two health care insurance programs were designed to provide services for the entire population, not just for the needy. A straight province-by-province 50-50 formula would have had at least two flaws from the federal viewpoint: (1) it would have put absolutely no brake on provincial spending: the federal government would have been committed to contribute half of every dollar spent, without limit, which could have tempted the provinces to be somewhat loose about their spending; and (2) it would give equal proportional relief to very unequally wealthy provinces. To meet these considerations, the cost-sharing under hospital insurance (1957) gave the 50-50 formula one twist, under medicare (1966, implemented 1968) another. Though no longer in force, these formulas usefully illustrate the nitty-gritty of federal-provincial financial relations.

The Former Hospital Insurance Formula

Each year, from the total cost of all covered hospital services, the cost per capita nation-wide and the cost per capita in each province were calculated. Each province received from the federal treasury an amount equal to one-quarter of the nation-wide per capita cost plus one-quarter of the province's own per capita cost, multiplied by the number of covered persons in the province. A province whose per capita costs were *below* the national average would have *more* than half of its total costs met by the federal government; and vice versa. This provided an incentive to the provinces to control their costs; also, it was expected to favour the poorer provinces partly because of the limits on what they could afford, and partly because they were expected to have lower labour costs.

As it turned out, poorer provinces did not always have lower per capita costs; sometimes the very factors that made a poorer province poorer—for example, a high proportion of older people, or of poor people with many health problems—could boost its per capita costs and thereby diminish its advantage under the formula. To illustrate the variation under the formula, in the last year of the overall 50-50 formula, the federal share of hospital costs varied across provinces from a low of 47 per cent to a high of 60 per cent.

The Former Medicare Formula

The hospital insurance cost-sharing formula called for detailed auditing by federal authorities of each province's expenditures. By 1966, the balance of forces in federal-provincial relations had tilted in favour of the provinces to such an extent that such federal surveillance of provincial expenditures was no longer acceptable. In the federal Medical Care Act, therefore, province-by-province expenditures were left out of the calculation; the federal government was authorized to pay to each participating province one-half of the system-wide per capita cost of insured services multiplied by the number of insured persons in the province. Costs were expected to vary so greatly from province to province that federal contributions would cover a good deal less than half the costs in the wealthier provinces and a good deal more than half in the poorer provinces. And so it came to pass: in the final year of this arrangement, the federal share of the costs of medicare in the provinces ranged from 41 per cent to 75 per cent.

ESTABLISHED PROGRAMS FINANCING/EQUALIZATION

Beginning in 1977–8, the modified 50-50 cost-sharing pattern was replaced by a different approach, Established Programs Financing, which (1) no longer tied the federal share to actual expenditures, and (2) was intended to produce greater equalization of the provinces' ability to afford health care and other programs.

The federal government was dissatisfied at being locked into paying half the large and increasing costs of programs over the administration of which it had little control. The federal government linked its financial concerns to an analysis that questioned whether continued increases in spending on hospital and medical care were making Canadians any healthier. The financial and the analytical arguments are combined to great effect in the 1974 federal Green Paper, *A New Perspective on the Health of Canadians.*

The provinces, for their part, argued that tax resources ought logically to be so shared that provinces could afford to exercise their jurisdictional responsibilities, without being tied down by the strings attached to intergovernmental transfers.

In the 1960s, even before medicare was installed nation-wide, the federal government offered an alternative kind of cost-sharing for hospitalization that included equalizing cash grants to the provinces (in addition to those that already existed) and a readjustment of taxes in the tax fields exploited by the two levels of government, notably personal and corporation income taxes. (In case the point has been forgotten, Canada's Constitution allows both provincial and federal governments to collect such taxes, and *all* provinces collect income taxes, Quebec more overtly than the others.) Such an approach would relieve Ottawa of the burden of meeting a fixed share of program expenditures, and at the same time would give provinces more leeway with respect to both revenues and expenditures.

Quebec, traditionally the most insistent of all the provinces on provincial autonomy, was the only province to accept Established Programs Financing, perhaps influenced by the fact that the terms were quite favourable to her. Other provinces were reluctant to change.

The original hospitalization and medicare agreements allowed either side to propose modifications with five years' notice. In the early 1970s the federal government forced the issue, gave the required notice of its intention to change the basis of its grants to the provinces, and accompanied its new proposals with a guarantee that no province would receive less in the first few years of the new regime than it would have received under the old 50-50 terms. Arduous intergovernmental negotiations resulted in the passage in 1977 of the rather gracelessly entitled Federal-Provincial Fiscal Arrangements and Federal Post-Secondary Education and Health Contributions Act, which became the vehicle for most federal sharing of the costs of provincially legislated programs. Conspicuously excluded was the Canada Assistance Plan, which still retained 50-50 sharing, province by province (until 1997).

Established Programs Financing was linked with equalization payments to the provinces. Federal transfers to the provinces, more generous to the less wealthy of them, were enshrined in the original British North America Act of 1867. In various forms, they have continued ever since.

As of 1977–8, the federal government began to make 'block grants' to the provinces sufficient to give all of them an equal total per capita transfer. The equalization payments, based on each province's fiscal capacity— its ability to raise taxes from its population—were folded into the block grants. The amount of the block grants was based on what the provinces had got in 1975–6 for health care and education, when the 50-50 principle was still in effect for health care, plus $20 per capita for Extended Health Care Services, a recognized emerging need. This total per capita transfer was originally to escalate in keeping with economic growth (as measured by the Gross National Product). The block grant was made up of three components: the above-described equalization (cash) payment (not paid to provinces of above-average wealth), a virtual (non-cash) transfer in the form of 'tax room', and a residual cash payment to make up the difference between these two and the total (equal) per capita transfer. Thus the block grants had three parts:

$$\text{Block grant} = \text{equalization grant (if any)} + \text{tax room} + \text{residual grant}$$

(i) *Equalization*: A province's fiscal capacity boils down to the income and wealth of its population, but the calculation of equalization payments is more complicated than that: it translates the economic activity of each province into the estimated yield of a hypothetical package of over thirty tax measures; the average of the yield of this tax package for five provinces (Ontario, Quebec, British Columbia, Saskatchewan, and Manitoba) is calculated; then, all provinces whose hypothetical tax yield is below this average are given cash transfers to bring them up to this average. Provinces whose hypothetical tax yields are above average receive no equalization payments. In most years, accordingly, Ontario, British Columbia, and Alberta have not received equalization grants.

(ii) *Tax room*: The federal government reduced, by a certain percentage, its collection of income taxes for its own use; this was expressed as the federal government giving up 'tax points' to the province. The province had to take the responsibility for collecting the revenue from its own taxpayers; but the 'tax room' meant that total provincial plus federal taxes paid by the people of the province would not increase. There is no equalizing potential in the transfer of 'tax points'. The dollar value of the 'tax room' thus opened to the province was calculated and was considered part of its annual block grant, even though, as provincial officials afterwards were prone to say, it was not a grant of funds, it was a transfer of taxes already being collected.

(iii) *Residual payment:* The final part of the block grant was the amount of cash necessary to bring the 'tax room' plus the equalization payment up to the level of the across-the-board per capita block grant which, to repeat, was to be equal in every province, and was to grow with GNP.

The federal government insisted that both the cash parts of the block grants and the conceded 'tax room' be accounted for and publicly acknowledged as federal contributions to meeting the expenditures of provinces in health care and education.

The history of EPF became more complicated in the late 1980s and early 1990s when, as an anti-inflationary measure, the federal government imposed certain limits on year-to-year increases in the block grants. The 'tax room' continued to grow in value; the total block grants either grew more slowly or were frozen. This meant that the cash part of the block grants diminished.

Provinces were still obliged to make access to hospital and medical care universal without financial barriers, to make eligibility fully portable throughout Canada, and to make administration publicly accountable. Arguably, the EPF mode of cost-sharing allowed the provinces greater freedom in allocating their funds in the health care field. They no longer had to show, as they previously had, how much they spent on hospital services, how much on medical services. As long as their programs respected the conditions of the federal cost-sharing legislation, they would receive the block grants to which they were entitled, and could use them as they thought wise. For some observers, the reverse of the coin is that federal administrative controls became more remote and attenuated. The federal government can still exert an influence, however, as is shown by its intervention through the Canada Health Act (1984), effectively suppressing user charges in hospitals and extra billing by doctors.

Lastly, the installation of EPF after 1977 was accompanied by the addition to the federal transfers of a $20 per capita annual amount to assist with the costs of 'complementary health services', meaning extended care in rest homes, home care, converted psychiatric care beds, and ambulatory care. This was done to make it easier for provinces to provide such services, coming to be seen as needed, where appropriate, in place of the relatively costlier conventional hospital care.

EPF and the Canada Assistance Plan both came to an end in 1996.

THE CANADA HEALTH AND SOCIAL TRANSFER

The 1995 federal government Budget was characterized by the National Council of Welfare as 'a giant step backward in Canadian social policy ... the worst social policy initiative undertaken by the federal government in more than a generation' (1995, pp. 1, 26), which at least conveys a sense of its impact. In the Budget the federal government announced that henceforth federal transfers in support of education, health care services, *and* public assistance—a new departure—would be wrapped up in a single transfer, labelled the Canada Health and Social Transfer. At the same time, the federal government projected reductions in the total transfers compared to those of recent years. Aside from questioning the propriety of making fundamental changes in social policy, without preliminaries, in the budget statement, commentators complained that the minister of finance gave only the sketchiest of indications of how the new CHST was to be calculated. The summary document of the Budget merely said: 'The new transfer will be provided through cash payments and tax points', and 'The Equalization program, which benefits the lower-income provinces, is untouched, and payments will continue to grow, ensuring that all provinces can provide comparable levels of service at comparable rates of taxation.' On that vague hint, one may hazard a guess that CHST will probably resemble EPF in its broad outlines. The most significant definite change is that the Canada Assistance Plan has been terminated, so that federal support of public assistance will now be based on the measured capacity of provinces to raise taxes, not on their welfare expenditures.

This attempt to explain the evolution of federal cost-sharing of social programs could end here, but it would be incomplete without mentioning the fact that justifiably caused more comment than the restructuring of the federal transfers subsequent to 1996, namely, the severe reductions in the amounts to be transferred over the coming years—annual reductions ranging from $2 billion to $3 billion. As stated above, even a slowdown in the growth or a freeze of the level of the total per capita block grant means that as the tax points portion continues to grow with GNP, the cash portion declines. The projected reductions in the block grant level make the decline in the cash transfer even steeper. Thus the federal government is helping to reduce its own deficit by reducing its transfers to the provinces; the provinces can only absorb the impact of the reductions either by raising taxes, never a popular option, or by reducing expenditures. As we have seen, they are doing their best to accomplish the latter.

When, in the 1990s, the federal government first limited the growth of EPF transfers and then froze them at the current level for a period of years, as part of its anti-inflation cost-cutting measures, the consequence was, as we have seen, that the cash transfers grew smaller as the tax points component grew larger. Many analysts quickly pointed to the ominous likelihood that the cash transfers would dwindle to zero by some foreseeable date, early in the twenty-first century. Only in late 1996 did the federal government modify its formulation of CHST to promise a minimum cash payment of about $11 billion a year to all the provinces combined; this assurance brought the greatest relief to the have-not provinces, for whom the more pessimistic projections had predicted little cash beyond their equalization payments.

BIBLIOGRAPHY

Bégin, Monique (1987). *Medicare: Canada's Right to Health*. Ottawa: Optimum.

Canada. Department of Finance (1993). *Federal Transfers to the Provinces*. A fairly lucid explanation of equalization payments, the Established Programs Financing system, and the Canada Assistance Plan. The Canada Health and Social Transfer still lay in the future when this document was published.

Canadian National Forum on Health. In 1994 the federal government created the National Forum on Health, something like a counterpart in the health field to the National Council of Welfare, an advisory body on welfare matters. As yet the Forum has not struck the critical, even combative tone that the NCW consistently adopts in its reviews of federal welfare initiatives. The Forum does provide a wealth of material through its Internet site, http://wwwnfh.hwc.ca/.

Institute for Research on Public Policy (1993). *Federal Transfers to the Provinces*. Montreal: The Institute. Expert discussions of many aspects—administrative, economic, political—of intergovernmental transfers. Dated 1993, this volume makes no references to the CHST.

Maslove, Allan M. (1995). *Time to Fold or Up the Ante: The Federal Role in Health Care*. Halifax: Dalhousie University Department of Economics Working Paper No. 95. In examining the pattern of federal support of provincial health care programs, Maslove explores alternative scenarios, none overly optimistic.

Mendelson, Michael (1995). *Looking for Mr. Good-Transfer: A Guide to the CHST Negotiations*. Ottawa: Caledon Institute of Social Policy. Formerly a highly placed Ontario civil servant, Mendelson describes part of the process leading up to the radical change in intergovernmental financial relations represented by the switch to the Canada Health and Social Transfer, and shows some estimates of the effect on the financial capacities of the provinces.

National Council of Welfare (1995). *The 1995 Budget and Block Funding*. The Council is principally concerned here with the effect of CHST on welfare.

Taylor, Malcolm (1988). *Health Insurance and Canadian Public Policy*. 2nd edn. Toronto: University of Toronto Press. See pp. 422–35.

Torjman, Sherri, ed. (1993). *Fiscal Federalism for the 21st Century*. Ottawa: Caledon Institute of Social Policy.

Useful Web Sites

Since the Internet is now widely used by students and others as a source of information, the following list of Web sites of resources in the social policy field will be helpful. Two words of caution:
1. Web site addresses are subject to change, especially those in the private realm.
2. All have their own ways of arranging and communicating information, and some are easier to use than others.

Each year, the *Canadian Almanac and Directory* provides a list of current Web sites and home pages in Canada. The student seeking information would do well to consult this list to verify exact addresses.

GOVERNMENT OF CANADA

Health Canada: http://www.ca/links/english.html
Health Canada Seniors Directorate:
    http://hpb1.hwc.ca/datahpsb/seniors/senpage.html
Human Resources Development Canada:
    http://www.hrdc-drhc.gc.ca/hrdc/menu-en.html
Finance Canada, Social Security: http://canada.9c.ca/main_e.html
Revenue Canada: http://www.revcan.ca/menue.html
Statistics Canada: http://www.statcan.ca
National Forum on Health: http://wwwnfh.hwc.ca/

PROVINCIAL GOVERNMENTS

Alberta
Department of Family and Social Services:
    http://www.gov.ab.ca/dept/fss.html
Department of Health:
    http://www.gov.ab.ca/dept/health.html
Alberta Research Council:
    http://www.arc.ab.ca

British Columbia
Ministry of Health and Ministry Responsible for Seniors:
    http://www.hlth.gov.bc.ca
Ministry of Labour:
    http://www.labour.gov.bc.ca/welcome.htm

Manitoba
Department of Health:
   http://www.gov.mb.ca/health/index.html
Department of Family Services:
   http://www.gov.mb.ca/fs/first/ffindex.html

New Brunswick
Department of Health and Community Services:
   http://www.gov.nb.ca/hcs/

Newfoundland
Department of Health:
   http://www.gov.nf.ca/health/starthel.htm
Department of Social Services:
   http://www.gov.nf.ca/doss/startdos.htm

Northwest Territories
Department of Health and Social Services:
   http://www.hlthss.gov.nt.ca/

Nova Scotia
Provincial government site:
   http://www.gov.ns.ca/

Ontario
Provincial government site:
   http://www.gov.on.ca/
Ministry of Community and Social Services:
   http://www.gov.on.ca/CSS/
Workers' Compensation Board:
   http://www.wcb.on.ca/

Prince Edward Island
Department of Health and Social Services:
   http://www.gov.pe.ca/hss/index.html

Quebec
Provincial government site:
   http://www.gouv.qc.ca/
Ministère de la santé et des services sociaux:
   http://www.gouv.qc.ca/français/minorg/msss/msss_intro.html
Ministère de la sécurité du revenu:
   http://www.gouv.qc.ca/français/minorg/msp/msp_intro.html

Saskatchewan
Department of Health:
 http://www.gov.sk.ca/govt/health/
Department of Social Services:
 http://www.gov.sk.ca/govt/socserv/

PRIVATE SOURCES

Caledon Institute of Social Policy: www.cyberplus.ca/~caledon/
Canadian Centre for Occupational Health and Safety:
 http://www.ccohs.ca
Canadian Centre for Policy Alternatives:
 http://www.policyalternatives.ca
Canadian Council on Social Development:
 http://www.achilles.net/~council/
Le Devoir (Montreal):
 http://www.ledevoir.com
Financial Post: http://www.canoe.ca/FP
Fraser Institute: http://www.fraserinstitute.ca
Halifax *Chronicle-Herald*: http://www.herald.ns.ca
C.D. Howe Institute: http://www.cdhowe.org
Maclean's Magazine: http://www.canoe.ca/macleans
Montreal *Gazette*: http://www.montrealgazette.com
National Advisory Council on Aging:
 http://hpbl.hwc.ca/datahpsb/seniors/senpage.html
Toronto *Globe and Mail*:
 http://www.TheGlobeAndMail.ca

Index